The Genus *Psychotria* (Rubiaceae) in the Philippine Archipelago

The Genus *Psychotria* (Rubiaceae) in the Philippine Archipelago

S.H. Sohmer and A.P. Davis

BRIT
2007

Sida, Botanical Miscellany
Number 27
ISBN-13 978-1-88978-15-7
ISSN 0833-1475

Founded by
Wm. F. Mahler and Barney L. Lipscomb, 1987
Editor: Barney Lipscomb
Botanical Research Institute of Texas
509 Pecan Street
Fort Worth, Texas 76102-4060, USA
Contributing Spanish Editor: Félix Llamas
Dpt. Biología Vegetal, Universidad de León
Campus de Vegazana s/n, 24071 León, Spain
Layout and Cover Design: rhorngraphics rlrlchorn@verizon.net

The Genus Psychotria (Rubiaceae) in the Philippine Archipelago
By Seymour H. Sohmer and Aaron P. Davis

Cover photograph: *Psychotria gitigensis* (photo credit, S.H. Sohmer)
The Genus Psychotria (Rubiaceae) in the Philippine Archipelago
is published with the support of:
The National Science Foundation (grant #BSR-8306982)
and the
Botanical Research Institute of Texas

Distribution of copies by:
Botanical Research Institute of Texas
509 Pecan Street
Fort Worth, Texas 76102-4060, USA
Telephone: 1 817 332 4441
Fax: 1 817 332 4112
Website: www.brit.org/sida
Email: sida@brit.org

BRIT PRESS
27

Sida, Botanical Miscellany, No. 27
30 March 2007

Printed in the United States of America

Table of Contents

New names and new combinations

Psychotria alabatensis Sohmer & A.P. Davis, sp. nov.—**142**
Psychotria alzapaniensis Sohmer & A.P. Davis, sp. nov.—**123**
Psychotria banahaensis var. **montalbanensis** Sohmer & A.P. Davis, var. nov.—**188**
Psychotria barkleyi Sohmer & A.P. Davis, nom. nov.—**26**
Psychotria bayombongensis Sohmer & A.P. Davis, sp. nov.—**203**
Psychotria boholensis Merr. ex Sohmer & A.P. Davis, sp. nov.—**131**
Psychotria casiguraensis Sohmer & A.P. Davis, sp. nov.—**208**
Psychotria castroi Merr. & Quisumb. ex Sohmer & A.P. Davis, sp. nov.—**98**
Psychotria catanduaniensis Sohmer & A.P. Davis, sp. nov.—**184**
Psychotria christarusselliae Sohmer & A.P. Davis, sp. nov.—**39**
Psychotria conglomeratiflora Sohmer & A.P. Davis, sp. nov.—**120**
Psychotria crassifolia var. **culionensis** Sohmer & A.P. Davis, var. nov.—**157**
Psychotria dingalanensis Sohmer & A.P. Davis, sp. nov.—**171**
Psychotria foxworthyi Sohmer & A.P. Davis, sp. nov.—**145**
Psychotria frakei Sohmer & A.P. Davis, sp. nov.—**147**
Psychotria galintanensis Quisumb. ex Sohmer & A.P. Davis, sp. nov.—**188**
Psychotria hilonghilongensis Sohmer & A.P. Davis, sp. nov.—**220**
Psychotria iridensis Sohmer & A.P. Davis, sp. nov.—**134**
Psychotria madulidii Sohmer & A.P. Davis, sp. nov.—**60**
Psychotria mariguidonensis Sohmer & A.P. Davis, nom. nov.—**208**
Psychotria mendozii Sohmer & A.P. Davis, sp. nov.—**214**
Psychotria merrittioides Sohmer & A.P. Davis, sp. nov.—**142**
Psychotria olsenii Sohmer & A.P. Davis, sp. nov.—**20**
Psychotria palimlimensis Sohmer & A.P. Davis, sp. nov.—**223**
Psychotria paloensis var. **bukidnonensis** Sohmer & A.P. Davis, var. nov.—**196**
Psychotria paloensis var. **ovatifolia** Sohmer & A.P. Davis, var. nov.—**197**
Psychotria paloensis var. **subelliptifolia** Sohmer & A.P. Davis, var. nov.—**199**
Psychotria paloensis var. **wenzeliana** Sohmer & A.P. Davis, var. nov.—**201**
Psychotria paracalensis Sohmer & A.P. Davis, sp. nov.—**20**
Psychotria patulinervia Merr. & Quisumb. ex Sohmer & A.P. Davis, sp. nov.—**206**
Psychotria pilosella subsp. **erythrotricha** (Elmer) Sohmer & A.P. Davis, comb. et stat. nov.—**109**
Psychotria pilosella subsp. **erythrotricha** var. **camarinensis** (Merr.) Sohmer & A.P. Davis, comb. et stat. nov.—**111**
Psychotria pilosella subsp. **pilosella** var. **samarensis** Sohmer & A.P. Davis, var. nov.—**108**
Psychotria pulgarensis Sohmer & A.P. Davis, nom. nov.—**103**
Psychotria quisumbingiana Sohmer & A.P. Davis, sp. nov.—**136**
Psychotria reynosoi Sohmer & A.P. Davis, sp. nov.—**191**
Psychotria rubiginosa Elmer var. **jacobsii** Sohmer & A.P. Davis, var. nov.—**129**
Psychotria sohotonensis Sohmer & A.P. Davis, sp. nov.—**228**
Psychotria subsessiliflora var. **carinata** (Quisumb. & Merr.) Sohmer & A.P. Davis, comb. et stat. nov.—**79**
Psychotria sulitii Merr. & Quisumb. ex Sohmer & A.P. Davis, sp. nov.—**100**
Psychotria surigaoensis Sohmer & A.P. Davis, sp. nov.—**100**
Psychotria tayabensis var. **castanea** (Merr.) Sohmer & A.P. Davis, comb. et stat. nov.—**82**
Psychotria tayabensis var. **euphlebia** (Merr.) Sohmer & A.P. Davis, comb. et stat. nov.—**84**
Psychotria tayabensis var. **fasciculiflora** (Merr.) Sohmer & A.P. Davis, comb. et stat. nov.—**84**
Psychotria tayabensis var. **pinagteponanensis** Sohmer & A.P. Davis, var. nov.—**88**
Psychotria yagawensis Sohmer & A.P. Davis, sp. nov.—**66**
Psychotria yenii Sohmer & A.P. Davis, sp. nov.—**68**

The Genus *Psychotria* (Rubiaceae) in the Philippine Archipelago

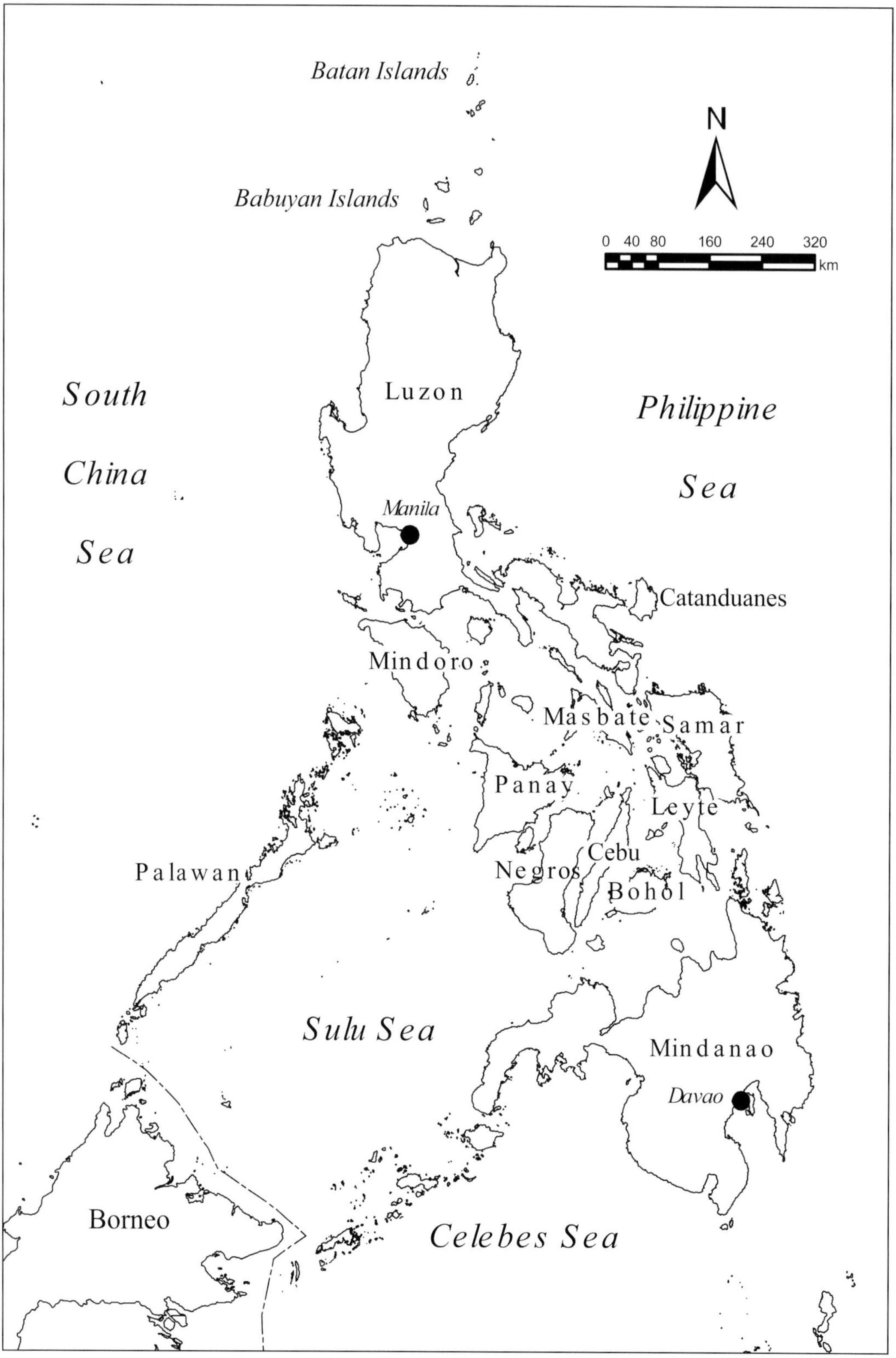
Batan Islands
Babuyan Islands
N
0 40 80 160 240 320
km
South China Sea
Luzon
Philippine Sea
Manila
Catanduanes
Mindoro
Masbate
Samar
Panay
Leyte
Cebu
Negros
Palawan
Bohol
Sulu Sea
Mindanao
Davao
Borneo
Celebes Sea

THE GENUS *PSYCHOTRIA* (RUBIACEAE) IN THE PHILIPPINE ARCHIPELAGO

S.H. Sohmer

Botanical Research Institute of Texas
509 Pecan Street
Fort Worth, Texas 76102-4060, U.S.A.

A.P. Davis

The Herbarium
The Royal Botanic Gardens, Kew
Surrey, TW9 3AB, UK

ABSTRACT

The genus *Psychotria* is probably the world's largest predominately woody genus (Davis et al. 2001), with perhaps around 2000 species (Sohmer 1998). It has a pantropical distribution and is often a very obvious and species rich component of humid, tropical forests, particularly in Africa, south-east Asia, and South America. The high number of species, coupled with bewildering systematic complexity, has resulted in a genus that is poorly known, particularly in south-east Asia. In this work we fill a gap in our knowledge of the genus by presenting a taxonomic revision of Philippine *Psychotria*.

RESUMEN

El género *Psychotria* es probablemente el que más especies leñosas tiene en el mundo (Davis et al. 2001), con unas 2000 especies (Sohmer 1998). Tiene una distribución pantropical, y es a menudo muy aparente y un componente rico en especies de los bosques tropicales húmedos, particularmente en África, sureste de Asia, y Sur América. El alto número de especies, junto con una complejidad sistemática desconcertante, ha dado como resultado un género que está pobremente conocido, particularmente en el sureste de Asia. Este trabajo viene a llenar un hueco en el conocimiento del género presentando una revisión de las *Psychotria* de Filipinas.

INTRODUCTION

The genus *Psychotria* is probably the world's largest predominately woody genus (Davis et al. 2001), with perhaps around 2000 species (Sohmer 1998). It has a pantropical distribution and is often a very obvious and species rich component of humid, tropical forests, particularly in Africa, south-east Asia, and South America. The high number of species, coupled with bewildering systematic complexity, has resulted in a genus that is poorly known, particularly in south-east Asia. In this work we fill a gap in our knowledge of the genus by presenting a taxonomic revision of Philippine *Psychotria*.

In this work we have recognized 112 species of Philippine *Psychotria*, including 17 infraspecific taxa; 32 species and nine varieties are new to science, and five new combinations are made.

Philippine *Psychotria* are distributed from sea level to over 3000 m elevation, generally in evergreen, humid forest. It is likely that nearly all species would have once existed as understorey components of undisturbed primary forest, with some capable of rapid recolonization after moderate to severe disturbance. In forest with a low canopy, such as those found in higher altitudes, *Psychotria* species may well form part of the canopy; climbing *Psychotria* species often emerge from the canopy. With the large loss of primary forest in the Philippine archipelago, and the more widespread occurrence of secondary forest, several species are now found almost exclusively in secondary vegetation. Within their various habitats, *Psychotria* species exist mainly as shrubs or small trees, with the notable exception of 22 climbing species. All Philippine *Psychotria* species are woody, even though a few may not reach many centimeters above the forest floor. In areas with remaining, good quality forest *Psychotria* may be abundant and sometimes species rich.

Despite the large number of species in the Philippine archipelago, many are now assumed to be extinct (see Koopowitz et al. 1998; Sohmer 2001; and see below), due to the drastic decline of Philippine forests over the past 20–40 years. Many of the species are known only from collections made early in the twentieth century by A.D.E. Elmer, and E.D. Merrill, and colleagues and staff of the latter. A significant number of the taxa have not been recollected since that era of Philippine botany ended in the late 1920s.

The fieldwork performed during the course of this revision (by S.H. Sohmer) was not as thorough as we would have liked, due to the fact that there were practical and political difficulties associated with getting into the forests that remained in the Philippines at the time. Areas of northern Luzon, as well as most of Mindanao, were essentially off-limits during the 1980s. Despite these difficulties, approximately six months were spent in the field, visiting most of the rest of the archipelago, including the Visayas and Palawan. On the basis of this fieldwork, it is possible to categorically state that little or nothing remains of the forests that Merrill, Elmer and others worked in during the early part of the twentieth century. We can further infer that many of the species known only in herbaria are either extinct or severely restricted in their present day distributions. Sohmer (2001), for example, estimates that more than 50% of known Philippine *Psychotria* species are extinct (and see below: Extinction and Extinction Threat in Philippine *Psychotria*). We should acknowledge the aforementioned workers who toiled, indefatigably 50-80 years ago, for most of the evidence that is available now in the form of preserved herbarium samples. These specimens serve as yet another reminder of the diversity of plant life we have lost, are losing, and are yet to lose, as well as of the principal value of that institution known as the herbarium.

This revision was carried out principally in the Herbarium Pacificum of Bishop Museum (BISH) by the first author (S.H. Sohmer) as well as during visits to other herbaria, including K, L, P, PNH, PR, and W (for abbreviations see Holmgren et al. 1990). The later work undertaken for this revision (by A.P. Davis) was carried out mainly at K, but also at BRIT, and L).

Significant historical collections of the genus exist at K, L, and P. The most significant in this regard is L, where interest has long been focused on the Malesian region, and where large and valuable collections of *Psychotria* from this region are held. A total of nearly four months were spent at L (by S.H. Sohmer) working in the hospitable environment provided by the staff of that herbarium. Specimens from these herbaria as well as from A, BM, GH, MO, and UC were borrowed for study at BISH. During the course of the revision we had at our disposal all of the significant collections of this genus, from all of the above cited herbaria.

Like so many other taxa in the Philippines the quality of what little information we have about these taxa is diminishing. It is for this reason alone that we have judged the publication of photos of the type specimens to be so important.

METHODOLOGY

This revision is based mainly on herbarium specimens, which were either studied in their respective herbaria or at BISH and K, as loans. The herbarium material of the following herbaria was consulted: A, BISH, BM, GH, K, L, MO, P, PNH, PR, UC, W (see Holmgren et al. 1990). This information was supplemented by fieldwork (S.H. Sohmer) and by literature review. Initially, all herbarium label data was stored on record cards but this was later transferred to an electronic database. During the course of the revision more than ca. 1800 collections, comprising ca. 3500 specimens, were studied.

For species description and keys, all measurements are based on herbarium material. Fresh or spirit-preserved material was used whenever available to gain a better understanding of form, structure and color, but measurements for descriptions and keys were taken from dry material only.

Pyrenes were prepared by immersing fruits in boiling water for three to five minutes, depending on the size and thickness of the fruit. The mesocarp and any other soft tissue was removed using a blunt scalpel blade and blunt needle. Preformed germination slits (PGSs) are visible with a hand lens or binocular microscope once the mesocarp has been removed, although slight pressure on the pyrene walls with tweezers was sometimes needed to reveal the PGSs. The endosperm was examined by cutting the pyrene in transverse section with a sharp scalpel. To test for an ethanol soluble pigment, a single seed, or one half of a seed, was placed into a test-tube and just covered with 70% ethanol for ca. 5 minutes, at room temperature. The resulting solution was held against a white piece of paper and the color assessed. On completion of the test, the seeds were dried and returned to their respective herbarium sheets. Pyrene observations were made using a Leica MZ95 stereomicroscope.

TAXONOMIC HISTORY OF PHILIPPINE *PSYCHOTRIA*: AN OVERVIEW

The taxonomic history of Philippine *Psychotria* began in 1829, when Chamisso and Schlechtendal (1829) described *P. philippensis* Cham. & Schltdl. and *Coffea luzoniensis* Cham. & Schltdl. The former is now recognized as *Scyphiphora hydrophyllacea* Gaertner (Merrill 1915a) and the latter as *P. luzoniensis* (Cham. & Schltdl.) F.-Vill. (Fernandez-Villar 1880). The descriptions are fairly thorough and were based on specimens collected by Chamisso on the Voyage of the Rurik in 1815-1818. *Coffea luzoniensis* was described at the same time that two other species of *Psychotria* were described from Hawai'I, which were also then considered to be in the genus *Coffea*.

In 1830, Bartling described five species of Philippine *Psychotria* based on collections made by Haenke, viz. *P. pauciflora* Bartl. ex DC., *P. ixoroides* Bartl. ex DC., *P. membranifolia* Bartl. ex DC., *P. linearis* Bartl. ex DC., and *P. manillensis* Bartl. ex DC. (Bartling 1830:522). In 1837 Blanco described a species he called *Paederia tacpo* Blanco (Blanco 1837), a taxon that Rolfe quite rightly transferred to *Psychotria* in 1886. However, this was the same taxon that Chamisso and Schlechtendal (1829) had already described as *Coffea luzoniensis*, and which Fernandez-Villar (1880) had already transferred to *Psychotria*, where it properly belongs. *Paederia tacpo* is thus a synonym of *Psychotria luzoniensis*.

In all this time and right up to the end of the nineteenth century, there were no resident botanists in the Philippines except Blanco, who apparently caused more harm than good as far as taxonomy is concerned. This would change dramatically with the coming of the twentieth century.

Elmer D. Merrill arrived in the Philippines in 1906 and began a career that was to last well over half a century and span more than half the globe as he moved from his first professional appointment with the Bureau of Science in Manila to the University of California, Berkeley Herbarium, and, finally, to the Arnold Arboretum. Merrill made one of the single greatest contributions to Philippine botany, which included considerable work on *Psychotria*. His tenure with the Botany Section of the Biological Laboratory of the Bureau of Science, which eventually became what is today the Philippine National Herbarium, was particularly significant. He apparently dedicated himself fully to the task of botanical exploration and publication. His collections and those made by native collectors, which he himself trained, formed the backbone of the herbarium prior to its destruction in World War II. Merrill promoted the collecting of duplicates and sending them out far and wide; enough material exists in other herbaria, therefore, to reconstitute a significant proportion of those early collections from an era that was certainly a golden age for Philippine botany. This golden age can never again be repeated because concomitant with the American occupation of the Archipelago came changes that would cause a huge population explosion and enormous development, leaving ever smaller amounts of the original vegetation. In connection with *Psychotria*, Merrill published 13 articles in the *Philippine Journal of Science* (Merrill 1906a & b, 1909, 1910a & b, 1913, 1915a & b, 1917, 1921, 1923, 1925, 1926) in which 59 new species of Philippine *Psychotria* were described, not counting new combinations or varieties. The *Philippine Journal of Science* began in 1906 with volume one and ended in 1926 with volume 29, when Merrill left the Philippines for California. During the course of working with the Philippine flora, particularly in the early years, a significant proportion of the material he received each year represented new taxa. For example, in 1906 he states: ' ... the percentage of undescribed species and interesting forms is very large, as is to be expected in a region botanically so little known...practically no work in some groups can be done in Manila, owing to a lack of literature;' (Merrill 1906b).

In 1915 Merrill (1915b) formally applied the generic name *Grumilea* Gaertn. to all Philippine *Psychotria* with a ruminate endosperm. He thereupon made 37 new combinations, which as the years progressed fell into disuse as the value of ruminate endosperm for generic differentiation became untenable.

The *Philippine Journal of Science* was Merrill's principal vehicle for botanical publication but it was by no means his only one. Towards the end of his tenure in the Philippines he summarized his years of botanical experience in a three volume work entitled *An Enumeration of Philippine Flowering Plants*. In this publication he listed the then known species of *Psychotria* (Merrill 1923), including synonyms, and he formally reestablished the genus *Psychotria* to include all those taxa he had either transferred to *Grumilea* (Merrill 1915b) or had described as new in that genus. Merrill lists 101 species and specifically excludes five others. To date this is the only summary of the species of *Psychotria* for the Philippines, and for Merrill it was most likely a catharsis as he did not do too much more with this genus in the Philippines, although he certainly continued to contribute to Philippine botany.

Another American botanist, A.D.E. Elmer, along with Merrill, stands apart from other botanists in his contributions to Philippine botany. He arrived in the Philippines at approximately the same time as Merrill in 1904 and worked for a year for the Bureau of Science. Within a short period, however, Elmer became a "free-lance" collector and the world's major herbaria are full of his duplicate collections. It is Elmer who, as an individual, contributed most to documenting the Philippine flora. The nature of his livelihood assured multiple sets of each collection. The taxa represented by Elmer's collections in the world's herbaria today are better represented than any others. Elmer, however, did not merely collect. He also created a publication called *Leaflets of Philippine Botany*, which he published serially for many years and whose main purpose was to serve as a vehicle to describe new species based almost entirely upon his collections. He charged a subscription rate of two cents per page until 1913 when the rate was raised to two and one-half cents per page and finally to three cents per page thereafter.

Elmer published 47 new species of *Psychotria* in the *Leaflets of Philippine Botany*, from 1906 to 1934, the first being on page 22 and the last on page 3,265.

When this golden age of Philippine botany ended in the late 1920s and early 1930s, so too did the most active period of work with *Psychotria*. The result of the above works was that some 100 species of *Psychotria* were known, but these lacked any systematic treatment other than the enumeration provided by Merrill in 1923. Quisumbing, a student of Merrill's, and the first Filipino botanist to take up systematic botany as a profession, published twice on Philippine *Psychotria*, once in 1928 and again in 1930, adding three new species of *Psychotria* to the literature.

TAXONOMIC CHARACTERISTICS OF PHILIPPINE *PSYCHOTRIA*

The characteristics used in this revision are principally morphological, and the most important are those that can be seen on herbarium specimens. As many of the taxa have not been recollected since the early part of the twentieth century, for some species there are many features that may never be known aside from what was reported in original descriptions and what may be seen on remnant herbarium specimens. It was therefore often difficult for us to follow a given morphological characteristic, such as stipule type, for all of the taxa. In the identification key we therefore used structures that were more likely to be present on the greatest number of specimens.

Habit

Philippine *Psychotria* species exist as shrubs, trees, or climbers. The climbing species are mostly true climbers, even though some are reported as scandent or scrambling shrubs. Where appropriate material has been collected, as well as through field observations (S.H. Sohmer), the occurrence of adventitious roots (anchoring the climbers to their host as they climb) has been recorded. It seems that most and perhaps all climbing Philippine *Psychotria* possess adventitious roots, although it is impossible to ascertain whether adventitious roots exist on those taxa for which this feature is absent on herbarium specimens. Quite often the adventitious roots are some distance from the part of the plant that is collected as the specimen. The climbing species climb on trees or rock outcrops, usually in moist, shady places. Most, or all, climbing species produce pendent branches, and at the apices of these branches the inflorescences are produced. On trees in dense forest the stimulus to branch is evidently provided by the impact of direct sunlight available to the plant once the canopy is reached. One may often see a profusion of branching and flowering on those individuals in or at a clearing caused by disturbances of the canopy. The occurrence of climbing species is probably much greater than collecting has indicated. The ability to recognize young plants often leads to the observation of several beginning their climb up a particular tree to join others who have already made it to the top. It would be an interesting study to see if the connection to the ground is ever eventually lost in older plants. Then they could be classified as true epiphytes. It seems more than mere coincidence that specialized epiphytic taxa such as *Hydnophytum* Jack sometimes resemble climbing *Psychotria* in their sterile condition, and it is possible that the myrmecophilous, epiphytic Psychotrieae (Hydnophytinae) such as *Hydnophytum* were derived from a climbing *Psychotria*-like ancestor.

The habits of the non-climbing species are shrubs, small trees, or trees. These definitions are used here rather broadly, as there is not much difference in overall habit between most *Psychotria* species that would be classified as shrubs from those that would be classified as trees, other than gross size. Most species are 1–5 m high, rarely up to 10m; there are a few records of trees up to 15 m but these have not been verified.

Hamilton (1985) has provided an interesting new view of branching patterns in neotropical *Psychotria* species and has shown that their branching patterns can be classified into three types. Old-world *Psychotria* species have not yet been looked at in this way, but it would be interesting to see how the branching patterns compare with other characteristics used to classify them.

Stipules

This study infers that stipule morphology could be important characteristic for classifying *Psychotria*, and in particular the way in which each stipule (within the stipule pair) is arranged in relation to the other. The problem is, of course, that stipules are often not found on the plant parts most likely to be collected, i.e. branches bearing flowers and/or fruit. Furthermore, to really comprehend the nature of the stipules in a given taxon, they have to be studied thoroughly in the field and perhaps followed by a developmental analysis in the laboratory. Based upon *in situ* observations in the Philippines (S.H. Sohmer), the stipules appear to be either valvate or imbricate. Valvate stipules are generally not very much longer than broad and they meet, margin-to-margin, like the two shells of a clam. The margins of valvate stipules may be joined and then the stipules become slightly to profoundly tubular. If the margins are joined only at the base the stipule pair appears slightly tubular but if joined for most or nearly all of its length the stipule become profoundly (cylindrical) tubular. This type of stipule may remain around the young shoot for some time before falling. Imbricate stipules, apparently the most common type in the Philippines, are usually much longer than broad, and the margins of one member of the pair are generally enveloped by the margins of the other. Imbricate stipules are generally narrow with long acuminate apices (before opening). We have not seen and positively identified any other stipular type in the Philippines, except for a tubular type (see above), which is apparently restricted to one species group (Linearis Species Group). It is possible that there are other types present in the Philippine Archipelago, because the stipules of some taxa are unknown at this time. In New Guinea, for example, there is, in addition to the above types (of which valvate is by far the most common) a calyptrate type with the stipules fused along one margin to form a hood-like structure over the juvenile plant parts, and which come away as one piece. There are variations of these types and a thorough study would not only clarify our understanding of this matter, but could also provide further characteristics for the infra-generic classification of *Psychotria*.

Stipules vary in terms of size, color, texture, and nature of the indumenta. These traits may help in the differentiation of species and other taxa. For the identification of some species and species groups (but excluding those recognised here) the apex of the stipule is significant, and specifically whether it is entire or cleft (divided). Infrequently the stipules may be semi-persistent to persistent, and this also serves to differentiate taxa, such as *P. rubiginosa* Elmer var. *jacobsii* Sohmer & A.P. Davis. Most *Psychotria* species worldwide have stipules that are caducous (Andersson 2002), which means they fall very early (e.g. they are only found on the uppermost nodes of the young stems).

Leaves

The leaves of Philippine *Psychotria* are very variable, particularly with respect to shape, size, venation and pubescence. Considerable variation is also manifest within species. One feature shared by many *Psychotria* species, including those from the Philippines, is that the secondary veins join near the margin to form an indistinct to distinct sub-marginal vein (brochidodromus). However, there are also species in which the veins do not join together (camptodromus).

In combination with inflorescence morphology, leaves provide the most practical means of differentiating taxa of *Psychotria* in the Philippines. The characteristics of these structures are almost always certain to be available. The size and shape of the leaf blades are important, as are their venation patterns, texture, and if present, hair type and coverage. Another important feature is the color of the leaf blades upon drying, i.e. when herbarium specimens are produced. In common with other *Psychotria* many Philippine *Psychotria* species have leaves that dry a pinkish or reddish color, although other colors are frequently evident. Differences in color are most likely related to heavy metal metabolism, which is apparently consistent over time and space in some taxa regardless of the manner in which the specimens were prepared. Colors are either a dark brick-red or brownish orange, grayish green or khaki-green to yellow-green. *Psychotria membranifolia* and *P. balabacensis* Merr. always have yellowish green leaf blades and twigs. A character apparently directly related to that of dried leaf color is tissue blackening, which occurs when plant parts, such as leaves, twigs, flowers and fruit, are fixed in alcohol or FAA fixative (formaldehyde/ethanol/acetic acid). The solution that these parts are fixed in also often turns black. In a preliminary systematic study of the occurrence of this phenomenon, Dr. Philip Rundel indicated (pers. comm.) that there seems to be some systematic significance in the occurrence of the tendency to accumulate heavy metals such as aluminium and the occurrence of this dark color.

The number of lateral veins per side of the leaf-blade is often a useful taxonomic and/or diagnostic character. Another important feature is the presence or absence of a tertiary venation pattern that, if present, is visible with a ×10 hand lens. The pattern is one in which the tertiary and quaternary veins form a reticulate pattern on the lower surfaces of the blades. Some taxa, such as *P. galintanensis* Sohmer & A.P. Davis, *P. pallidifolia* Merr., *P. weberi* Merr., and others always have it, while a taxon like *P. banahaensis* Elmer, in the same species group, never does.

One of the features unique for two species within Philippine *Psychotria* (*P. papillata* (Merr.) Merr. and *P. pilosella* Elmer), is the occurrence of hairs on the upper surface of the leaf blades.

Inflorescence

The inflorescence position of all species is terminal, although sometimes growth of the shoot occurs during floral and fruit development and then the inflorescence becomes overtopped. The overtopped condition sometimes becomes fixed in some taxa, but this does not appear to occur in any of the Philippine species. Overtopped inflorescences should not be confused with pseudoaxillary inflorescences, which are very rare or perhaps absent in *Psychotria* (A. Davis unpubl. data).

The inflorescence is an important structure from a taxonomic perspective in Philippine *Psychotria*. Five of the nine species groups recognized for Philippine *Psychotria* are generally defined by the nature of inflorescence structure. The most significant character is whether the inflorescence is pedunculate or sessile. A pedunculate inflorescence has a single main axis subtending the rest of the inflorescence parts (e.g., the axes (inflorescence branches), and flowers)—this main axis is the peduncle (see Fig. 1j). Sessile inflorescences lack a peduncle. In this revision the use of the terms pedunculate and sessile are not explicitly used in the descriptions, as they are likely to cause confusion. A truly sessile inflorescences (i.e., lacking a peduncle) may not appear 'sessile' to many users of this work, as the flowers and inflorescence parts still possess other axes, e.g., inflorescence branches. This is exemplified in Fig. 1: the inflorescences in a, b, c, and k are sessile in the true botanical sense, because they lack a peduncle, but a, c and k may be considered by some users of this work as pedunculate. In the descriptions we have used the terms monochotomous and trichotomous to describe most inflorescence types. Monochotomous inflorescences (see Fig. 1j) are in the true sense pedunculate, which is indicated in the taxon descriptions in parentheses. Trichotomous inflorescences (Fig. 1c, 1k) lack a peduncle, but have three major inflorescences axes arising from the apex of the shoot. In most cases this differentiation appears to be fixed within species and some species groups; in the few cases where it does not, the respective taxa appear more than once in the identification key. In some taxa the inflorescence may have five or more main axes arising from the shoot apex, a pentachotomous arrangement is shown in Fig. 1b).

Following on from the trichotomous arrangement, species may have inflorescences that are further reduced in the number and length of inflorescence axes and branches (e.g., Fig. 1c) or the inflorescence may be very simple (much reduced) and lack inflorescences axes and branches altogether (e.g., Fig. 1a). One species group (Pilosella Species Group) is delimited by inflorescence that are usually trichotomous but the flowers are sessile at the ends of the inflorescence branches (i.e. lacking pedicels, or where the pedicels are very small making the flowers appear to be in a capitulum, e.g., Fig. 1c). There are two species groups (Linearis Species Group and Tayabensis Species Group) whose primary differentiation from the rest of their congeners in the Philippines is the presence of a very reduced inflorescence, lacking inflorescence axes, and inflorescence branches.

When inflorescence axes and branches are present, or well developed, bracts and bracteoles are usually present. In some cases bracts and bracteoles may be useful for taxonomic purposes and particularly the margins of these structures, which can be entire, fimbriate, or serrate, depending on the species.

Flowers

Like most *Psychotria*, the flowers of the Philippine species are heterostylous, with long styled (pin) and short styled (thrum) flowers. We have attempted, where possible, to recognize this condition

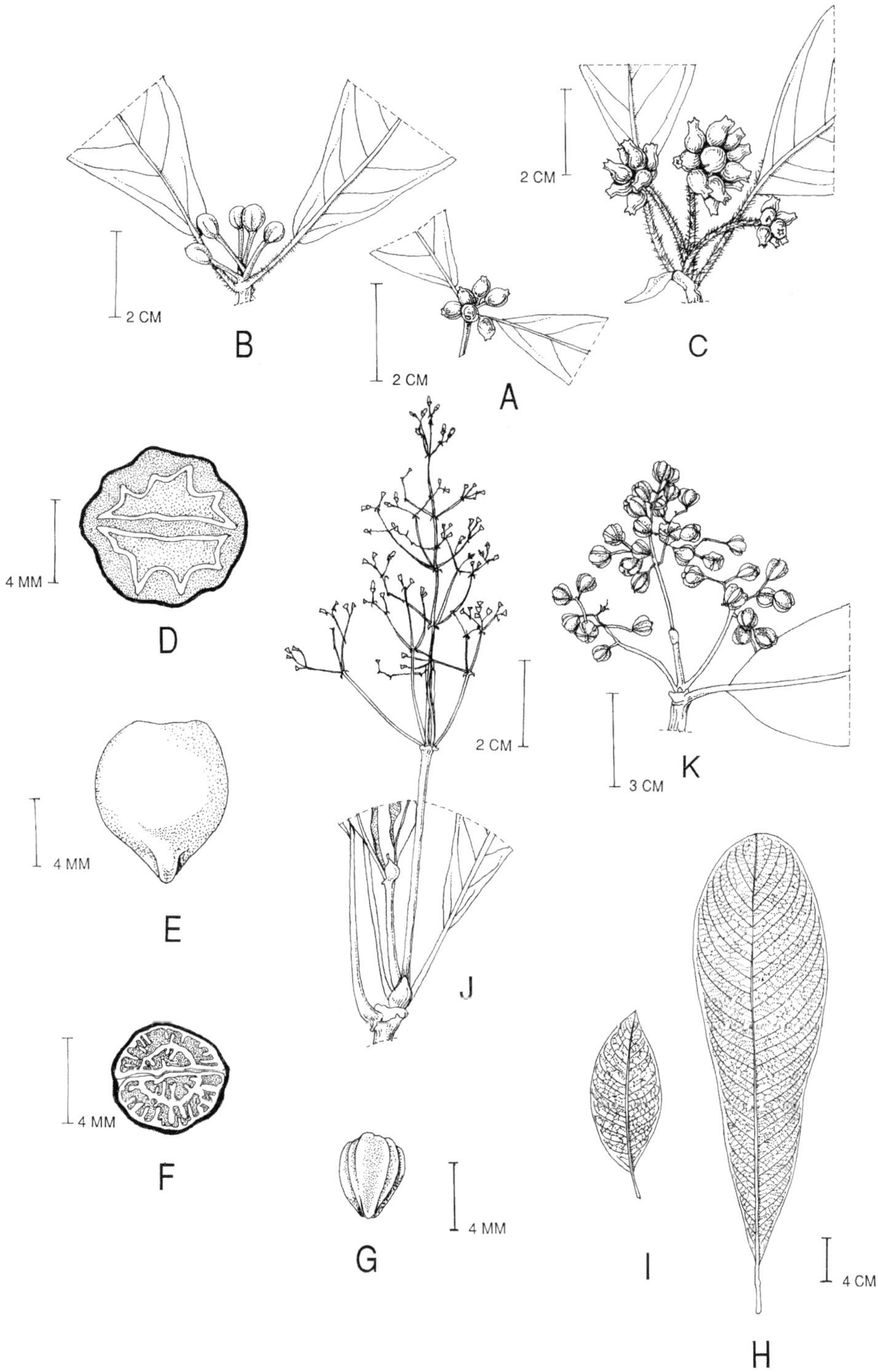

Fig. 1. Taxonomic characters helpful in delimiting species groups in Philippine *Psychotria*. **A,** inflorescence sessile with flowers ± sessile (fruit shown): *P. nagapatensis* (*Ramos BS 33193*) (A); **B,** inflorescence sessile with individual flowers pedicellate (fruit shown): *P. madulidii* (*Edaño BS 78408*) (A); **C,** inflorescence trichotomous (sessile), flowers sessile (fruit shown) and relatively large: *P. pilosella* subsp. *pilosella* var. *samarensis* (*Sulit PNH 14346*) (A); **D,** endosperm smooth: *P. boholensis* (*Ramos BS 43201*) (PNH); **E,** pyrene wall smooth: *P. boholensis* (*Ramos BS 43201*) (PNH); **F,** endosperm ruminate: *P. oconnori* (*Edaño BS 78400*) (A); **G,** pyrene wall with ribbing: *P. galintanensis* (*Ramos & Edaño BS 48899*) (UC); **H,** leaf blade large, usually with more than 20 lateral veins per side: *P. conglomeratiflora* (*Sulit PNH 6306*) (PNH); **I,** leaf blade small with less than 20 lateral veins per side: *P. frakei* (*Frake PNH 38038*) (A); **J,** inflorescence monochotomous (pedunculate): *P. rubiginosa* var. *jacobsii* (*Jacobs 7847*) (L); **K,** inflorescence (infructescence shown) trichotomous (sessile): *P. catanduaniensis* (*Ramos & Edaño BS 75589*) (NY).

and where possible to indicate whether the measurements for floral parts in the descriptions are from pin or thrum individuals. Flower parts vary from, 4-6-merous, and rarely up to 8-merous, but most species have 4 or 5 merous flowers. The flowers are either sessile or pedicellate. The pedicel, if present, is usually between 1-3 mm.

The calyx and the hypanthium are usually indistinguishable from one another and so we have cited measurements for those parts together (as calyx (incl. hypanthium)), and a separate measurement for the calyx lobes. The calyx (incl. hypanthium) is usually ca. 1 mm long and wide, rarely more than 2 mm long and wide, although in one taxon (*P. paloensis* Elmer var. *bukidnonensis* Sohmer & A.P. Davis), it is over 4 mm wide. At the apex of the calyx there is a very short (e.g. ca. 1 mm long) tubular limb, to which the calyx lobes are attached. Sometimes the calyx lobes are more or less obsolete, and the limb is repand (i.e. a slightly uneven to undulate margin).

The corolla of most species is white, which seems to be the most common color for *Psychotria*. Otherwise the corolla may be white with hints of other colors, or pale yellow and pale green. The following colors have been recorded for Philippine *Psychotria*: light cream, yellowish white, greenish to greenish white, yellowish green, or greenish yellow. The corolla tube varies in length from less than 1mm, up to 9 mm, but is usually less than 5 mm long, and most often 2-3 mm long. The shape of the corolla tube is variable but is normally short-tubular, usually widening towards the apex, or in some species it is more or less campanulate. The external surface is generally glabrous, although there are exceptions where the tube is pubescent (e.g. *P. ramosii* Sohmer & A.P. Davis). In most species the internal surface of the corolla contains a ring of relatively long, somewhat crisped hairs at the throat, or somewhere between the mid-point of the corolla length and just below the throat, depending on whether the flower is long or short styled. The corolla lobes are usually as long as the tube, but this varies among taxa. The texture of the corolla, thick or thin, has not been systematically studied as the flowers of many taxa are unknown to us.

The stamens are located within the corolla, with the filaments fixed within the ring of corolla throat hairs (if present). If the ring of corolla throat hairs is absent, the filaments are fixed to the corolla either just below the apex of the throat or above the mid-point of the corolla tube length. Sometimes the filaments are inserted below the mid-point, but to our knowledge never occur near or at the base of the corolla. The stamens are mostly basifixed, or fixed slightly above the base (supra-basifixed). The anthers are generally ellipsoid to linear, and usually 1-3 mm long. The filaments are generally 1-3 mm long, although the length varies considerably depending upon whether stamens are long or short styled flowers. The style varies from 1 mm to more than 4 mm long, again depending upon whether it is from a short or long styled flower; the style has two lobes at the apex.

Without a great deal of time spent in the field studying a few, or at most several species, it is impossible to come to a meaningful understanding of the floral biology in Philippine *Psychotria*. Fieldwork for this study (by S.H. Sohmer) was invariably undertaken during the fruiting season, and populations visited were almost all under a great deal of stress in increasingly disturbed habitat with often only a few individuals available for study.

Fruit

Philippine *Psychotria* have small fleshy, indehiscent fruits, containing two pyrenes (Fig. 2). The pyrene walls are composed of endocarp tissue and so the structure is botanically known as a drupe. Each pyrene contains a single seed. This fruit type is the same as most other species of *Psychotria*. The size of the fruit varies within given parameters, generally from between 4-5 mm wide, and up to ca. 12 mm in length. The shape may be globose, obovoid, turbinate or ellipsoid. The turbinate fruit may be associated with pyrenes that have a pronounced 'foot' connecting them to the base of the fruit. The color of the ripe fruit is generally red, orange or black in the terrestrial, non-climbing Philippine *Psychotria* species, and white in the climbing taxa. Fruits are usually glabrous, but sometimes pubescent, and they are smooth if the pyrenes they contain do not have ridges or ribs on the outer (dorsal/abaxial) surface. Otherwise, if the pyrenes have ridges or ribs on their backs, the fruit will show this externally when dry (e.g. on herbarium specimens). The distinctly ribbed or ridged fruits present on herbarium specimens often provide a good spot character for the identification of *Psychotria*.

In outline shape the pyrenes are usually more or less obovate but may also be circular to nar-

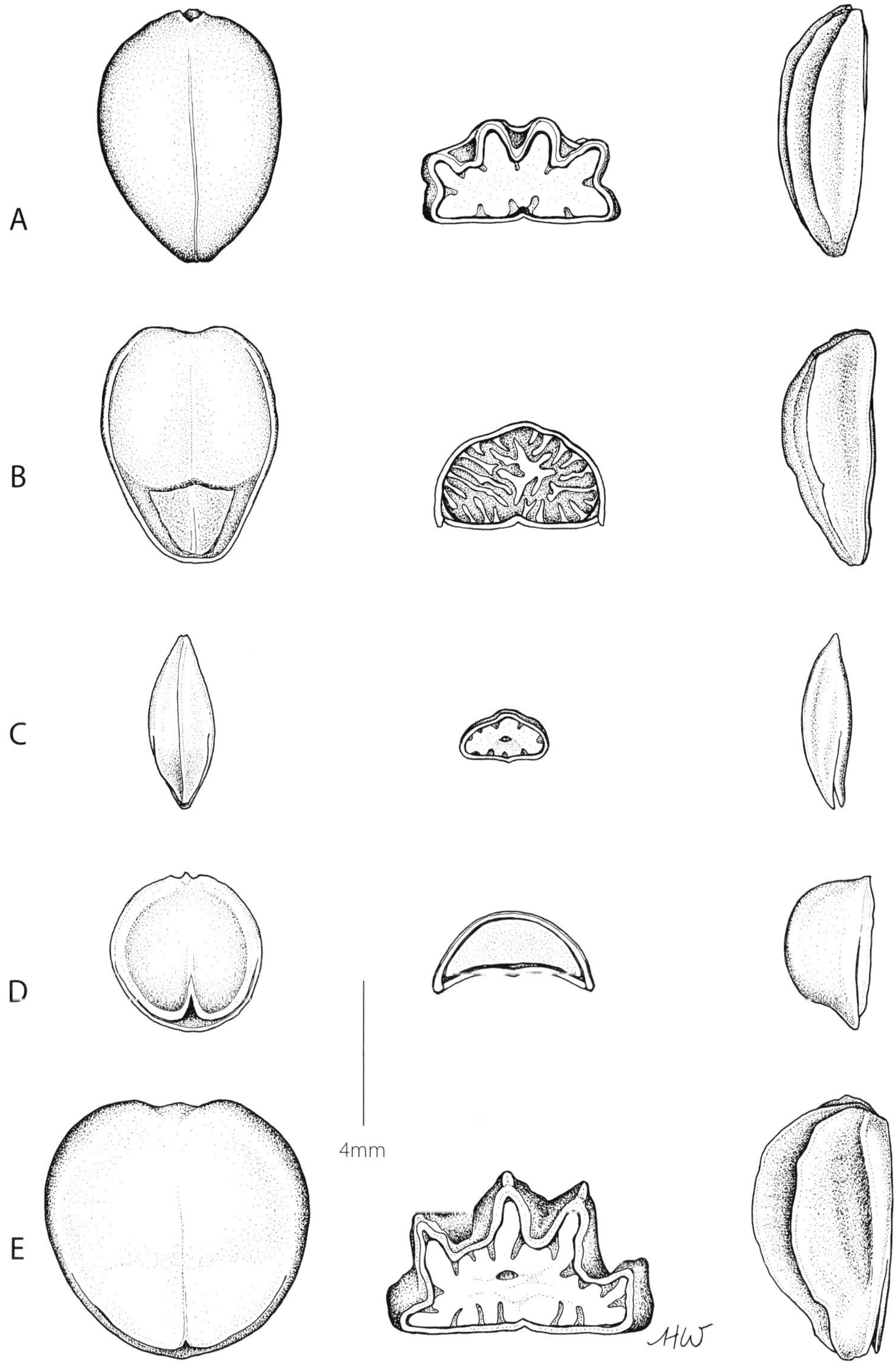

Fig. 2. Pyrene and seed characters in different species of Philippine *Psychotria*, showing details of pyrene shape, preformed germination slits (PGSs), and endosperm. From left to righ: basal view of pyrene; transverse section (T.S.) through pyrene; side view of pyrene. A, *P. manillensis* (*Weber 1561*)—pyrene lacking PGSs, endosperm slightly ruminate (small invaginations on lower surface and between ribs); B, *P. papillata (McGregor BS 43850)*—marginal PGSs from base to near apex, endosperm distinctly ruminate; C, *P. obscurinervia* (*Elmer 18294*)—marginal PGSs from base to middle of pyrene, endosperm slightly ruminate (small invaginations on lower and upper surfaces); D, *P. membranifolia* (*Gaerlan & Chavez PPI 26351*)—marginal PGSs from base to middle of pyrene and with a very short (cleft) median PGS, endosperm entire; E, *P. reynosoi* (*Barbon, Romero & Fuentes PPI 13019*)—marginal PGSs ± basal, endosperm ruminate. Scale bar = 4 mm.

rowly elliptic. The apex is generally rounded and the base acute. In transverse section the inner surface (ventral/adaxial) is more or less flat, although in the Membranifolia Species Group it is concave; the outer surface (dorsal/abaxial) of all species is convex (usually semicircular). The outer surface of the pyrene may be smooth, ribbed or ridged and is of taxonomic importance. For example, in the Luzoniensis Species Group several species can be separated on the basis of strongly ridged pyrenes.

Preformed germination slits (PGSs) are of prime systematic importance for *Psychotria* and the tribe Psychotrieae in general (Petit 1964; Robbrecht 1988, 1989; Nepokroeff et al. 1999; Andersson 2001; Davis et al. 2001; Piesschaert 2001; Andersson 2002). Presence versus absence and position of PGSs provide most of the systematic data. Nearly all Philippine *Psychotria* possess PGSs, which are of the same kind as the type species of *Psychotria*, *P. asiatica* L. (Davis et al. 2001). The PGSs of Philippine *Psychotria* are located on the inner (adaxial) surface of the pyrene, running along the inside edges of the margins, from the base of the pyrene for 1/2 to 2/3 of pyrene's length, although in some cases they continue all the way to the apex. When the pyrene is mature the inner (adaxial) part of the pyrene can be easily lifted upwards with a pair of fine forceps, like the lid of a box, exposing the seed. Only two species of Philippine species lack PGSs, viz. *P. manillensis* and *P. barkleyi* Sohmer & A.P. Davis, although we have not, of course, been able to sample those species that are only known from flowering specimens.

The seeds are encased within the pyrene wall (endocarp) and are roughly the same shape as the pyrenes but are obviously smaller. In all species examined by us the endosperm and seed coat is a darkish color (e.g. brown or red-brown), apart from those in the Membranifolia Species Group (*P. balabacensis* and *P. membranifolia*), which are always pale colored. As with preformed germination slits (PGSs), systematic significance has been given to the presence vs. absence of an ethanol soluble seed coat (Petit 1964; Piesscaheart 2001; Andersson 2002). It is easy to ascertain whether a seed has an ethanol soluble pigment within the seed coat (Seed Coat Pigment (SCP) positive), by immersing a single seed in ethanol (50–70%) for ca. 5 minutes, at room temperature. If the ethanol turns light to dark red an SCP has been detected (SCP positive). Of the samples examined for the absence of an ethanol soluble SCP, most were positive. A notable exception is the Membranifolia Species Group, which tested as SCP negative (ethanol solution clear); these two species also have entire endosperm (see below).

Like most *Psychotria* species, nearly all Philippine species possess ruminate endosperm (see Fig. 1d). The rumination is manifest as shallow to deep invaginations. In species with non ruminate endosperm the seed surface is entire (smooth), although the seed may still be furrowed or ridged like the pyrene (see Fig. 1d). The Membranifolia Species Group is, in part, characterized on the presence of non ruminate (entire) endosperm. Non ruminate endosperm can be a good key characteristic for the separation of species, and can be used with other characters for the circumscription of species and species groups (e.g. the Membranifoia Species Group). It is not, however, a character that can be used on its own for the characterization of genera. Likewise, the presence of ruminate endosperm is not a character of principal systematic value. The genus *Grumilea* was based almost entirely on the presence of ruminate endosperm, but its variability in Philippine *Psychotria* (and *Psychotria* in general) led to the relatively early demise of this genus.

COMMENTS ON THE SYSTEMATIC POSITION OF PHILIPPINE *PSYCHOTRIA*

Considerable study has been focused on the problems concerning the systematics of the tribe Psychotrieae and in particular the delimitation of *Psychotria* (e.g. Andersson & Rova 1999; Nepokroeff, Bremer & Systma 1999; Andersson 2001; Piesschaert 2001; Davis et al. 2001; Andersson 2002). There has been a suggestion that *Psychotria* should become more inclusive, by adding more species from other Psychotrieae genera (Nepokroeff et al. 1999), whereas others have suggested a mixture of addition and division (Andersson 2002). For this revision the pertinent question is: do Philippine *Psychotria* species really belong in *Psychotria*? The answer is, categorically, yes. Work on typification has provided a clear definition of *Psychotria* in the very strictest sense (Davis et al. 2001), and Philippine *Psychotria* falls within this narrow circumscription. It is unlikely that Philippine *Psychotria* will be relocated in another genus after further research.

A definitive characterization of *Psychotria* is difficult as there are very many species (c. 2000) and morphological exceptions are frequent, although the following characterization works in most cases. *Psychotria* may be defined by the possession of the following characters: stipules usually caducous (falling); inflorescences terminal; corolla usually small (< 5 mm long); anthers inserted within the ring of corolla throat hairs (if hairs present); pyrenes usually with ribs or ridges on the outer (dorsal/abaxial) surface; pyrenes with two marginal preformed germination slits (PGSs); seed coat usually soluble in 70% ethanol (SCP positive); and endosperm usually ruminate. The assumption that marginal preformed germination slits (PGSs) are rare in *Psychotria*, and that the absence of this feature may be used as a defining character for the genus (Andersson 2002) is not supported by examination of either the type species (*P. asiatica*; see Davis et al. 2001) or other species from Asia (including the Philippines) and Australasia (A.P. Davis, unpubl. data)

SPECIES GROUPS IN PHILIPPINE *PSYCHOTRIA*

Based on gross morphology alone, Philippine *Psychotria* can be divided into nine species groups (Table 1). These nine groups provide a practical means for subdividing Philippine *Psychotria* into more manageable units, so that any single species can be placed into a smaller group of species and then more easily identified to species, subspecies or variety. Without a full systematic study of *Psychotria* it would be impossible to say whether any of the species groups represent natural (monophyletic) units, and for this reason we have not given them formal taxonomic status.

The principal characteristics for the species groups involve habit features of the leaves, stipules, nature of the inflorescence, flowers, and fruits (including pyrene characters). Table 2 provides a summary of the relevant characters and their distribution, and gives the number of species presently recognized in each species group. A short discussion on each of the species groups is given below, and a general description is given before each species group section in the main body of the text.

I. Ixoroides Species Group (p. 19)

The Ixoroides Species Group is the most easily characterized, as they are all climbers. Of the species studied in the field (S.H. Sohmer) all of them climbed by means of adventitious roots. Some species known only from herbarium material are reported as being scandent or climbing shrubs. Further details on the nature of climbing in Philippine *Psychotria* are given above in the section above on habit.

It seems that all species of the Ixoroides Species Group have white fruits, although the fruit color of some species is as yet unknown. To our knowledge no other species of Philippine *Psychotria* has white fruits except those in the Ixoroides Species Group. There is a single report of white fruit for *P. boholensis* Merr. ex Sohmer & A.P. Davis (Gitingensis Species Group) but this record is very dubious because other collections of this species have orange or red fruit. The predominant color found in the other eight groups is red. Other species are recorded as having yellow, orange, purple or black fruits, although yellow and orange fruits quite often turn red at maturity.

Within the Ixoroides Species Group there appear to be two subgroups: those with trichotomous inflorescences (three main axes at the apex of the vegetative shoot, the middle of these the longest) and those with monochotomous (pedunculate) inflorescences where there is a single axis (the peduncle) at the apex of the shoot.

The monochotomous subgroup is by far the larger of the two subgroups with 16 of the 22 species usually having this type of inflorescence. The monochotomous subgroup is exemplified by *P. ixoroides*, which is (or was, in some cases) found on nearly all the principal islands of the archipelago. The basic morphology of the species in this subgroup is relatively uniform.

The trichotomous subgroup appears to be morphologically more diverse, even though there are fewer species. Two of the species, *P. cuernosensis* Elmer and *P. elliptifolia* Elmer, are distinguished from the rest by short contracted inflorescences. All of the others in this subgroup have large, spreading, diffuse inflorescences. Overall habit distinguishes *P. diffusa* Merr. from species such as *P. ramosissima* Elmer. Whereas the former is small, climbs at most a few meters and can, depending on the environment, be more shrubby or more vine-like, and the latter is robust and apparently can climb to the tops of large trees where large tangled masses of branches (bearing the inflorescence) are produced.

TABLE 1. Species groups in Philippine *Psychotria*. Taxa are arranged alphabetically within their respective species groups. The *Taxon number* refers to the number used for each species as used in the main body of text.

Taxon	Taxon number	Taxon	Taxon number
	I. Ixoroides Species Group		
P. barkleyi	6	*P. lianoides*	19
P. cadigensis.	3	*P. mindoroensis*	13
P. christarusselliae	14	*P. obscurinervia*	12
P. cuernosensis	17	*P. olsenii*	1
P. diffusa	22	*P. paracalensis*	2
P. diffusa var. *agusanensis*	22c	*P. paucinervia*	11
P. diffusa var. *cervina*	22b	*P. piperi*	5
P. diffusa var. *diffusa*	22a	*P. radicans*	8
P. elliptifolia	18	*P. ramosissima*	20
P. elmeriana	15	*P. rigidaefolia*	7
P. iwahigensis	10	*P. sarcocarpa*	9
P. ixoroides	16	*P. sibuyanensis*	4
P. lanaensis	21		
	II. Linearis Species Group		
P. cephalophora	25	*P. longipedicellata*	26
P. cuneata	27	*P. madulidii*	24
P. linearis	23	*P. yagawensis*	28
	III. Tayabensis Species Group		
P. amaracarpoides	32	*P. tayabensis* var. *castanea*	36b
P. nagapatensis	34	*P. tayabensis* var. *euphlebia*	36c
P. papillata	29	*P. tayabensis* var. *fasciculiflora*	36d
P. subcucullata	31	*P. tayabensis* var. *pinagteponanensis*	36e
P. subsessiliflora	35	*P. tayabensis* var. *tayabensis*	36a
P. subsessiliflora var. *carinata*	35b	*P. urdanetensis*	33
P. subsessiliflora var. *subsessiliflora*	35a	*P. yenii*	30
P. tayabensis	36		
	IV. Pillosela Species Group		
P. arborescens	41	*P. pilosella* subsp. *pilosella*	48a
P. bataanensis	38	*P. pilosella* subsp. *pilosella* var. *pilosella*	48aa
P. castroi	43	*P. pilosella* subsp. *pilosella* var. *samarensis*	48ab
P. elliptilimba	39	*P. pulgarensis*	46
P. lancilimba	47	*P. pygmaea*	40
P. pilosella	48	*P. scaberula*	37
P. pilosella subsp. *erythrotricha*	48b	*P. sulitii*	45
P. pilosella subsp. *erythrotricha* var. *camarinensis*	48bb	*P. surigaoensis*	44
P. pilosella subsp. *erythrotricha* var. *erythrotricha*	48ba	*P. wenzelii*	42
	V. Membranifolia Species Group		
P. balabacensis	50	*P. membranifolia*	51
	VI. Magnifolia Species Group		
P. alzapaniensis	54	*P. conglomeratiflora*	52
P. amplissima	55	*P. magnifolia*	53
	VII. Gitingensis Species Group		
P. alabatensis	63	*P. gitingensis*	72
P. boholensis	58	*P. iridensis*	59
P. cardiophylla	73	*P. manillensis*	62
P. crassifolia	71	*P. merrittioides*	64
P. crassifolia var. *crassifolia*	71a	*P. mindanaensis*	68
P. crassifolia var. *culionensis*	71b	*P. palawanensis*	60
P. foxworthyi	65	*P. panayensis*	66
P. frakei	67	*P. phanerophlebia*	57

Table 1. continued

Taxon	Taxon number	Taxon	Taxon number
	VII. Gitingensis Species Group		
P. pyramidata	69	*P. rubiginosa* var. *jacobsii*	56b
P. quisumbingiana	61	*P. rubiginosa* var. *rubignosa*	56a
P. rubiginosa	56	*P. sorsogonensis*	70
	VIII. Luzoniensis Species Group		
P. banahaensis	87	*P. negrosensis*	94
P. banahaensis var. *banahaensis*	87a	*P. nitens*	99
P. banahaensis var. *montalbanensis*	87b	*P. pallidifolia*	89
P. bayombongensis	93	*P. paloensis*	91
P. bontocensis	75	*P. paloensis* var. *bukidnonensis*	91b
P. capizensis	78	*P. paloensis* var. *ovatifolia*	91c
P. casiguraensis	96	*P. paloensis* var. *paloensis*	91a
P. catanduaniensis	86	*P. paloensis* var. *subelliptifolia*	91d
P. cordatula	74	*P. paloensis* var. *wenzeliana*	91e
P. dingalanensis	79	*P. patulinervia*	95
P. fenicis	82	*P. pinnatinervia*	84
P. galintanensis	88	*P. plumeriaefolia*	98
P. hilonghilongensis	103	*P. pubilimba*	85
P. lagunensis	80	*P. ramosii*	77
P. longissima	83	*P. reynosoi*	90
P. luconiensis	92	*P. velutina*	81
P. mariguidonensis	97	*P. weberi*	102
P. mendozii	100	*P. yatesii*	101
P. merrittii	76		
	IX. Subalpina Species Group		
P. acuminatissima	108	*P. palimlimensis*	106
P. gracilipes	107	*P. pauciflora*	111
P. ilocana	104	*P. sohotonensis*	109
P. loheri	112	*P. subalpina*	110
P. macgregorii	105		

II. Linearis Species Group (p. 58)

This group of seven species is easily recognized by its much reduced inflorescences, which lack a peduncle and any discernable inflorescence axes or branches, and consists only of individual flowers and their pedicels. The Linearis Species Group differs from the Tayabensis Species Group in that the flowers are usually pedicellate (i.e., possessing pedicels).

Psychotria linearis Bartl. ex DC. is certainly the most common species of the Linearis Species Group. Members of the group are distinguishable by the size and shape of the leaf blades, and particularly the shape of the blade at the base. Most of the taxa are pubescent, some extremely so. Tubular stipules, where the margins are fused to form a tube, are also frequent here, and a consistent feature of *P. linearis*. This may provide a further characteristic for distinguishing the Linearis Species Group, but as many specimens lack stipules we have not been able to fully evaluate this character.

III. Tayabensis Species Group (p. 66)

Like the Linearis Species Group, this group of eight species has very reduced inflorescences. The inflo-rescence consists only of individual flowers, lacking a peduncle and any discernable axes or branches. Unlike the Linearis Species Group, the inflorescences of the Tayabensis also lack pedicels, although sometimes the fruits become pedicellate as they mature. The stipules are never tubular. The most common species in this group is *P. tayabensis* Elmer, which is very widely distributed and consists of five varieties. The variation within this taxon centres on the size and venation of the leaf blades.

Within the Tayabensis Species Group morphological variation is extensive, particularly when

Table 2. Summary of characteristics delimiting species groups in Philippine *Psychotria*. Key: + = character present ; - = character absent; +/- = equivocal.

	Species Group	Climbing habit	Dry branchlet & leaf color yell.-gr.	Stipules	Leaf blade usually < 3 cm wide	Leaf blade >30 cm long	Leaf blade membran.	Inflores-cence parts greatly reduced	Inflores-cence type	Flowers
I	Ixoroides	+	+/-	valv.	+/-	-	-	-	mon./tric.	ped.
II	Linearis	-	-	valv., rarely imbr.	+/-	-	-	+	—	ped.
III	Tayabensis	-	-	valv., rarely calyp.	+/-	-	-	+	—	sess.
IV	Pilosella	-	-	valv.	+/-	-	-	-	tric./mon.	sess., rarely ped.
V	Membranifolia	-	+	valv.	-	-	+	-	mon.	pedicel.
VI	Magnifolia	-	-	imbr.	-	+	-	-	mon./tric.	ped.
VII	Gitingensis	-	-	valv./imbr.	+/-	+/-	-	-	mon./rarely tric.	ped., rarely sess.
VIII	Luzoniensis	-	+/-	imbr./valv.	+/-	+/-	-	-	tric. (& >), rarely mon.	ped.
IX	Subalpina	-	-	valv./valv.-. imbr./imbr	+	-	-	-	tric. (& >)	ped.

Table 2. (continued)

	Species Group	Fruits white	Inner (adaxial) surface of pyrenes	Ethanol soluble/ SCP present	Endosperm ruminate (+) entire (-)/	No. of species
I	Ixoroides	+	flat	+/rarely -	+/-	22
II	Linearis	-	flat	+/rarely -	+/-	6
III	Tayabensis	-	flat	+/rarely -	+	8
IV	Pilosella	-	flat	+/rarely -	+/rarely -	13
V	Membranifolia	-	concave	-	-	2
VI	Magnifolia	-	flat	+/rarely -	+	4
VII	Gitingensis	-	flat	+/rarely -	+	17
VIII	Luzoniensis	-	flat	+/rarely -	+/-	30
IX	Subalpina	-	flat	+/rarely -	+	10

Abbreviations: **calyp.** = calyptrate; **imbr.** = imbricate; **gr.** = green; **membran.** = membranaceous; **mon.** = monochotomous; **ped.** = pedicellate; **sess.** = sessile; **tric.** = trichotomous; **valv.** = valvate; **yell.** = yellow.

compared to some of the other species groups. The commonality of inflorescence morphology and the presence of other features gives this group a practical value, although it is doubtful that this group has a shared common ancestry. *Psychotria amaracarpoides* (Merr.) Merr. has a solitary fruit, and its inflorescence probably consists of a single flower, although flowers are as yet unknown for this species.

IV. Pilosella Species Group (p. 88)

The Piloesella Species Group comprises 13 species and is primarily characterized by trichotomous inflorescences bearing sessile flowers at the apices of the primary inflorescence axes (main branches). It may well be a polyphyletic group. This group is exemplified by *P. pilosella*, herein considered as comprising two subspecies with two varieties in each. The presence or absence of pubescence is a major characteristic in this group and roughly divides it into two parts.

V. Membranifolia Species Group (p. 114)

This is the smallest group in Philippine *Psychotria*, containing only two species: *P. membranifolia* and *P. balabacensis*. It seems to represent part of a distinct group of species, which is potentially monophyletic if species from other parts of Asia and perhaps elsewhere are included. This group of *Psychotria* is represented by considerably more diversity in New Guinea, for example. *Psychotria membranifolia* is widespread and found throughout the Malesian area but *P. balabacensis* is endemic to Palawan. However, further research may show that *P. balabacensis* is conspecific with one or more species belonging to this group outside the Philippine archipelago. The morphological variation of the two Philippine species is not fully understood at the present time.

The main characteristics for the Membranifolia Species Group (and other closely allied species from other parts of Asia) are: 1) leaves membranaceous or sometimes chartaceous; 2) leaves and young twigs usually yellow-green in color, when dry (as opposed to the gray-green, olivaceous green, brown, chocolate-brown, dark orange-brown, or reddish brown colors of most other *Psychotria*. 3) pyrenes usually with a smooth outer (abaxial/dorsal) wall, although sometimes ribbed or strongly and sharply ridged; 4) pyrenes with a concave inner (adaxial/ventral) surface, the centre having a weak longitudinal ridge (i.e. the inner surface is more or less, weakly M-shaped; 5) endosperm non ruminate (entire); 6) seed coat lacking a soluble seed coat pigment in 70% ethanol (SCP negative). Another basic phytochemical property is that when specimens of this group are preserved in alcohol they never turn black or discolor the solution, whereas many other *Psychotria* turn black or blackish in a fixing solution (such as 70% alcohol or FAA). This feature often readily permits identification of herbarium specimens collected in alcohol.

Species potentially related to the Membranifolia Species Group, occurring in western parts of Malesia through to New Guinea, the Philippines and the Pacific, require extensive systematic study. We need to understand the relationships between the constituent species and elucidate their systematic position within *Psychotria* and perhaps the Psychotrieae.

VI. Magnifolia Species Group (p. 119)

The Magnifolia Species Group comprises four species and is characterized by large stipules, large leaves, and frequently complex inflorescence structure, although they can be viewed as large-leaved representatives of the Luzoniensis Species Group (see below). The stipules are usually around 3–6 cm long but are sometimes up to 10 cm long. Most leaf blades in the group are over 30 cm long with more than 20 lateral (secondary) veins per side. One of the species, *P. conglomeratiflora* Sohmer & A.P. Davis, has an inflorescence with the original trichotomous branching pattern converted to a pseudo-monochotomous condition via fusion of the three previously distinct axes. The most commonly represented species in herbaria is *P. amplissima* Merr. *Psychotria magnifolia* Merr. has inflorescences with flowers clustered directly at the nodes of the primary axes.

VII. Gitingensis Species Group (p. 125)

The 17 species in this group are characterized by monochotomous (pedunculate) inflorescences, although in some species the inflorescence is rarely trichotomous. Within the group the nature of the stipules appears (insofar as they are known) to allow the species to be grouped. One subgroup is characterized by valvate stipules, and the other by imbricate stipules. Stipular features are potentially very informative but are not available on all specimens. More field work is necessary to complete our knowledge of stipule morphology for this group.

VIII. Luzoniensis Species Group (p. 162)

With 30 species, this is the largest group in Philippine *Psychotria*. Most of the species possess trichotomous inflorescences and pedicellate flowers. Around 20 species of this group form a rather uniform morphological assemblage, which can be separated into two smaller subgroups of unequal size on the basis of their pyrenes. In one subgroup of species the pyrenes are prominently ridged on the back, and in the other they are smooth or irregularly ribbed. The former group is exemplified by *P. banahaensis*, and the latter by *P. luzoniensis*.

Psychotria luzoniensis is by far the most common Philippine *Psychotria* species in herbaria. Whether this means *P. luzoniensis*is is indeed the most ubiquitous species in the Philippines or that it is (or was) found in habitats more accessible to collectors, is not known. We suspect that it is a combination of both, as the species is found in relatively disturbed situations and in secondary forest. The species morphologically grouped around *P. luzoniensis* are the most difficult to separate from each other. Dry leaf blade color as well as the presence or absence of prominent tertiary and quaternary venation patterns on the lower leaf surface are important characteristics (easily examined with a ×10 hand lens).

IX. Subalpina Species Group (p. 219)
This group of 10 species is similar to the Luzoniensis Species Group in that it has trichotomous inflorescences, although the leaves of the Subalpina Species Group are significantly smaller. Species within this group appear to be taxa adapted either to higher elevation wet forests or to more exposed, drier habitats, such as limestone outcrops and ridges. They possess leaf blades that are usually less than 3 cm wide and less than 10 cm long. We suspect that the species owe more of their uniformity to ecological adaptation than to common or close ancestry.

SPECIES DISTRIBUTION

Nearly all of the taxa treated in this revision are endemic to the Philippines. The exceptions are: *P. crassifolia* Miq., which is found in Borneo; *P. membranifolia*, recorded from New Guinea (Sohmer 1988); and *P. manillensis*, which is also found in Japan (Nansei-shoto [Ryuku Islands]) and Taiwan. It also likely that *P. gitingensis* Elmer is resident in Borneo. Taxonomic revision of *Psychotria* in SE Asia, and particularly Borneo, may possibly identify other species that are presently endemics of the Philippines. The most likely candidates are species occurring throughout the Philippines, including species of secondary forest and those in the southern most parts of the Philippine archipelago.

EXTINCTION AND EXTINCTION THREAT IN PHILIPPINE *PSYCHOTRIA*

In this work we provide provisional conservation assessments for all species and varieties using the criteria of the IUCN Red List Categories (IUCN 2001). We used the extent of occurrence (EOO) as the baseline measurement for all conservation assessments. EOO was approximated for each taxon using a minimum convex polygon to cover the inferred sites of present occurrence. Larger scale discontinuities and/or disjunctions within the overall distribution were excluded, for example sea and areas obviously devoid of forest cover. When producing the conservation assessments we considered two main factors: (1) that the Philippines have been largely deforested and (2) that most *Psychotria* species exist only in primary vegetation (A. Davis, S.H. Sohmer, pers. obs.). Myers et al. (2000) estimate that remaining primary vegetation in the Philippines amounts to 3%, although the amount of old growth forest (in which most species of *Psychotria* exist) is undoubtedly far lower. Bantayan (2001), for example, estimates that old growth forest is down to ca. 800,000 hectares. The situation is unlikely to improve: continuing decline can be projected for the area and extent of species ranges, quality of habitat, and number of populations and individuals for all plants in the Philippines due to deforestation and changing land use.

Three further factors were taken into consideration when considering taxa that may be extinct. (1) Lowland primary forest has all but gone in the Philippines except for Palawan and a few small forest remnants in poor condition elsewhere. We therefore used a limit of below 500 m to quantify lowland forest. (2) Pre 1925 was used as a cut-off date to define an era when the Philippines was more than two-thirds forested. (3) A single collection (botanical specimen) was used as the lowest possible value for a taxon range. The first two of these criteria are deliberately cautious: 700 m or slightly higher may be a more realistic limit for 'lowland' forest (e.g. Whitmore 1990), and in reality the Philippines may have been two-thirds forested up to 1945 (IUCN 1992). However, we believe that it is better to be cautious when assessing extinction, as the discovery of a single individual removes a taxon from the Extinct category, for example. The above criteria were used in the following manner: species known only from one collection; not collected after 1925; and recorded below (or assumed to be below) 500 m altitude are assessed as Extinct. We believe this gives a robust estimate of 'reasonable doubt that the last individual has died' (IUCN, 2001). Taxa failing to fulfil just one of the above criteria, within reasonable limits (i.e. not collected after 1945, known from above 500 m altitude, and known from two or three collections) and otherwise fulfilling the IUCN criteria for Critically Endangered, were assessed as Critically Endangered/possibly Extinct. We believe that taxa in this intermediate category have a reasonable chance of being extinct.

Information on areas of remaining primary forest and good condition secondary forest was taken from broad-spectrum literature sources (e.g. IUCN 1992; Davis et al. 1995; see also http://www.nationalgeographic.com/wildworld/terrestrial.html), some specific literature related to animals (e.g. Danielsen & Treadway 2004), low resolution satellite imagery, the Internet (for information on specific protected areas), and personal observation (S.H. Sohmer).

For all other extinction threat categories we recorded (e.g. from herbarium labels) whether a species is able to survive in secondary forest, as opposed to most Philippine *Psychotria* which only exist in good to reasonable quality primary forest. We assume that any remaining forest remnant, regardless of quality, could potentially contain *Psychotria* species able to exist in secondary forest.

On the basis of the provisional IUCN extinct threat assessments produced we estimate that 10 (ca. 9%) species are Extinct (EX), 34 (ca. 30%) are Critically Endangered (CR), 19 (ca. 17%) are Endangered (EN), 20 (18%) are Vulnerable (VU), 21 (19%) are Near Threatened (NT), and eight (ca. 7%) are Least Concern. The total number of species either extinct or within a threat category (CR, EN, VU, NT) is ca. 74%. The 34 species in the Critically Endangered category includes 21 species (ca. 19%) of those we have assessed as Critically Endangered/possibly Extinct (see above and main text). If one assumes that the species we have as Critically Endangered/possibly Extinct are actually extinct, then the number of Extinct species could be as high ca. 28%. It is our feeling that ca. 9% is a conservative estimate for extinction in Philippine *Psychotria*, whereas ca. 28% is likely to be a more realistic figure for extinction.

Our estimate of 0–28% for extinction in Philippine *Psychotria* species is lower than that given by Sohmer (2001) of ca. 50%, although the latter figure could be valid as an upper limit or worst case scenario. Koopowitz et al. (1998) used an earlier version of this manuscript (before 1990) to study extinction in the Philippines by using *Psychotria* as the main data source. Depending on the deforestation scenario adopted, Koopowitz et al. give extinction rates of between 24 and 56% for the Philippine flora.

It is of course difficult to accurately measure extinction of species or species groups without extensive fieldwork. However, a recent plant collecting initiative may provide some independent confirmation of high species loss in Philippine *Psychotria*. The Philippine Plant Inventory project (PPI) was initiated in 1990, and in five years 17,049 specimens were collected (Madulid & Sohmer 1997); by the end of the project ca. 20,000 were collected (Sohmer pers. comm.). We do not know exactly how many specimens of *Psychotria* were collected during the PPI project but we have examined a total 255 herbarium specimens from this collecting series. The PPI project collected 43 of the 112 Philippine *Psychotria* species, in a ten year period between 1990 and 1997; 68 species were not recollected. Of the 25 species that were only known from one site (i.e. only known from the type specimen, see above) before 1990, only two species (*P. sarcocarpa* Merr., *P. iwahigensis* Elmer) have been rediscovered during the course PPI project. This leaves 23 species that have not been rediscovered since the first and only original collection.

It is our hope that the provisional conservation assessments given here will achieve two objectives. Firstly, it will give an overview of extinction and extinction threat for Philippine *Psychotria* and perhaps other flowering plants that exist predominately in primary forest and have narrow distributions (e.g. see Koopowitz et al. 1989). Secondly, it will highlight those species of *Psychotria* for which conservation effort is most greatly needed, such as targeting species for *in situ* and *ex situ* conservation.

Based on all the above data (including Sohmer 2001 and Koopowitz et al. 1998) the outlook for the survival of most Philippine *Psychotria*, and many other naturally occurring plants species in the Philippines species, does not look very good.

THE GENUS *PSYCHOTRIA* (RUBIACEAE) IN THE PHILIPPINE ARCHIPELAGO

Psychotria L., Syst. Nat. ed. 10, 2:929. 1759. nom. cons. TYPE SPECIES: *Psychotris asiatica* L. *Grumilea* Gaertn., Fruct. Sem. Pl. 1:138 (1788). TYPE SPECIES: *Grumilea nigra* Gaertn.

Shrubs, ***small trees***, ***trees*** or ***climbers***; raphides present. ***Young stems*** (branchlets) usually terete, smooth to longitudinally rugulose. ***Stipules*** caducous, or infrequently semi-persistent to ± persistent, falling to leave a ± naked scar or scar with colleters (hair- or bristle-like), imbricate, valvate, or rarely tubular, ± broadly ovate to narrowly triangular, or lanceolate, usually without surface features and lacking appendages, apex entire (usually acute to obtuse), or cleft (divided or bifid). ***Leaves*** petiolate, chartaceous to coriaceous, rarely membranaceous, midrib and secondary veins usually prominent, secondary veins ascending at an angle of 30°–60°, usually curved and often distinctly curved near the margin and usually joining to form a distinct marginal veins (brochidodromus) or secondary veins not joined (camptodromus), tertiary venation obscure to prominent, sometimes invisible. ***Inflorescences*** terminal, sometimes overtopped by the growth of the shoot, pedunculate (monochotomous) or sessile (trichotomous to many-chotomous), paniculate to simple (if simple often greatly reduced, with short axes and branches, or these lacking, and few-flowered); bracts at bracteoles usually present, bracts at branching points of inflorescence, usually triangular, margins entire or rarely serrate, bracteoles usually subtending flower(s)/fruit(s), usually ovate to triangular, margins entire or rarely serrate. ***Flowers*** heterostylous, with short styled (thrum-eyed) or long styled (pin-eyed) flowers, 4–6(–8)-merous, erect, sessile or pedicellate. ***Calyx*** (incl. hypanthium) ± campanulate to ± obconical; calyx limb tubular, usually ca. 1 mm long, usually with lobes, or limb repand (lobes obsolete); calyx lobes usually deltate to triangular. ***Corolla*** usually white, whitish or pale green/yellow, external surface usually glabrous; corolla tube usually short-tubular but widening towards the apex, sometimes ± campanulate, thin to quite fleshy; internal surface usually hairy, hairs restricted to a ring below the throat, either emergent and prominent on the outside or concealed (positioned further down the corolla); corolla lobes distinctly valvate, becoming reflexed, usually deltate to triangular, often thickened and fleshy at the apex. *Stamens* fixed just below the corolla lobe sinuses (short styled), or within the throat (long styled), always fixed within the internal ring of hairs (if present); filaments 1–3 mm long; anthers ellipsoid to linear, usually 1–3 mm long. ***Disc*** hemi-subglobose, entire or slightly 5-lobed. ***Style*** filiform; stigma 2-lobed. ***Fruit*** a drupe, globose, obovoid, turbinate or ellipsoid, succulent (when fresh), smooth or prominently longitudinally ribbed (when dry), normally glabrous, sessile to pedicellate, mature fruits usually red, orange or black (for non-climbing species), or white (climbing species); calyx usually persistent, obscure to prominent. ***Pyrenes*** ± circular to narrowly elliptic in outline view, apex rounded, base acute to broadly acute, in transverse section ± semicircular, inner (ventral/adaxial) surface flat or rarely concave, outer (dorsal/abaxial) surface convex and either distinctly ribbed/ridged or smooth; preformed germination slits (PGSs) present or sometimes absent, if present running along the margins of the pyrene, from the base to 1/2 to 2/3 of the total length of each pyrene. ***Seeds*** ± the same shape and size as the pyrene but slightly smaller, seed coat pigment usually soluble in 70% ethanol (producing a red solution), or sometimes insoluble (producing a clear solution); endosperm usually ruminate, rumination present on the ad- and abaxial surfaces, invaginations 0.5–0.7 mm deep, or endosperm entire (non ruminate).

KEY TO THE SPECIES GROUPS OF PHILIPPINE *PSYCHOTRIA*

1. Climbers or scandent shrubs, often climbing by means of adventitious roots; fruits (mature) white ____ **I. Ixoroides Species Group**
1. Shrubs or trees, never climbing or scandent; fruits (mature) usually orange or red and other colors except white.
    2. Inflorescences and/or flowers sessile: either (1) inflorescences sessile with sessile flowers (Fig. 1A, fruit shown), (2) inflorescences sessile with individual flowers pedicellate (Fig. 1B, fruit shown), or (3) inflorescences with short axes/branches and flowers sessile at branch apices (Fig. 1C, fruit shown).
        3. Inflorescences unbranched; flowers pedicellate ____ **II. Linearis Species Group**
        3. Inflorescences branched or unbranched; flowers sessile.
            4. Inflorescences unbranched; flowers sessile, rarely becoming pedicellate during fruit development ____ **III. Tayabensis Species Group**
            4. Inflorescences branched (trichotomous or rarely pentachotomous); flowers and fruit sessile ____ **IV. Pilosella Species Group**
    2. Inflorescences with distinct peduncles, axes or branches, flowers sessile or pedicellate.
        5. Leaves membranaceous or chartaceous usually drying yellow-green; young stems usually drying yellow-green; endosperm entire (non ruminate; see Fig. 1D), always pale colored ____ **V. Membranifolia Species Group**
        5. Leaves chartaceous to coriaceous rarely drying a yellow-green; young stems rarely drying yellow-green; endosperm usually ruminate (Fig. 1F), always dark colored.

6. Leaf blades usually > 30 cm long, lateral veins usually > 20 per side (Fig. 1H); stipules often large and conspicuous, 3–6(–10) cm long ____ **VI. Magnifolia Species Group**
6. Leaf blades usually < 20 cm long, lateral veins usually < 20 per side (Fig. 1I); stipules usually < 5 cm long.
   7. Inflorescence usually monochotomous (pedunculate; see Fig. 1J) ____ **VII. Gitingensis Species Group**
   7. Inflorescence usually trichotomous (not pedunculate; Fig. 1K).
      8. Leaf blades usually > 3 cm wide ____ **VIII. Luzoniensis Species Group**
      8. Leaf blades usually < 3 cm wide ____ **IX. Subalpina Species Group**

I.—Ixoroides Species Group

Climbers, usually climbing by means of adventitious roots. Stipules valvate. Leaf blades less or more than 3 cm wide, never more than 20 cm long, usually less than 15 cm long; chartaceous to subcoriaceous, rarely coriaceous. Inflorescences with axes and branches, ***mainly monochotomous (pedunculate) but some species trichotomous.*** Flowers pedicellate. ***Fruits white***; pyrene inner (adaxial) surface flat; seed producing a red color when placed in 70% ethanol, due to a soluble seed coat pigment (SCP positive), rarely SCP negative; endosperm ruminate, sometimes entire, usually dark. Comprising 22 species.

Note.—Text in ***bold italics*** denotes features useful for the placement of taxa within the respective species group.

KEY TO THE SPECIES OF THE IXOROIDES SPECIES GROUP

1. Inflorescences monochotomous: with 1 principal axis (the peduncle) arising from the apex of the shoot.
   2. Leaf blades mostly > 8 cm long.
      3. Leaf blades always truncate or rounded at base ____ **1. P. olsenii**
      3. Leaf blades narrowly acute, acute, cuneate, obtuse or sometimes rounded at base.
         4. Fruits disposed along infructescence branches in pairs; ultimate branching pattern paired-opposite ____ **2. P. paracalensis**
         4. Fruits not disposed along infructescence branches in pairs, ultimate branches of infructescence always cymose.
            5. Leaf apex abruptly and sharply long acuminate, 1/5–1/4 length of blade ____ **3. P. cadigensis**
            5. Leaf apex rounded, obtuse, acute, cuneate, or acuminate to alternate, but never abruptly and sharply long acuminate.
               6. Leaf base acute to rounded, sometimes slightly attenuate ____ **4. P. sibuyanensis**
               6. Leaf base blades acute, cuneate, obtuse or rounded.
                  7. Leaf base obtuse or rounded ____ **5. P. piperi**
                  7. Leaf base acute or cuneate, rarely tending towards rounded.
                     8. Inflorescence peduncle (4.5–12.5(–14) cm, usually > 7 cm long ____ **6. P. barkleyi**
                     8. Inflorescence peduncle < 7 cm long.
                        9. Leaf blades coriaceous, often drying orange-brown ____ **7. P. rigidaefolia**
                        9. Leaf blades subcoriaceous, usually drying grayish green ____ **8. P. radicans**
   2. Leaf blades mostly < 8 cm long.
      10. Plants pubescent on all parts or at least on midrib below, hairs curled or crisped.
         11. Plants with dense crisped hairs on all parts, including calyx and corolla ____ **9. P. sarcocarpa**
         11. Plants with crisped hairs only on leaf blades below (at least on midrib), or rarely on young stems.
            12. Leaf blades > 5 cm long ____ **10. P. iwahigensis**
            12. Leaf blades usually < 5 cm long ____ **17. P. cuernosensis**
      10. Plants glabrous or only puberulous, hairs straight.
         13. Inflorescence internodes usually longer on the primary axis, first internode usually > 5 cm long, the lowest branches usually > 5 cm long, spreading.
            14. Inflorescence peduncle usually (1.7–)3.5–5 cm long; leaf apex abruptly acuminate ____ **11. P. paucinervia**
            14. Inflorescence peduncle usually 5–8 cm long; leaf apex obtuse or abruptly acute ____ **12. P. obscurinervia**
         13. Inflorescence internodes usually shorter on the primary axis, first internode usually <2 cm long, the lowest branches usually < 5 cm long, ascending.
            15. Leaf blades obtuse, leaf base cordate, rounded, or at least obtuse ____ **13. P. mindoroensis**
            15. Leaf blades acuminate, leaf base acute to obtuse, never rounded or cordate.
               16. Inflorescences contracted, with probably < 10 flowers, < 3 cm long, peduncle 0.2–1.2 cm long ____ **14. P. christarusselliae**
               16. Inflorescences not contracted, with > 10 flowers, > 3 cm long, peduncle 1–5.1 cm long.
                  17. Bract and bracteole margins distinctly fimbriate ____ **17. P. cuernosensis**
                  17. Bract and bracteole margins entire.
                     18. Leaf blades 2.5–3.5 cm long ____ **15. P. elmeriana**

18. Leaf blades 4–10.9 cm long.
    19. Fruits smooth (outer (dorsal) surface of each pyrene smooth) ____ **21. P. lanaensis**
    19. Fruits ribbed (outer (dorsal) surface of each pyrene ribbed) ____ **16. P. ixoroides**

1. Inflorescences trichotomous: with 3 principal axes (inflorescence branches) arising from the apex of shoot.
  20. Inflorescences relatively contracted, the principal axis < 3 cm long and rarely with < 70 flowers, and usually with far fewer.
    21. Stipules subcoriaceous, apex usually with 2 aristate tips; corolla tubes 2–4(–6) mm long, villous at throat ____ **17. P. cuernosensis**
    21. Stipules subcoriaceous or distinctly chartaceous (papery), apex usually with 1 aristate tip; corolla tubes ca. 1 mm long, glabrous at throat ____ **18. P. elliptifolia**
  20. Inflorescences diffuse, broadly spreading, the principal axis > 3 cm long (usually > 5 cm long) and ≥ 30 cm long, and usually with > 100 flowers, and usually with many more.
    22. Leaf blades not conduplicate in the field; prinicpal inflorescence axis usually < 15 cm long.
      23. Bracts leaf-like, narowly elliptic, (0.7–)2.5–5 cm long, margins entire; fruits ellipsoid or narrowly ellipsoid, 3–5 mm long, striated ____ **22. P. diffusa**
      23. Bracts deltate to triangular, 0.8–1.5 mm long, margins distinctly serrate; fruits obovoid, 5–7 mm long, strongly ribbed ____ **4. P. sibuyanensis**
    22. Leaf blades often conduplicate in the field; principal inflorescence axis usually > 20 cm long.
      24. Leaf base truncate or subcordate ____ **19. P. lianoides**
      24. Leaf base acute to obtuse ____ **20. P. ramosissima**

1. Psychotria olsenii Sohmer & A.P. Davis, sp. nov. (**Fig. 3**). TYPE: PHILIPPINES. TAWI TAWI: Tawi Tawi Prov.: Tarawakan, rainforest at 50 m, 14 Nov 1961, *Olsen 758* (HOLOTYPE: L).

Ut videtus liana est, stipulis valvatis glabris 2-2.5 mm longis apice acuto integro, foliorum laminis subcoriaceis glabris 2.9 × 6.7-4.4 × 9.5 cm latis et longis nervis lateralibus 6-7 in dimidio quoque apice acuto basi truncata vel rotundata, petiolis 0.6-0.7 cm longis, infructescentia monochotoma glabra 3.3-6 cm longa rhachidi cum 2-3 nodis et ramulis oppositis, fructibus costatis obovoidis 7 mm longis, pedicellis brevibus, pyrenis ignotis.

Climber, height/length not recorded, adventitious roots not recorded; young stems glabrous. Stipules valvate, broadly ovate to deltate, 2–2.5 mm long, glabrous; apex broadly acute to acute. Leaves: petioles 0.6–0.9 cm long, glabrous; leaf blades ovate, 6.7–9.5 × 2.9–4.4 cm, subcoriaceous, glabrous; secondary veins 6 or 7 pairs; tertiary venation obscure; base truncate or rounded; apex acute. Inflorescence/infructescence monochotomous, 3.3–6 cm long, glabrous; principal axis (peduncle) 1.2–4.5 cm long, with 2–3 primary nodes and opposite branching at each node; bracts ± deltate, ca. 0.6 mm long, margins entire; bracteoles unknown. Flowers unknown. Fruit ± ellipsoid, ca. 7 mm long, ribbed, color unknown, glabrous; calyx limb and lobes persistent but obscure; pedicels ca. 1.5 mm long; pyrenes unknown.

PARATYPES: None—known only from the type collection.

Distribution.—Known only from Tawi Tawi Island.

Ecology.—Forest; at low altitude (e.g. 50 m).

Phenology records.—Flowering: unknown; fruiting: November.

Preliminary conservation assessment.—Critically Endangered (**CR B1ab(iii)**), or possibly Extinct (**EX**). Only collected once in 1961.

Distinguishing characteristics.—Leave blades ovate; leaf base truncate or rounded.

Remarks.—Known only from the type collection. The short peduncle and round to truncate leaf-base separates this taxon, as presently conceptualized, from *P. rigidaefolia* (Elmer) Merr. Further collecting may show these characteristics are too variable for the maintenance of *P. olsenii*.

2. Psychotria paracalensis Sohmer & A.P. Davis, sp. nov. (**Fig. 4**). TYPE: PHILIPPINES. LUZON. Camarines Prov.: Paracale, Nov-Dec 1918, *Ramos & Edaño BS 33697* (HOLOTYPE: A; ISOTYPES: BM, P).

Laminis subcoriaceis glabris oblongo-ellipticis 4.6 × 10.5-6.1 × 13.7 cm latis et longis nervis lateralibus 7-9 in dimidio quoque apice obtuso vel rotundato interdum cum puncto abrupto breviori basi acuta vel obtusa, infructescentia monochotoma glabra 14.5-25.5 cm longa, pedunculo 8-9.1 cm longo, nodis primariis 2-3 cum ramis oppositis, fructibis secus ramas oppositis laevibus obovoideo-globosis 8-9 mm longis licet sessilibus, pyrenis in dorso laevibus, endospermio ruminato.

Climber, height/length not recorded, adventitious roots not recorded; young stems glabrous. Stipules unknown. Leaves: petioles 2.1–3.6 cm long, glabrous; leaf blades oblong-elliptic, 10.5–13.7 × 4.6–6.1 cm, subcoriaceous, glabrous; secondary veins 7–9 pairs; tertiary venation manifest to obscure; base

Fig. 3. *Psychotria olsenii* (*Olsen 758*) (L, holotype). **A,** habit of branch.

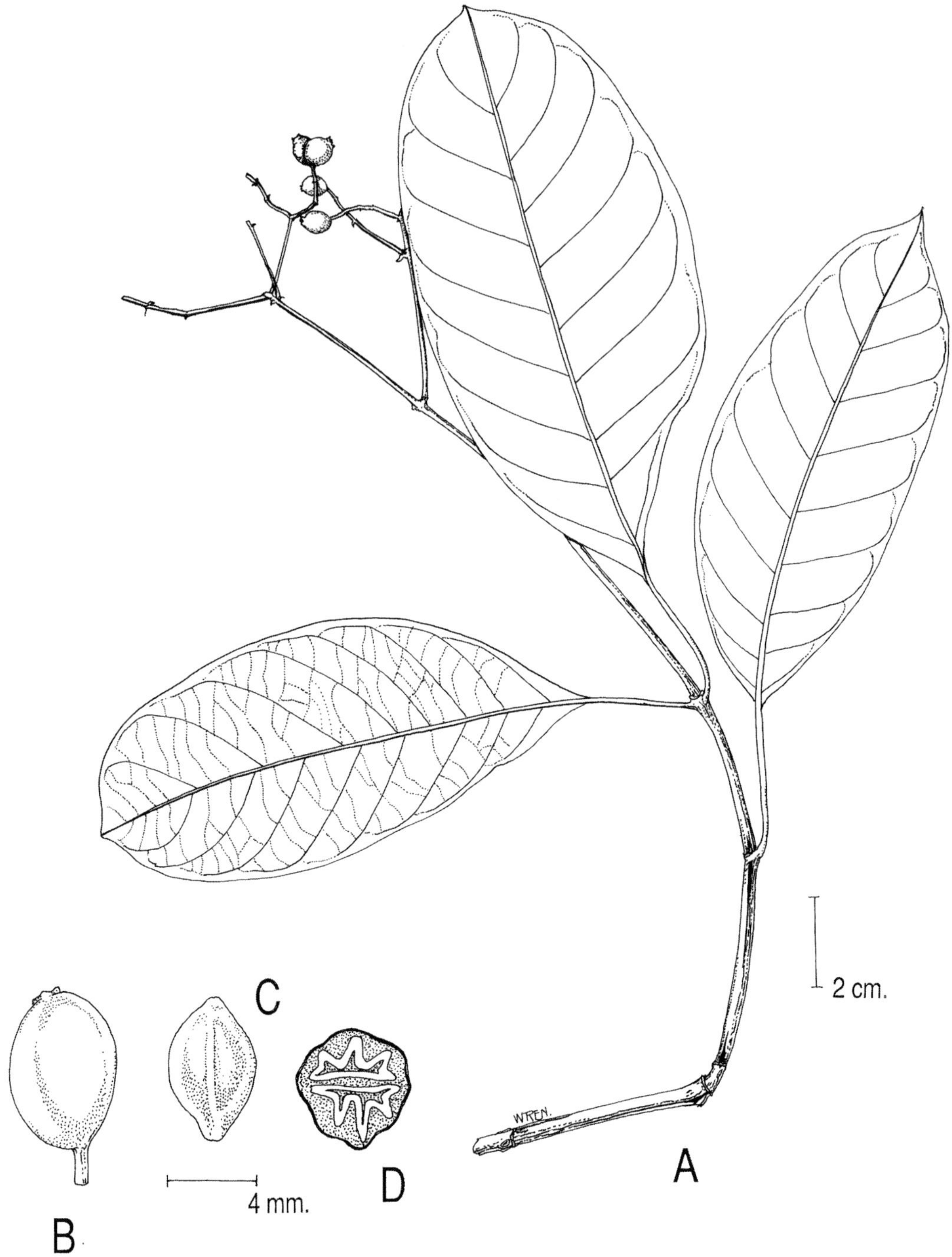

FIG. 4. *Psychotria paracalensis* (*Ramos & Edaño BS 33697*) (A, holotype). **A**, habit of fruiting branch; **B**, fruit; **C**, pyrene; **D**, fruit in cross section.

acute to obtuse; apex obtuse or round, sometimes apiculate. Inflorescence/infructescence monochotomous, 14.5–25.5 cm long, glabrous; principal axis (peduncle) 8–9.1 cm long, with 2–3 primary nodes and opposite branching at each node, fruit arranged along branches in pairs; bracts unknown; bracteoles unknown. Flowers unknown. Fruit obovoid to ± globose, 8–9 mm long, ± smooth, color unknown, glabrous; calyx limb and lobes persistent but not obvious; pedicels unknown; pyrenes smooth on outer surface; endosperm ruminate.

PARATYPES: None—known only from the type collection.

Distribution.—Known only from Paracale, Luzon.

Ecology.—Unknown.

Phenology records.—Flowering: unknown; fruiting: November, December.

Preliminary conservation assessment.—Extinct (**EX**). Only collected once in 1918.

Distinguishing characteristics.—Leaves broad, oblong-elliptic; fruit apparently sessile and borne in pairs along the infructescence axes.

Remarks.—We are a little uneasy about the lack of material available with which to characterize this new species. The evidence indicates a flowering and subsequent fruiting pattern in which a node possesses two sessile flowers opposite each other, translating directly into the manner in which the fruit are borne. There is also some indication that flowering and fruiting are continuous and that the inflorescence axes continue to grow up to a point. If subsequent collecting of this taxon is possible and it becomes clear that the evidence does not support our conclusion based on the type collection, this taxon would need to be synonymized with either *P. piperi* or *P. rigidaefolia* (Elmer) Merr.

3. Psychotria cadigensis Merr., Philipp. J. Sci. 12:175. 1917; Merr., Enum. Philipp. Fl. Pl. 3:553. 1923. (**Fig. 5**). TYPE: PHILIPPINES. LUZON: Quezon Prov.: Mt. Cadig, in forests at 350–500 m, Dec 1916, *Yates BS 25533* (LECTOTYPE: BM, designated here; ISOLECTOTYPE: K, P).

Climber or scandent shrub, height/length not recorded, adventitious roots not recorded; young stems glabrous. Stipules valvate, ± deltate, ca. 3 mm long, glabrous; apex truncate. Leaves: petioles 0.6–1.1 cm long, glabrous; leaf blades oblong-elliptic to oblong-ovate, 5.6–9.1 × 1.7–3.2 cm, chartaceous, glabrous; secondary veins 7–9 pairs; tertiary venation obscure to invisible; base acute; apex abruptly long-acuminate. Inflorescence/infructescence monochotomous, 13–16.5 cm long, glabrous; principal axis (peduncle) 4.5–5.2 cm long, with 2–3 primary nodes and opposite branching at each node, secondary branches branching 2–3 more times, ultimate branches terminating in cymules; bracts ± narrowly triangular, ca. 1.2 mm long, margins entire; bracteoles ± deltate, ca. 1 mm long. Flowers very poorly known; calyx (incl. hypanthium) ca. 1.2 mm long, glabrous; calyx lobes obsolete (limb rotate); corolla and other flower parts unknown. Fruit poorly known, ca. 5 mm long, white, glabrous, otherwise unknown; pyrenes unknown.

Other specimens examined: Known only from the type collection.

Distribution.—Known only from Mt. Cadig, Luzon.

Ecology.—Forest; 350–500 m.

Phenology records.—Flowering: December; fruiting: December.

Preliminary conservation assessment.—Extinct (**EX**). Only collected once in 1916.

Distinguishing characteristics.—Leaf blades chartaceous, with apices long-acuminate.

Remarks.—This species has not been recollected since the original material was collected. The collection designated by Merrill as the type (*BS 25484*) has not been found by us. We have therefore designated the syntype as a lectotype.

4. Psychotria sibuyanensis Elmer, Leafl. Philipp. Bot. 3:1028. 1911; Merr., Enum. Philipp. Fl. Pl. 3:563. 1923. (**Fig. 6**). TYPE: PHILIPPINES. SIBUYAN: Romblon Prov.: Mt. Giting-giting, forested ridge at ca. 700 m, Apr 1910, *Elmer 12355* (LECTOTYPE: BM, designated here; ISOLECTOTYPES: BISH, GH, K, L, MO, P, US, W).

Climber, 2–5 m long, with adventitious roots; young stems glabrous. Stipules valvate, triangular to narrowly triangular, 2–6 mm long, ± glabrous; apex acute to abruptly short-aristate. Leaves: petioles 0.5–1.9 cm long, glabrous; leaf blades narrowly elliptic to lanceolate, 6.5–13 × 1.7–3.4 cm, chartaceous to subcoriaceous, glabrous; secondary veins 7–12 pairs; tertiary venation obscure to invisible; base acute to rounded, sometimes slightly attenuate; apex narrowly acute. Inflorescence/infructescence monochotomous or rarely trichotomous, 5–11.3 cm long, glabrous to puberulous; principal axis (usaully the peduncle) 3.5–6.7 cm long, with 2–3 primary nodes and opposite branching at each node, secondary branches branching 1 or 2 more times, ultimate branches terminating in rather dense cymules; bracts leaf-like, narowly elliptic, (0.7–)2.5–5 cm long, margins entire; bracteoles ± deltate to linear, 1–3.8 mm long. Flowers poorly known (immature), pedicels short or ± absent, otherwise

Fig. 5. *Psychotria cadigensis* (*Yates BS 25533*) (BM, lectotype).

Fig. 6. ***Psychotria sibuyanensis*** **(*Elmer 12355*) (BM, lectotype).**

unknown. Fruit obovoid, 5–7 mm long, strongly ribbed, white, glabrous; calyx limb and lobes persistent, usually with distinct deltate to triangular lobes; pedicelss 1–3.5 mm long; pyrenes with pronounced ribbing on outer surface; endosperm ruminate.

Other specimens examined: **Luzon. Cagayan Prov.:** Claveria, Barangay Kilkiling, Taggat Watershead Area, *Garcia, Fuentes & Romero PPI 18724* (BRIT); Claveria, Kilkiling, Mt. Taggat, *Reynoso, Fuentes & Majaducon PPI 11971* (BRIT); Barangay San Juan, Santa Praxedes, *Garcia, Fuentes & Romero PPI 18457* (BRIT). **Sibuyan: Romblon Prov.:** Mt. Guiting-Guiting [Giting-Giting], *Reynoso,*

Fuentes & Rivers PPI 14080 (BRIT); Magdiwang, Barrio Hawasan, *Stone et al. PPI 6702* (BRIT, K); trail from Sitio Agsalay, Barrio Jaosan to Punong River, *Sohmer 12372* (K); Punong River, near trail for Barrion San Jaosan, *Sohmer 12436* (K). **Panay. Capiz Prov.:** Jamindan, *Ramos & Edaño BS 30977* (A, BM, K, P). **Antique Prov.:** Mt. Balingaso, 10 km north-east of Libertad, *Sands & Vendivil SANDS 3269* (K). **Mindanao. Surigao del Norte Prov.:** Dinagat Island, Cambinliw, Santiago, Loreto, *Gaerlan, Sagcal & Fernando PPI 4570* (BRIT, K); Dinagat Island, Anahaw, Kaantojan, *Gaerlan & Fernando PPI 19126* (BRIT). **Surigao Prov.:** Twagaau Tubud Placer, *Wenzel 2856* (A, K, UC); Mawpayang Tubud Placer, *Wenzel 2901* (UC).

Distribution.—Luzon, Sibuyan, Panay and Mindanao.

Ecology.—Primary forest, including forest with dipterocarps and dipterocarp forest, also in open and exposed places, such as ridges; 150–700 m.

Phenology records.—Flowering: January to May; fruiting April to September.

Preliminary conservation assessment.—Near Threatened (**NT**). Last collected in 1995.

Distinguishing characteristics.—Leaves narrowly elliptic to lanceolate, base acuminate to attenuate.

Remarks.—In general appearance *P. sibuyanensis* is close to *P. ixoroides* but they can be easily separated because the leaves and fruits of the former are larger.

5. Psychotria piperi Merr., Philipp. J. Sci. 17:457. 1921; Merr., Enum. Philipp. Fl. Pl. 3:561. 1923. (**Fig. 7**).
TYPE: PHILIPPINES. MINDANAO. Surigao del Norte Prov.: Hinatuan, 16 May 1911, *Piper BS 513* (not seen).

Climber, 2–4 m long, with adventitious roots; young stems glabrous. Stipules unknown. Leaves: petioles 1.6–2.2(–3.6) cm long, glabrous; leaf blades broadly elliptic to broadly oblong, 11–12.4 × 6.9–7.5 (–8.1) cm, subcoriaceous to coriaceous, glabrous; secondary veins 7–10(–12) pairs; tertiary venation manifest to prominent; base obtuse to rounded; apex rounded with a short acute tip. Inflorescence/Infructescense monochotomous, 8–11.5 cm long, glabrous to minutely puberulous; principal axis (peduncle) 4.2–4.7(–11.5) cm long, with 2–3 primary nodes and opposite branching at each of the secondary branches, which also possess opposite branching, terminating in cymules; bracts ovate to triangular, 0.9–2 cm long, margins entire; bracteoles broadly ovate to triangular, 0.9–1.4 mm long. Flowers poorly known (immature), pedicels 1.5–1.9 mm long; calyx (incl. hypanthium) ca. 1.5 mm long, glabrous; calyx lobes obscure, ± obsolete; corolla and other flower parts unknown. Fruit obovoid, 9–10 mm, smooth to slightly ribbed, colour unknown, glabrous; calyx limb and lobes persistent; pedicels 1–2 mm long; pyrenes smooth on outer surface; endosperm ruminate.

Other specimens examined: **Samar. Western Samar Prov.:** Mt. Calbiga, *Sulit PNH 6397* (A, PNH); Loquilocon, *Wright, Sulit & Conese PNH 6154* (K); Paranas, Campo Uno, Central Samar, *Reynoso & Majaducon PPI 24066* (BRIT). **Mindanao. Surigao del Norte Prov.:** Dinagat Island, Mabini, Tabajon, *Gaerlan et al. PPI 16887* (K); Dinagat Island, without further data, *Ramos & Pascasio BS 35188* (K). **Agusan del Sur Prov.:** No. 1 ERDS Experimental Forest, Mambalili, *Gaerlan, Fuentes & Majaducon PPI 23152* (BRIT).

Distribution.—Samar and Dinagat Island.

Ecology.—Dipterocarp forest, and mixed dipterocarp forest (both primary forest types); 100–400 m.

Phenology records.—Flowering: January; fruiting: April, May.

Preliminary conservation assessment.—Vulnerable (**VU B1ab(iii)**). Last collected in 1996.

Distinguishing characteristics.—Fruits relatively large; leaves broadly elliptic or oblong.

Remarks.—*Psychotria piperi* is similar to *P. barkleyi*, but differs on the basis of its generally shorter petioles, narrower leaves, obtuse to rounded leaf bases and shorter inflorescence peduncle (see descriptions for these species).

6. Psychotria barkleyi Sohmer & A.P. Davis, nom. nov. (**Fig. 8**). *Psychotria longipedunculata* Elmer, Leafl. Philipp. Bot. 3:1027. 1911. *Grumilea longipedunculata* (Elmer) Merr. Merr., Philipp. J. Sci. 10:133. 1915. Non *Psychotria longipedunculata* (Gardener) Müll. Arg., Mart., Fl. Bras. 6:237. 1881. TYPE: PHILIPPINES. SIBUYAN: Romblon Prov.: Mt. Giting-giting [Mt. Giting-giting], along Patos River in very humid, alluvial forest at 250 m, May 1910, *Elmer 12532[a]* (LECTOTYPE: US, designated here; ISOLECTOTYPES: A, BISH, BM, GH, K, L, MO, NY, W).

Climber, height/length not recorded, adventitious roots not recorded; young stems glabrous. Stipules valvate, ± triangular, ca. 4 mm long, glabrous; apex acute. Leaves: petioles (1.1–)2.2–4.5 cm long, glabrous; leaf blades elliptic-oblong to elliptic-ovate, 9.3–15.3 × 4–6.2 cm, chartaceous to subcoriaceous, glabrous; secondary veins 7–9 pairs; tertiary venation manifest to weak; base acute or cuneate, rarely almost rounded; apex short-acuminate or acute. Inflorescence/infructescence monochotomous, 12–28 cm long, glabrous; principal axis (peduncle) 4.5–12.5(–14) cm long, with 3–4 primary nodes and opposite branching at each node, secondary branches branching 2–3 more times and terminating in

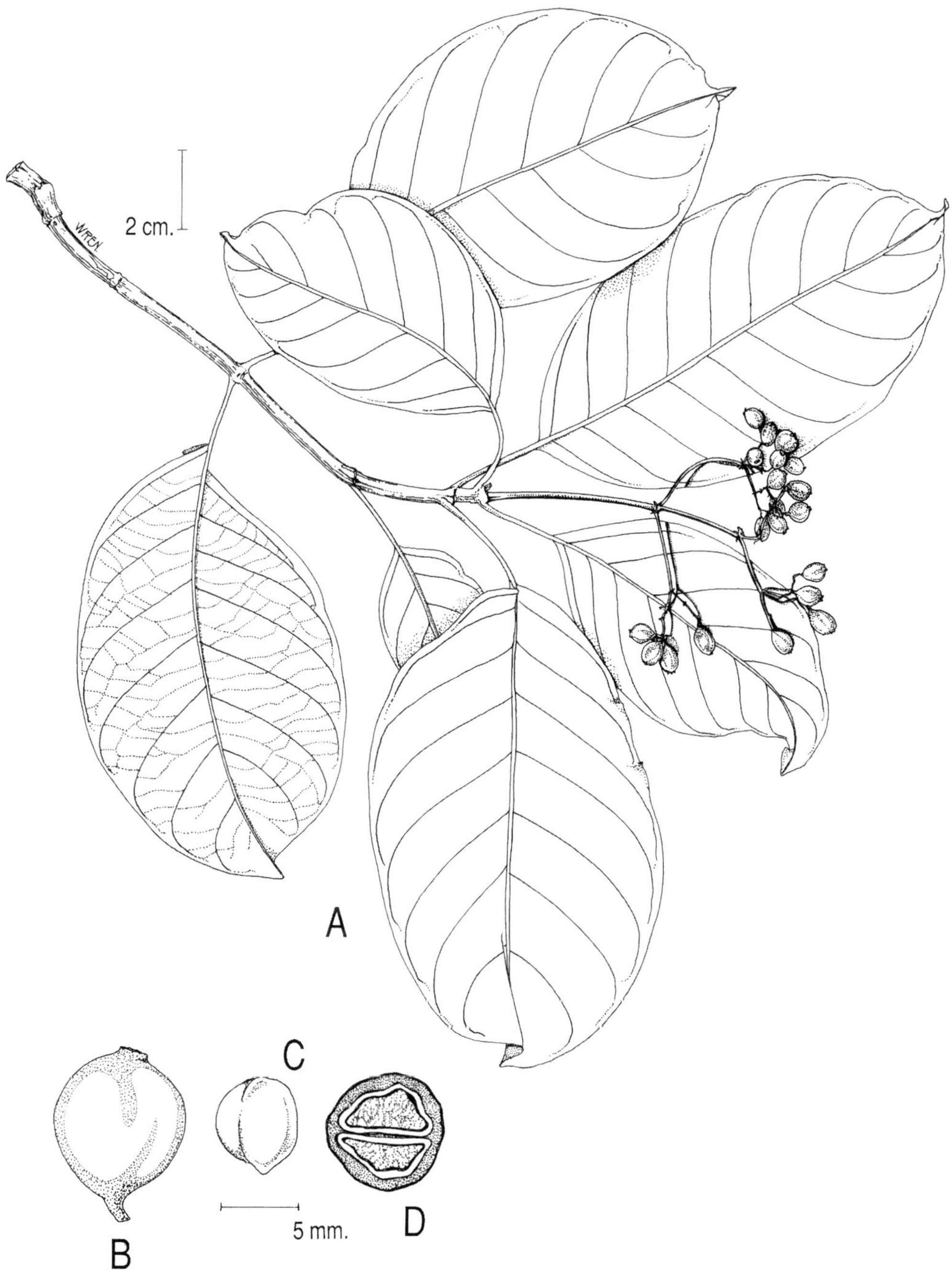

Fig. 7. *Psychotria piperi* (*Sulit PNH 6397*). **A,** habit of fruiting branch; **B,** fruit; **C,** pyrene; **D,** fruit in cross section.

cymules; bracts ± narrowly triangular, (1.5–)2–2.7 mm long, margins entire; bracteoles deltate to cusp-like, 0.5–1.2 mm long. Flowers poorly known (buds only); calyx (incl. hypanthium) ca. 1.7 mm long, glabrous; calyx lobes very shallowly triangular, ca. 0.2 mm long, or ± obsolete; corolla: color unknown; corolla tube ca. 3.5 mm long, external surface glabrous, otherwise unknown. Fruit obovoid, 6–8 mm long, white, smooth, glabrous; calyx limb and lobes persistent but obscure; pedicels 0.5–2.5 mm long; pyrenes smooth on outer surface; endosperm ruminate.

Paratypes: **Luzon. Camarines Norte Prov.:** Bicol National Park (NE), *Stone et al. PPI 3713* (BRIT), *Stone et al. PPI 3817* (BRIT).

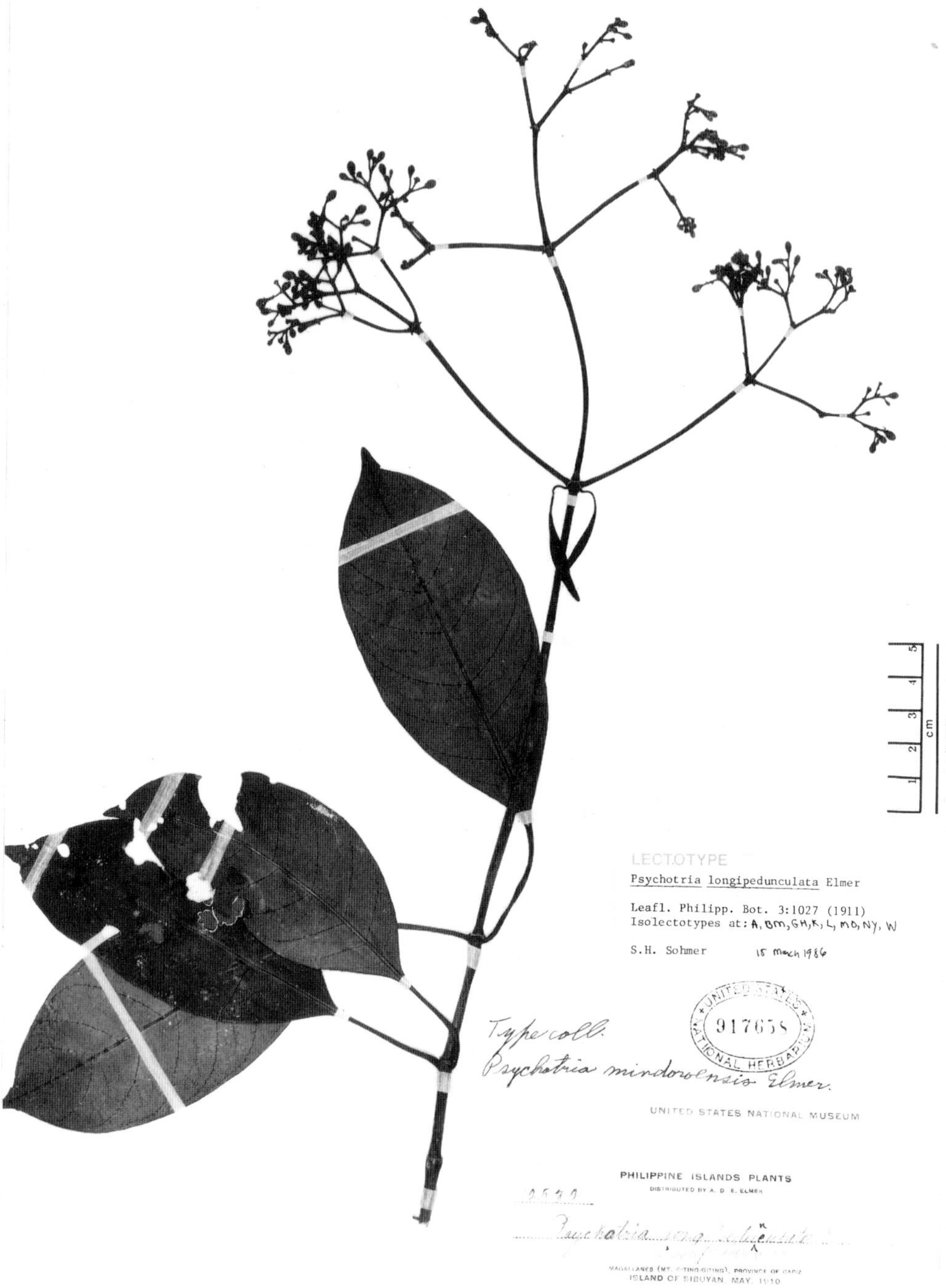

Fig. 8. *Psychotria barkleyi* (*Elmer 12532[a]*) (US, lectotype).

Camarines Sur Prov.: Kamugong River, *Edaño BS 75889* (NY). **Sibuyan. Romblon Prov.:** Mt. Guiting-guiting, *Reynoso, Fuentes & Rivers PPI 14087* (BRIT).

Distribution.—Sibuyan and Luzon.

Phenology records.—Flowering: May; fruiting: November, December.

Ecology.—Primary and disturbed primary, humid forest, including mixed dipterocarp forest; 200–500 m.

Preliminary conservation assessment.—Vulnerable (**VU B1ab(iii)**). Last collected in 1994.

Distinguishing characteristics.—Inflorescence very long with peduncles usually more than 15 cm long.

Remarks.—We have renamed *P. longipedunculata*, as *P. barkleyi*. This species is named in honor and memory of the late Ted Barkley, a Research Associate at the Botanical Research Institute of Texas (BRIT).

The type specimen of *P. longipedunculata* (*Elmer 12532*) is also cited by Elmer as the type of *P. mindoroensis* (Elmer 1911:1027, 1032). Elmer states that: '...only a fragment of this was secured from the same tree upon which *P. longipedunculata* Elm. was discovered.' As these two types are obviously different plants, we have indicated this here by citing the types as *Elmer 12532[a]* (*P. longipedunculata*) and *Elmer 12532[b]* (*P. mindoroensis*).

The main characteristic separating *P. barkleyi* from *P. obscurinervia* are the larger leaf blades of the former species.

7. Psychotria rigidaefolia (Elmer) Merr., Philipp. J. Sci. 10:128. 1915; Merr., Enum. Philipp. Fl. Pl. 3:562. 1923. *Randia rigidaefolia* Elmer, Leafl. Philipp. Bot. 1:352. 1908. TYPE: PHILIPPINES. LUZON: Tayabas Prov.: in humid forests on the road between Lucban and Sampaloc, *Elmer 9118* (not seen).

Climber, 3–5 m long, adventitious roots not recorded; young stems glabrous. Stipules unknown. Leaves: petioles 1.4–3.4 cm long, glabrous; leaf blades broadly elliptic to narrowly lanceolate, 5.9–15 × 2.4–4.3 cm, coriaceous, glabrous; secondary veins 6–9 pairs; tertiary venation manifest to rather prominent; base acute, or obtuse to rounded; apex acuminate to short-acuminate. Inflorescence/infructescence monochotomous, 6.5–12 cm long, glabrous; principal axis (peduncle) 2.9–6 cm long, with 2–3 primary nodes and opposite or sometimes verticillate branching at each node, secondary branches branching 1–2 times, terminating in cymules; bracts absent (?caducous); bracteoles deltate to broadly ovate, 0.7–1.3 mm long. Flowers unknown. Fruit globose, 5–6 mm long, smooth or slightly rugose (in dry state), white, glabrous; calyx limb and lobes persistent; pedicels 2–4 mm long; pyrenes smooth on outer surface; endosperm ruminate.

Other specimens examined: **Luzon. Camarines Sur Prov.:** Mt. Potianay, *Edaño BS 75956* (A, NY, PNH). **Catanduanes. Catanduanes Prov.:** Mt. Abucay, *Ramos & Edaño BS 75216* (NY). **Catanduanes Prov.:** without further data: *Ramos BS 30564* (NY, P). **Samar. Western Samar Prov.:** Mt. Calbiga, *Sulit & Conese PNH 6419* (A, L). **Leyte. Leyte Prov.:** Biliran, *McGregor BS 18730* (A); Mt. Suiro, *Sulit* PNH *21581* (L, PNH), *Sulit PNH 21639* (L); Dagami, M. *Ramos BS 15296* (K).

Distribution.—Luzon, Samar, Catanduanes and Leyte.

Ecology.—Humid forest; low elevations.

Phenology records.—Flowering: unknown; fruiting: August.

Preliminary conservation assessment.—Vulnerable (**VU B1ab(iii)**). Last collected in 1954.

Distinguishing characteristics.—Leaves stiffly coriaceous, often drying a conspicuous orange-brown color.

Remarks.—In the absence of type material the identity of Elmer's taxon is not clear. Most of the specimens examined had already been identified as *P. rigidaefolia* and they appear to constitute a taxon that can be separated in the herbarium from other climbing taxa. Merrill (1923) cited a number of new collections as belonging to this taxon, and one of these is cited here (*Ramos BS 30564*). However, the only other collection Merrill listed, *Elmer 12532*, was cited by Elmer (1911) as the type collection for *P. longipedunculata* (now called *P. barkleyi*). These two taxa resemble each other. The identity of *P. rigidaefolia*, therefore, remains slightly unclear.

8. Psychotria radicans (Merr.) Merr., Enum. Philipp. Fl. Pl. 3:562. 1923. (**Fig. 9**). *Grumilea radicans* Merr., Philipp. J. Bot. 10:139. 1915. TYPE: PHILIPPINES. LEYTE: Leyte Prov.: Buenavista, near Jaro, in forests at ca. 500 m, 10 Oct 1914, *Wenzel 1193* (LECTOTYPE: BM, designated here; ISOLECTOTYPE: MO).

Climber or scandent shrub, ca. 3 m high, adventitious roots not recorded; young stems glabrous. Stipules valvate, triangular, 3–4 mm long, glabrous; apex acute to shortly aristate. Leaves: petioles 1.4–2.1 cm long, glabrous; leaf blades oblong-elliptic to oblong-lanceolate, 8–13 × 2.4–4 cm, subcoriaceous, glabrous; secondary veins 6–8 pairs; tertiary venation obscure to invisible; base acute; apex acuminate. Inflorescence/infructescence monochotomous, 7–11 cm long, glabrous; principal axis

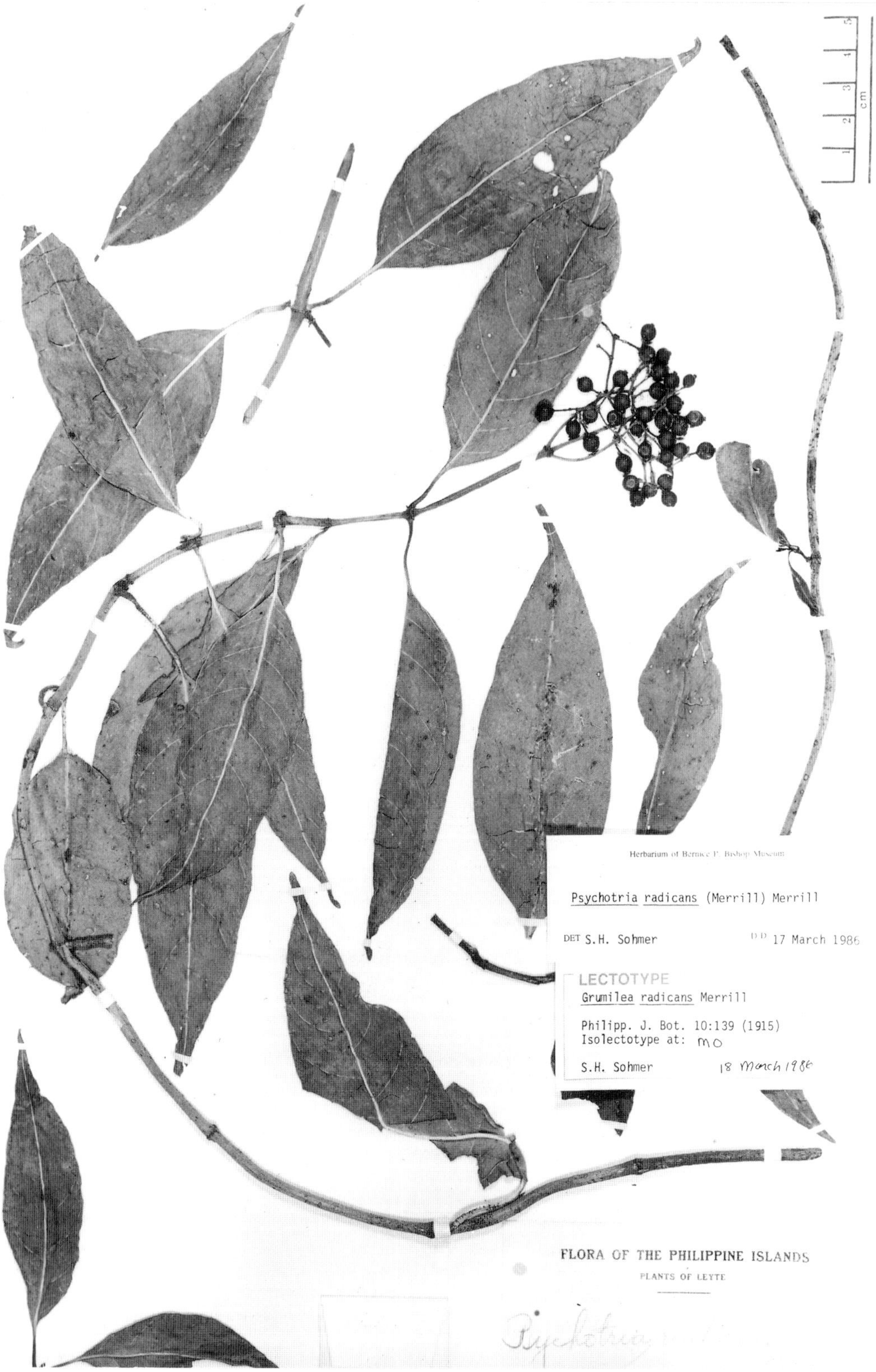

FIG. 9. *Psychotria radicans* (*Wenzel 1193*) (BM, lectotype of *Grumilea radicans*).

(peduncle) 4.5–5 cm long, with 2–3 primary nodes and opposite branching at each node, ultimate branches terminating in cymules; bracts ± narrowly triangular, ca. 1 mm long, margins entire; bracteoles ± deltate, ca. 0.7 mm long. Flowers unknown. Fruit ± globose, 5–6 mm long, smooth, color unknown, glabrous; calyx limb and lobes persistent; pedicels ca. 1.2 mm long; pyrenes smooth on outer surface; endosperm ruminate.

Other specimens examined: **Samar. Western Samar Prov.:** Basey, *Fontanoza FB 31142* (NY). **Leyte. Without further data:** *Wenzel 754* (BM, MO).

Distribution.—Leyte and Samar.

Ecology.—Forest; 500–1000 m.

Phenology records.—Flowering: unknown; fruiting: October.

Preliminary conservation assessment.—Vulnerable (**VU B1ab(iii)**). Last collected in 1929.

Distinguishing characteristics.—Scandent shrub or climber; leaves oblong-elliptic to oblong-lanceolate, and usually drying grayish green as opposed to reddish brown; fruits smooth (when dry), pyrenes smooth on the outer (dorsal) surface.

Remarks.—*Psychotria radicans* superficially resembles *P. ixoroides*, but the fruits of *P. radicans* are much larger and are smooth (when dry); the fruits of *P. ixoroides* are ribbed. *Psychotria radicans* has been collected on Biliran Island according to Merrill (1923). We are not completely certain that the collection from Samar belongs here.

9. Psychotria sarcocarpa Merr., Philipp. J. Sci. 8:53. 1913; Merr., Enum. Philipp. Fl. Pl. 3:563. 1923. (**Fig. 10**). TYPE: PHILIPPINES. MINDANAO: Zamboanga del Sur Prov.: Sax River Mountains, in wet, dense forest along small streams at 650–1000 m, 27 Nov 1911, *Merrill 8077* (LECTOTYPE: US, designated here; ISOLECTOTYPES: A, BM, K, L, P).

Climber, ca. 3 m long, adventitious roots not recorded; young stems pubescent to densely pubescent. Stipules valvate, ovate to narrowly triangular, 5–11 mm long, pubescent; apex acute. Leaves: petioles 0.5–0.7 cm long, pubescent; leaf blades oval-elliptic, 3.1–4.7 × 1.7–2.8 cm, chartaceous, glabrous above, pubescent (with crisped or curled hairs) below; secondary veins 5–7 pairs; tertiary venation inconspicuous to invisible; base obtuse or acute; apex rounded to obtuse, sometimes with the slightest very short acuminate tip. Inflorescence/infructescence monochotomous, ca. 4 cm long, densely pubescent; principal axis (peduncle) 0.6–1 cm long, apparently with 1 primary node and verticillate branching; bracts absent (?caducous); bracteoles ± narrowly triangular, ca. 0.6 mm long. Flowers poorly known; calyx (incl. hypanthium) 3–4 mm long, pubescent to densely pubescent; calyx lobes ± deltate, ca. 1 mm long; corolla: color unknown; corolla tube ca. 2.5 mm long, external surface pubescent, otherwise unknown; corolla lobes ca. 2.5 mm long, otherwise unknown; stamens unknown; style unknown. Fruit ellipsoid, 8–12 mm long, ribbed, white, pubescent; calyx limb and lobes persistent; pedicels 5–10 mm long; pyrenes ribbed on outer surface; endosperm ruminate.

Other specimens examined: **Mindanao: Basilan Prov.:** Basilan Island, Udiongan, Ayala, *Gaerlan & Ferdinand PPI 26171* (BRIT, K).

Distribution.—Known only from the Sax River Mountains and Basilan Island, Mindanao.

Ecology.—Forest, including disturbed dipterocarp forest; recorded from wet, dense forest along small streams; 600–1000 m.

Phenology records.—Flowering: unknown; fruiting: June, November, December.

Preliminary conservation assessment.—Vulnerable (**VU B1ab(iii)**). Last collected in 1996.

Distinguishing characteristics.—Most plant parts densely pubescent, including the calyx and corolla; leaves short, ovate-elliptic.

Remarks.—Merrill (1913) averred that his species was similar to *P. ovalis*, 'but with smaller, fewer-nerved leaves and very different fruit.' More material and field study of this somewhat enigmatic taxon is needed. The nature of the inflorescence is not well understood.

10. Psychotria iwahigensis Elmer, Leafl. Philipp. Bot. 4:1351. 1912; Merr., Enum. Philipp. Fl. Pl. 3:556. 1923. (**Fig. 11**). TYPE: PHILIPPINES. PALAWAN: Palawan Prov.: Mt. Pulgar, in forests at ca. 100 m, Apr 1911, *Elmer 13052* (LECTOTYPE: NY, designated here; ISOLECTOTYPES: BM, L).

Climber, height/length not recorded, adventitious roots not recorded; young stems puberulous to ± pubescent. Stipules valvate, triangular, 3–5 mm long, puberulous to pubescent; apex acute or with a

Fig. 10. *Psychotria sarcocarpa* (*Merrill 8077*) (US, lectotype).

Fig. 11. *Psychotria iwahigensis* (*Elmer 13052*) (NY, lectotype).

long aristate point. Leaves: petioles 0.6–0.9 cm long, puberulous to ± pubescent; leaf blades elliptic to elliptic-oblong, or elliptic obovate, 5.5–9.7 × 2.4–3.4 cm, chartaceous, glabrous above, puberulous to pubescent (hairs crisped), often pubescent on midrib; secondary veins 7–10 pairs; tertiary venation manifest to obscure; base acute to cuneate; apex narrowly acute to broadly acute. Inflorescence/infructescence monochotomous, 2.3–7.2 cm, puberulous; principal axis (peduncle) 1–4.5 cm long, with 2–3 primary nodes and verticillate branching at the first primary node and opposite branching at those above, the branches themselves branched oppositely; bracts triangular, 0.7–1 mm long, margins entire; bracteoles deltate to triangular, 0.3–0.8 mm long. Flowers poorly known (immature), pedicels 1–2 mm long; calyx (incl. hypanthium) cinereous, otherwise unknown; corolla: color unknown; corolla tube ca. 5 mm long, external surface puberulous, otherwise unknown; corolla lobes ca. 2.5 mm long; stamens: unknown; style unknown. Fruit ellipsoid to ellipsoid-obovoid, 4.1–5.4 mm long, distinctly ribbed, color unknown, glabrous; calyx limb and lobes persistent but inconspicuous; pedicels 1.6–2 mm long; pyrenes distinctly ribbed on outer surface; endosperm unknown (immature).

Other specimens examined: **Palawan.** Mt. Cleopatra Ranges, So. Calabayog, Barangay Tanabag, Puerto Princessa, *Reynoso & Majaducon PPI 24292* (BRIT, K); Rizal, Ransang, Balen-Balen, *Gaerlan, Romero & Sagcal PPI 13322* (BRIT).

Distribution.—Known only from Palawan.

Ecology.—Humid forest; low latitude (e.g., ca. 100 m).

Phenology records.—Flowering: unknown; fruiting: April, June.

Preliminary conservation assessment.—Vulnerable (**VU B1ab(iii)**). Last collected in 1996.

Distinguishing characteristics.—Leaves pubescent below with crisped hairs at least on the midrib, leaf blades over 5 cm long; inflorescence monochotomous.

Remarks.—Until recently this species was poorly known but new collections made under the Philippine Plant Inventory project (PPI), have allowed us to make a more complete description for this species.

11. Psychotria paucinervia Merr., Philipp. J. Sci. 12:176. 1917; Merr., Enum. Philipp. Fl. Pl. 3:561. 1923. (**Fig. 12**). TYPE: PHILIPPINES. LUZON: Abra Prov.: Mt. Posooy, near summit of mountain in forests, 4 Feb 1917, *Ramos BS 27055* (LECTOTYPE: US, designated here; ISOLECTOTYPE: K).

Scandent shrub, height/length not recorded, adventitious roots not recorded; young stems glabrous. Stipules valvate, ± ovate, 6–7 mm long, glabrous; apex acute. Leaves: petioles 0.7–1.1 cm long, glabrous; leaf blades obovate-elliptic, 4.5–7.1 × 1.9–3.4 cm, subcoriaceous to coriceous, glabrous; secondary veins 5–6 pairs; tertiary venation inconspicuous to invisible; base acute to obtuse; apex abruptly acuminate. Inflorescence/infructescence monochotomous, (5–)11–14 cm long, glabrous; principal axis (peduncle) (1.7–)3.5–5 cm long, with 2–3 primary nodes and opposite branching at each node, secondary branches branching 1 or 2 more times, terminating in cymules; bracts triangular with a distinct apical seta, 1.4–2 mm long, margins entire; bracteoles like the bracts, 1–1.5 mm long. Flowers ± unknown (very immature). Fruit ± unknown, 5–7 mm long, white, glabrous, otherwise unknown; pyrenes unknown.

Other specimen examined: **Luzon. Cagayan Prov.:** Bauan, Mt. Tabuan, *Ramos BS 77087* (A).

Distribution.—Northern Luzon.

Ecology.—Forest; below 1000 m.

Phenology records.—Flowering: unknown; fruiting: February.

Preliminary conservation assessment.—Critically Endangered (**CR B1ab(iii)**), or possibly Extinct (**EX**). Last collected in 1929.

Distinguishing characteristics.—Scandent shrub; inflorescence with opposite branches, peduncle 5 cm long or less.

Remarks.—Very little is known about this taxon. It has been collected only once since Merrill first described it. Flowers are unknown, and at this time the fruit is also unknown, although they were available when Merrill described this species. In overall appearance this taxon closely resembles *P. diffusa* Merr.

12. Psychotria obscurinervia Merr., Philipp. J. Sci. 17:455. 1921; Merr., Enum. Philipp. Fl. Pl. 3:560. 1923.

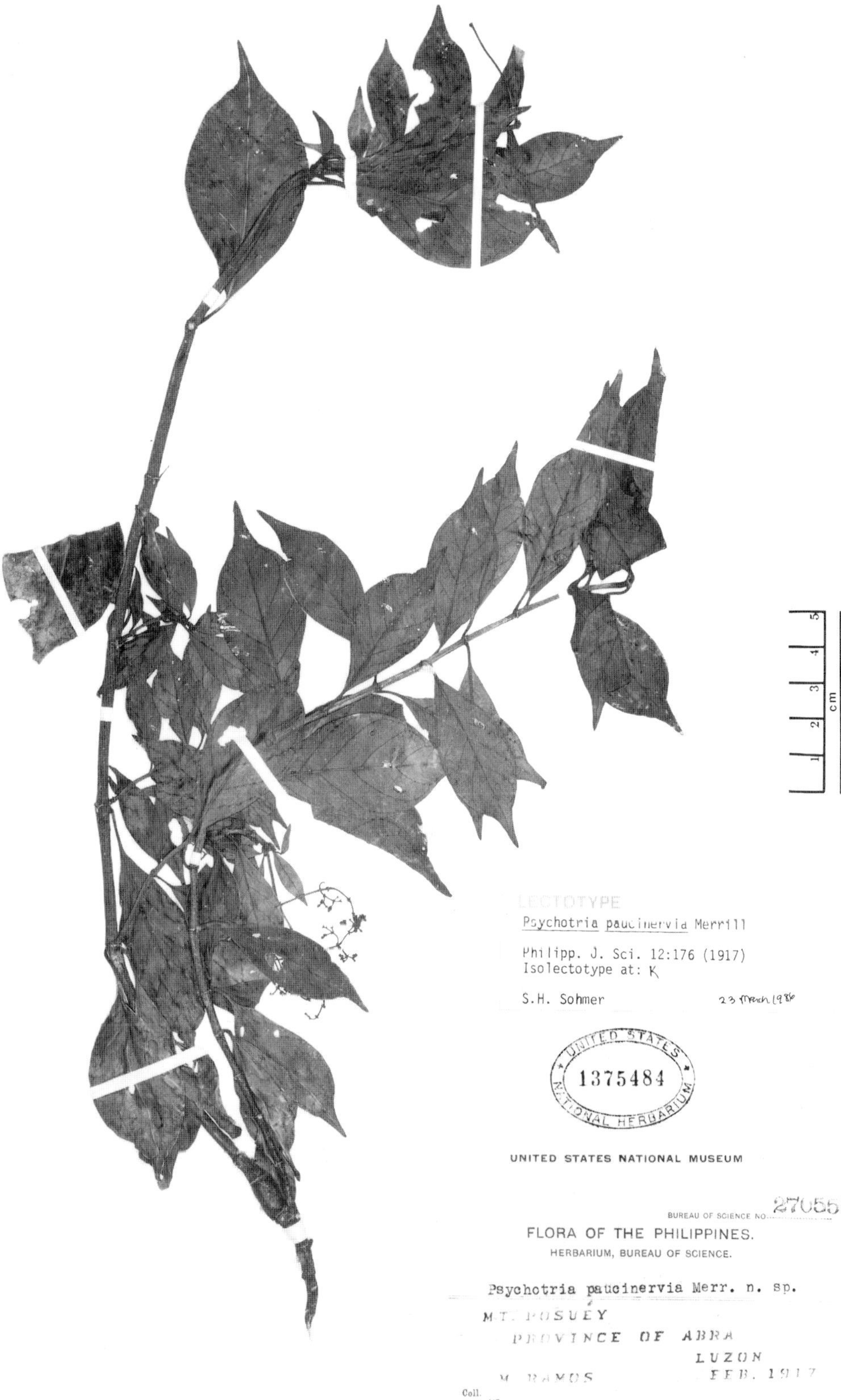

Fig. 12. *Psychotria paucinervia* (*Ramos BS 27055*) (US, lectotype).

(**Fig. 13**). TYPE: PHILIPPINES. LUZON: Ilocos Norte Prov.: between Bangui and Claveria in low elevation forest, 30 Aug 1918, *Ramos BS 33077* (LECTOTYPE: A, designated here; ISOLECTOTYPES: BM, K, P, US).

Psychotria vulcanica Elmer, Leafl. Philipp. Bot. 9:3264. 1934. TYPE: PHILIPPINES. LUZON: Sorsogon Prov.: Mt. Bulusan, forest at ca. 200 m, Oct 1915, *Elmer 14431* (LECTOTYPE: A, designated here; ISOLECTOTYPES: BISH, GH, K, L, MO, NY, P, UC, US).

Climber or scandant shrub, height/length not recorded; adventitious roots not recorded; young stems glabrous. Stipules valvate, ± ovate, 3.5–4.3 mm long, glabrous; apex subtruncate or obtuse. Leaves: petioles 0.6–1 cm long, glabrous; leaf blades oblong-elliptic to obovate, 4.5–8.6 × 1.4–3.8 cm, chartaceous, glabrous; secondary veins 4–7 pairs; tertiary venation inconspicuous to invisible; base acute to obtuse; apex obtuse to acute. Inflorescence/infructescence monochotomous, 12.5–21.5 cm long, glabrous; principal axis (peduncle) 5–8 cm long, with 2–4 primary nodes and opposite branching at each node, the branches themselves branched 2–3 times in the same pattern, flowers numerous; bracts absent (?caducous); bracteoles deltate, cup-like, 0.6–1.1 mm long. Flowers 4-merous, pedicels 1.5–3 mm long; calyx (incl. hypanthium) 1.5 mm long, glabrous; calyx lobes shallowly triangular, ca. 0.5 mm long; corolla: color unknown; corolla tube 2.5 mm long (short styled), external surface glabrous, villous at throat within; corolla lobes ca. 1.5 mm long; stamens: filaments ca. 1 mm long; anthers ca. 1 mm long; style ca. 1 mm long, lobes ca. 0.4 mm long. Fruit narrowly ellipsoid, 4–5 mm long, obscurely striated to ± smooth, white, glabrous; calyx limb and lobes persistent; pedicels 1.7–3 mm long; pyrenes smooth on outer surface; endosperm ruminate.

Other specimens examined: **Luzon. Laguna Prov.:** Mt. Maquiling, *Elmer 17504* (A, BM, K, L, MO, NY, P, UC, W), *Elmer 18294* (A, BM, K, L, MO, NY, P, UC, US), *Servinas BS 16883* (BM, K, L, NY, US). **Sorsogon Prov.:** without precise locality: *Elmer 16564* (A, BISH, BM, GH, L, MO, NY, P, UC, US).

Distribution.—Luzon.

Ecology.—Forest; low elevation (e.g. ca. 20 m).

Phenology records.—Flowering: June, July; fruiting: June, July, October, November.

Preliminary conservation assessment.—Extinct (**EX**). Last collected in 1918.

Distinguishing characteristics.—Inflorescence monochotomous, large and spreading, peduncle 5 cm long or more; all plant parts glabrous.

Remarks.—*Psychotria obscurinervia* is evidently a robust plant in nature but has not been collected since the early part of this century; the last collection seen was made in 1918 We suspect that this is due to the fact that this species appears to be restricted to low elevation forests and these have nearly all disappeared on Luzon. Many of the collections appear on herbarium sheets with few or no leaves. It bears some similarity to *P. barkleyi* but the average leaf blade size in *P. obscurinervia* is smaller.

13. Psychotria mindoroensis Elmer, Leafl. Philip. Bot. 3:1031. 1911; Merr., Enum. Philipp. Fl. Pl. 3:559. 1923. (**Fig. 14**). TYPE: PHILIPPINES. SIBUYAN: Romblon Prov.: Mt. Giting-giting, along Patos River in very humid, alluvial forest at 250 m, May 1910, *Elmer 12532[b]* (PNH, destroyed). NEOTYPE: PHILIPPINES. MINDORO: Mindoro Oriental Prov.: Baco River, Apr–May 1905, *McGregor 312* (NEOTYPE: NY, designated here; ISONEOTYPES: K, US).

Climber, height/length not recorded, with adventitious roots, young stems glabrous. Stipules valvate, ovate to ± obovate, 1.9–4 mm long, glabrous; apex acute/apiculate or with two aristate points. Leaves: petioles 0.3–0.6(–1) cm long, glabrous; leaf blades ovate to ovate-oblong, 4.4–9(–12) × 1.4–4.4(–4.8) cm, chartaceous to subcoriaceous, glabrous; secondary veins 4–8 pairs; tertiary venation inconspicuous; base cordate, rounded, or at least obtuse; apex acute to acuminate. Inflorescence/infructescence monochotomous, 3–7.7 cm long, glabrous to puberulous; principal axis (peduncle) 1.5–5.5 cm long, with 2–3 primary nodes and opposite branching at each node, secondary branches branching 1 or 2 more times, ultimate branches terminating in cymules or individual flowers/fruits; bracts absent (?caducous); bracteoles deltate to broadly ovate, 0.3–0.5 mm long. Flowers 5-merous, pedicels ca. 1.5 mm long; calyx (incl. hypanthium) ca. 2.5 mm long, glabrous; calyx lobes reduced, ± obsolete; corolla: white; corolla tube 1.5–2 mm long, external surface glabrous, villous at throat within; corolla lobes ca. 2 mm long (long styled); stamens: filaments ca. 1 mm long; anthers ca. 1 mm long (long styled); style 1.5–2 mm long, lobes ca. 0.5 mm long. Fruit ellipsoid to obovoid, 5–6 mm long, distinctly ribbed, white, glabrous; calyx limb and lobes persistent; pedicels 2–3.2 mm long; pyrenes ribbed on outer surface; endosperm entire.

Fig. 13. *Psychotria obscurinervia* (*Ramos BS 33077*) (A, lectotype).

Fig. 14. *Psychotria mindoroensis* (*McGregor 312*) (US, syntype).

Other specimens examined: **Luzon. Quezon Prov.:** Baler, Disalit River, *Quisumbing PNH 2053* (A); Umiray, *Ramos & Edaño BS 29001* (P, US); Malbog, Guinayangan, *Oro FB 30696* (NY); Polillo Island, Lukutan, *Castro & Anonuevo PNH 6548* (A); Quezon National Park, *Reynoso et al. PPI 3887* (BRIT). **Laguna Prov.:** without further data: *McGregor BS 22957* (K, P). **Albay Prov.:** SE slopes of Mt. Malinao, Taqui River, San Bernado, Tiwi Side, *Reynoso, Sagcal & Fernando PPI 21266* (BRIT). **Sorsogon Prov.:** Irosin (Mt. Bulusan), *Elmer 15242* (A, BM, K, L, MO, NY, P, UC, US). **Camarines Sur Prov.:** Barangay Caputatan, Caramoan, *Reynoso, Sagcal & Fuentes PPI 14829* (BRIT). **Sorsogon Prov.:** without further data: *Ramos BS 23460* (A, NY, P, US). **Luzon. Without further data:** *unknown collector 97* (W). **Catanduanes. Catanduanes Prov.:** Mt. Nagpakdit, *Ramos & Edaño BS 75438* (NY). **Mindoro. Mindoro Oriental Prov.:** Mt. Halcon, *Ramos & Edaño BS 40690* (A, P, US); San Teodoro, Sibuang River Camp, *Ridsdale 842* (BISH). **Mindoro Occidental Prov.:** Mt. Iglit-Baco National Park, Sitio Sapaton, Barangay Poypoy, Calintaan, *Barbon & Garcia PPI 22578* (BRIT). **Panay. Capiz Prov.:** Jamindan, *Ramos & Edaño BS 31103* (US), *BS 31331* (BM, K, L, NY). **Leyte. Without further data:** *Wenzel 279* (A, US). **Bohol. Bohol Prov.:** Batuan River, *Ramos BS 43169* (A, BM, UC). **Mindanao. Agusan Prov.:** Butuan, Guinabasan, *Mendoza PNH 42119* (L). **Surigao del Norte Prov.:** without further data: *Wenzel 2539* (A, K, MO). **Surigao Prov.:** Mauub Placer, *Wenzel 3303* (GH, K, MO, NY, UC). **Misamis Prov.:** without further data: *Quimpo FB 30223* (NY, UC). **Zamboanga del Norte Prov.:** Fulalh, *Frake PNH 38081* (A, L); Muyar, *Frake PNH 38102* (A, PNH); Dipelawon, *Frake PNH 38283* (A, L). **Lanao Prov.:** Davis Falls, *Zwickey 569* (A). **Basilan Prov.:** Basilan Island, Udiongan, Ayala, *Gaerlan & Ferdinand PPI 26177* (BRIT). **Palawan.** Iwahig River, *Merrill 777* (NY).

Distribution.—Quite widespread, in Luzon, Catanduanes, Mindoro, Panay, Leyte, Bohol and Mindanao.

Ecology.—Wet forest, including disturbed dipterocarp forest and secondary forest; some collections reported on limestone, sea level to 850 m.

Phenology records.—Flowering: January, May to June, August; fruiting: February, June, July, October.

Preliminary conservation assessment.—Near Threatened (**NT**). Last collected in 1995

Distinguishing characteristics.—Glabrous throughout; petioles short,and in some cases leaves nearly sessile, leaf bases rounded, cordate, or at least obtuse; inflorescence nodes relatively close together (primary internode less than 2 cm long) and the lowest branches less than 5 cm long, ascending.

Remarks.—It is quite certain that the type specimen of *P. mindoroensis* was destroyed during the destruction of the Philippine National Herbarium (PNH) during World War II. Elmer (1911:1032) states that the type specimen for *P. mindoroensis* (*Elmer 12532[b]*) is 'only a fragment' and so it is unlikely that there are any duplicates. We have not located *Elmer 12532[b]*, despite searching through the many herbaria in which Elmer specimens are held. We have seen the specimen *Elmer 12532[a]*, the type specimen of *P. longipedunculata*, for example (see above). It seems, therefore, that the only known collection *Elmer 12532[b]* was once held at PNH. Elmer (loc. cit) states that R.C. McGregor discovered *P. mindoroensis* in 1905, along the Baco river, Mindoro. We believe that this record is represented by the specimen *McGreggor 312*, which we have selected as the neotype.

14. Psychotria christarusselliae Sohmer & A.P. Davis, sp. nov. (**Fig. 15**). Type: PHILIPPINES. Bohol, without precise locality, *Ramos BS 43230* (holotype: P; isotypes: A, BM, K).

Liana vel frutex scandeus est, novellis puberulis et radices adventes ferrentibus, stipulis valvatis ovatis 1.5–3 mm longis apice acuto vel acuminato integro, laminis coriaceis glabris lineari-ellipticis vel lineari-lanceolatis 0.4 × 2.3–1 × 5.5 cm latis et longis nervis lateralibus 5–7 in dimidio quoque nervis tertiis conspicuis apice acuto basi acuta vel obtusa, petiolis glabris 1–4 mm longis, inflorescentia monochotoma rhachidi cum pilis multis brevioribus velatis condensatis 1.5–2.2 cm longis 1–2 nodis, ramulis oppositis, pedunculo 2–12 mm longo, floribus paucis minus 10, ramulis et cymulis cum bracteis et bracteolis ovatis 1 mm longis sustentibus, fructibus globosis 5 mm in diametro.

Climber or scandent shrub, height/length not recorded, with adventitious roots; young stems puberulous. Stipules valvate, ovate, 1.5–3 mm long, glabrous; apex broadly acute to acuminate. Leaves: petioles 0.1–0.4 cm long, glabrous; leaf blades linear-elliptic to linear-lanceolate, 2.3–5.5 × 0.4–1 cm, coriaceous, glabrous; secondary veins 5–7 pairs; tertiary venation inconspicuous; base acute to obtuse; apex acute. Inflorescence/infructescence monochotomous, 1.5–2.2 cm long, axes puberulous to pubescent; principal axis (peduncle) 0.2–1.2 cm long, with 1–2 nodes and opposite branching at each node, secondary branches branching once more time, ultimate branches terminating in 1–2 flowers/fruits (few-flowered, < 10 flowers per inflorescence); bracts narrowly triangular to linear, ca. 0.8 mm long, margins entire; bracteoles narrowly triangular, ca. 0.6 mm long. Flowers unknown. Fruit poorly known, globose, ca. 5 mm in diam., otherwise unknown; pyrenes unknown.

Paratypes: None—known only from the type collection.

Distribution.—Known only from Bohol.

Ecology.—Unknown.

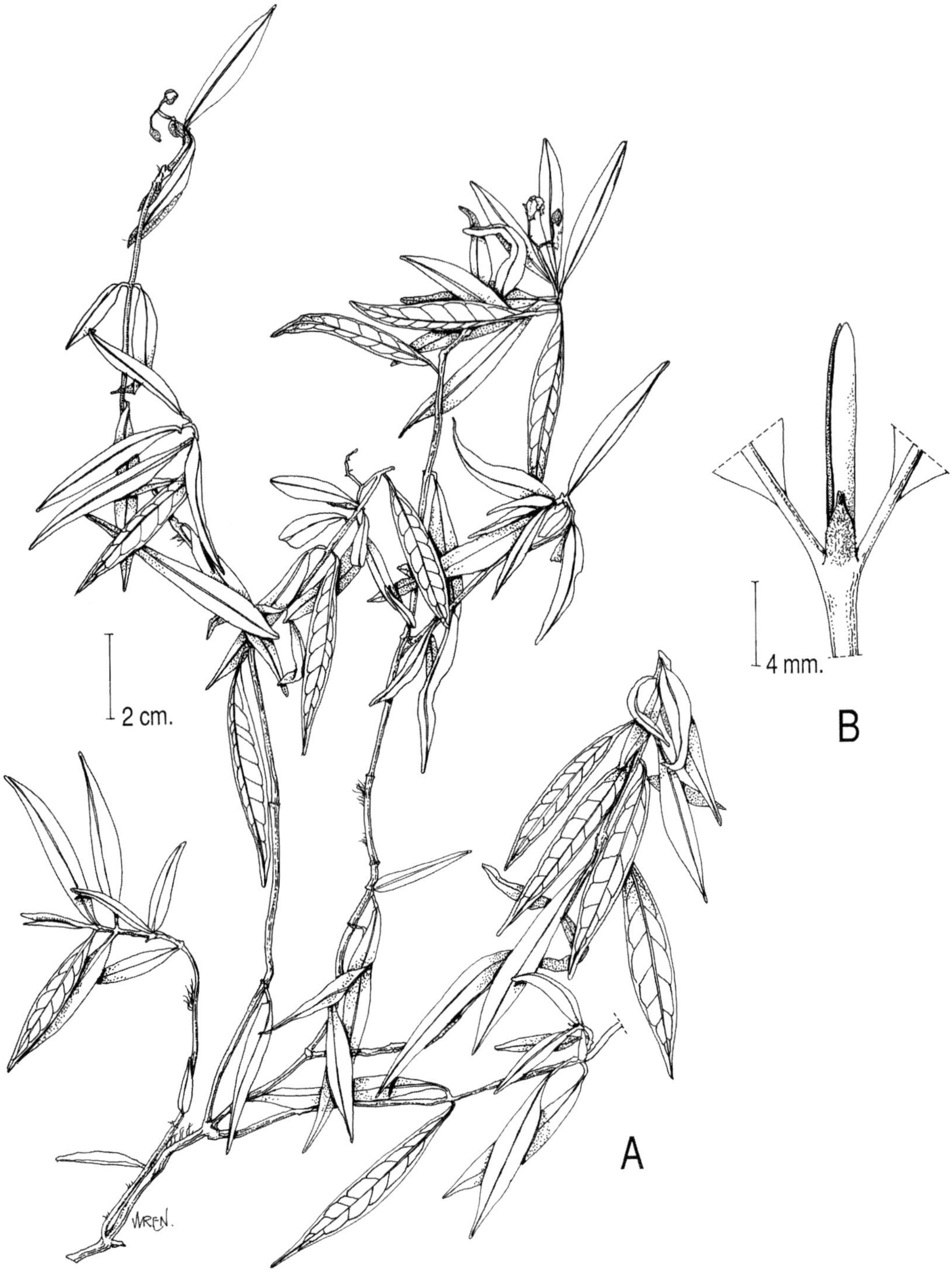

FIG. 15. *Psychotria christarusselliae*. **A,** habit (*Ramos BS 43230*) (P, holotype); **B,** stipule (*Ramos BS 43230*)(A, isotype).

Phenology records.—Flowering: unknown; fruiting: August to October.

Preliminary conservation assessment.—Critically Endangered (**CR b1ab(iii)**), or possibly Extinct (**EX**). Last collected in 1923.

Distinguishing characteristics.—Leaves small, narrow; inflorescence monochotomous, contracted, and less than 3 cm long.

Remarks.—*Psychotria christarusselliae* is named after Christa Russell, who was a primary research assistant during the earlier part of this work.

This taxon does not resemble any other species of climbing Philippine *Psychotria*. The inflorescence separates this taxon from the narrow-leafed variations of *P. ixoroides* Bartl. ex DC.

The specimen *Olsen 246* (L) is a juvenile form of what would be a climbing species of *Psychotria*, which seems close to *P. christarusselliae*. However, the lack of visible lateral veins has precluded us from formally citing it here. Another potential collection that may belong here is *Steiner PNH 37697* (PNH) from Bulusan in Sorsogon Province. The taxon is named for Christa Russell whose assistance was instrumental in advancing this revision of Philippine *Psychotria*.

15. Psychotria elmeriana Hochr., Candollea 5:270. 1934, as 'elmerana.' (**Fig. 16**). Type: as for *P. epiphytica* Elmer. *Psychotria epiphytica* Elmer, Leafl. Philipp. Bot. 5:1890. 1913. Non *Psychotria epiphytica* Krause, Verh. Bot. Vereins Prov. Brandenburg 50:108. 1909. Type: PHILIPPINES. Mindanao: Agusan Prov.: Cabadbaran (Mt. Urdaneta), wet, stony, forested slopes at ca. 1800 m, Oct 1912, *Elmer 14199* (lectotype: A, designated here; isolectotypes: BISH, BM, GH, K, L, MO, NY, P, UC, US, W).

Climber or ?epiphyte, height/length not recorded, with adventitious roots; young stems glabrous. Stipules valvate, ± deltate to triangular, 2–4 mm long, glabrous; apex obtuse to rounded, or acute. Leaves: petioles 0.3–0.6 cm long, glabrous; leaf blades elliptic to elliptic-ovate or elliptic-obovate, 2.5–3.5 × 0.9–1.6 cm, subcoriaceous, glabrous; secondary veins 3–4 pairs; tertiary venation invisible; base obtuse to acute; apex acute to acuminate. Inflorescence/infructescence monochotomous, 4–6.1 cm long, glabrous; principal axis (peduncle) 1–3.4 cm long, with 2–3 primary nodes and opposite branching at each node, secondary branches branching 1 or 2 more times, ultimate branches terminating in cymules; bracts leaf-like, narowly elliptic-oblong, ca. 1.7 cm long (but often caducous), margins entire; bracteoles triangular, ca. 1.5–2.5 mm long. Flowers unknown. Fruit ellipsoid, ca. 4.5 mm long, smooth, white, glabrous; calyx limb and lobes persistent; pedicels ca. 0.5 mm long, or ± absent (subsessile); pyrenes smooth on outer surface; endosperm ruminate.

Other specimens examined: **Mindoro. Mindoro Oriental Prov.:** Laylay, 10 km from San Teodoro, *Sohmer 12297* (BISH, PNH).

Distribution.—Known from Mt. Urdaneta, Mindanao, and possibly from Mindoro.

Ecology.—Forest; ca. 1800 m.

Phenology records.—Flowering: unknown; fruiting: October, November.

Preliminary conservation assessment.—Vulnerable (**VU B1ab(iii)**). Last collected in 1985.

Distinguishing characteristics.—Infructescence monochotomous, the primary internodes less than 2 cm long, the lowest branches less than 5 cm long, ascending, not spreading; leaf blades 2.5–3.5 cm long.

Remarks.—Elmer (1913) reported this species as an epiphyte but we believe that it is a climber.

16. Psychotria ixoroides Bartl. ex DC., Prodr. 4:522. 1830. Merr., Enum. Philipp. Fl. Pl. 3:556. 1923. (**Fig. 17**). Type: PHILIPPINES. Luzon: without precise locality, *Haenke s.n.* (holotype: PR).

Climber or scandent shrub, height/length not recorded, with adventitious roots; young shoots glabrous or very rarely puberulous. Stipules valvate, deltate or ovate, 2–6 mm long, glabrous; apex acute to infrequently acuminate, infrequently with a pair of acuminate tips. Leaves: petioles 0.3–1.9 cm long, glabrous; leaf blades narrowly to broadly elliptic, oblanceolate or narrowly lanceolate, 4–8 (–10.9) × 1.3–3.7 cm, subcoriaceous to coriaceous, glabrous; secondary veins 5–9 pairs; tertiary venation manifest to prominent; base cuneate; apex acute or acuminate. Inflorescence/infructescence monochotomous, 1.5–7.3 cm long, glabrous or sometimes puberulous; principal axis (peduncle) 1–5.1 cm long, with 2–3 primary nodes, usually with verticillate branching at each node, secondary branches branching 1 or 2 more times, ultimate branches terminating in cymules; bracts leaf-like, ± narowly elliptic, 0.6–3.5 cm long, but sometimes absent (?caducous), margins entire; bracteoles triangular to narrowly triangular, 0.3–2.5 mm long. Flowers 5-merous, pedicels 0.5–2 mm long; calyx (incl. hypanthium) 1.5–2 mm long, glabrous or minutely puberulous; calyx lobes tooth-like to deltate 0.1–0.1 mm long; corolla: white; corolla tube ca. 2 mm long, external surface puberulous, villous at throat within; corolla lobes ca. 1.2 mm long; stamens: filaments ca. 1 mm long (short styled); anthers ca. 1 mm long; style ca. 2 mm long, lobes unknown. Fruit obovoid-ellipsoid, 4–6 mm long, ribbed,

Fig. 16. *Psychotria elmeriana* (*Elmer 14199*) (A, lectotype for *Psychotria epiphytica*).

Fig. 17. *Psychotria ixoroides* (*Haenke s.n.*) (PR, holotype).

white, glabrous; calyx limb and lobes persistent; pedicels 0.5–2 mm long; pyrenes ribbed on outer surface; endosperm mostly ruminate (entire in one species examined).

Other specimens examined: **Luzon. Ilocos Norte Prov.:** Burgos, *Ramos BS 32819* (A, BM, K P, US). **Isabela Prov.:** Palanan Bay, *Escritor BS 21181* (US). **Pangasinan Prov.:** Mt. San Isidro, Labrador, *Fenix BS 30001* (A, BM, P). **Nueva Ecija Prov.:** Mt. Umingan, *Ramos & Edaño BS 26313* (A, US). **Bulacan Prov.:** Angat, *Ramos & Edaño BS 34085* (A, K, UC); Angat Watershed Area, Barangay San Lorenzo, Norzagaray, *Garcia et al. PPI 15151* (BRIT), *Garcia et al. PPI 15197* (BRIT, K). **Rizal Prov.:** Mt. Susong-Dalaga, *Ramos & Edaño BS 29310* (A, P, US); Mt. Irid, *Ramos & Edaño BS 48464* (NY, UC); San Isidro, *Ramos BS 336* (US); **Rizal Prov.:** without further data: *Ramos BS 2674* (L). **Bataan Prov.:** Mt. Mariveles, *Merrill 3253* (BM), *3887* (BM, K), *Whitford 253* (K, NY); **Bataan Prov.:** without further data: *Topacio FB 19086* (BISH, US), *Topaico BS 20027* (K). **Camarines Sur Prov.:** Lupi, Sooc, BNP, *Reynoso, Romero & Fuentes PPI 1459* (BRIT); Barangay Guijalo, *Reynoso, Sagcal & Fuentes PPI 14621* (BRIT); *Reynoso, Sagcal & Fuentes PPI 14626* (BRIT); Mt. Asog, Iriga, Barbon, *Romero & Fuentes PPI 8462* (BRIT, K). **Quezon Prov.:** Mt. Cadig, *Yates BS 25437* (A); Tagkawayan, *Mendoza PNH 97728* (L); Tagkawayan, Barangay Bagong Silang II, *Barbon, Garcia & Farnando PPI 12310* (K); Quezon National Park, *Sohmer 12105* (BISH); Quezon National Park, *Castro, Barbon & Garcia PPI 22174* (BRIT); Polillo City, *Fox PNH 9183* (A); Polillo Island, without further data, *Robinson BS 9056* (L, P), *Salvoza FB 29692* (UC); Alabat Island, Quezon National Park, Malikboy, Pagbilao, Pantayen to Silio River trial, *Gaerlan & Romero PPI 23672* (BRIT); Alabat Island, without further data, *Ramos & Edaño BS 47979* (NY, UC). **Sorsogon Prov.:** Mt. Kapung Kayan, Bulusan Lake, *Sulit & Gojar PNH 2687* (A, L); Mt. Bulusan National Park, Barangay Inlagadian, Casiguran, *Garcia & Fernando PPI 22826* (BRIT); Albay, *Vidal 3078* (A, K); Albay, Castilla, *Vidal 3021* (K), *Cuming 1321* (K). **Luzon. Without further data:** *Haenke s.n.* (MO), *Vidal 402* (A, K). **Catanduanes. Catanduanes Prov.:** Simamla, *Ramos & Edaño BS 75546* (NY). **Mindoro. Mindoro Oriental Prov.:** Inland from San Teodoro, at and above Subaan Reserve, *Ridsdale 1167* (K); Puerto Galera, *Bartlett 13737* (A), *Ramos BS 46357* (NY, P, UC). **Mindoro Occidental Prov.:** Paluan, *Ramos BS 39547* (A, GH, P, UC). **Mindoro. Without further data**: *Merritt FB 12073* (US). **Eastern Samar Prov.:** East of Station, Sitio Cantinoc, Barangay 1, Balangiga, *Gaerlan & Chavez PPI 26328* (BRIT). **Samar. Western Samar Prov.:** Paranas, Campo Uno, Central Samar, *Reynoso & Majaducon PPI 24069* (BRIT, K), *Reynoso & Majaducon PPI 24176* (BRIT). **Leyte. Leyte Prov.:** Palo, *Elmer 7112* (K, NY); Mt. Top, Dagami, *Frohne PNH 35688* (A, L); Biliran Island, Mt. Suiro, *Sulit PNH 21504* (L, PNH, SING); Biliran Island, Libtong, Naval, *Barbon, Alvarez & Fuentes PPI 8560* (BRIT); Bailong, *Gaerlan, Sagcal & Romero PPI 10491* (BRIT, K). **Panay. Aklan Prov.:** Libacao, *Martelino & Edaño BS 35385* (NY); Antique Prov.: lower portion of Mt. Baloy, Barangay San Agustin, Valderama, *Garcia & Romero PPI 15715* (BRIT), *Garcia & Romero PPI 15751* (BRIT). **Camotes/Cebu. Cebu Prov.:** Alcoy, Manguera Forest, *Madulid & Reynoso PPI 27372* (BRIT, K). **Mindanao. Surigao Prov.:** Sukailang, *Mendoza PNH 10247* (A, L); Hinatuan, *Piper BS 502* (L); **Surigao del Norte Prov.:** Siagaro Island, Poneas Islet, Sitio Dabo, Sunkoi Hill, *Stone, Barbon & Garcia PPI 9438* (BRIT); Siargao Island, without further data, *Ramos & Pascasio BS 34920* (A); Dinagat Island, Mabini, Tubajon, *Gaerlan et al. PPI 16893* (BRIT); Dinagat Island, without further data, *Ramos & Convocar BS 83977* (A). **Surigao del Sur Prov.:** Hinatuan, *Piper BS 502* (L). **Surigao Prov.:** without further data: *Mallonga FB 26267* (A), *Ramos & Pascasio BS 34483* (NY), *BS 34551* (A, UC). **Davao Prov.:** Mt. Galintan, *Ramos & Edaño BS 48916* (BM, NY, UC). **Palawan. Palawan Prov.:** Puerto Princesa, *McGregor BS 45958* (UC); Mt. Iraan, Aborlan, *Sulit PNH 12546* (A, L); Mt. Victoria, *Edaño PNH 14207* (A, L); Barangay Bundog, near Sumindap River, *Pipoly & Romero et al. PPI 38268* (BRIT); Barangay Bundog, Sumindap River, *Pipoly & Romero et al. PPI 38248* (BRIT), *Pipoly & Romero et al. PPI 38268* (BRIT, K); Rizal, Ransang, Madubis, *Gaerlan, Romero & Sagcal PPI 13298* (BRIT, K); Mt. Mantalingahan, Brooke's Point, *Edaño PNH 28* (A, PNH), *PNH 31* (A); San Antonio Bay, *Merrill 837* (K, NY, US); Pasig Bay, *Kondo & Edaño PNH 36546* (BISH, PNH); Taytay, Lake Manguao, *Stone & Reynoso et al. PPI 290* (BRIT, K); Balabac Island, Panaan, *Ramos & Edaño BS 49871* (NY, UC).

Distribution.—A ubiquitous species found on nearly all the main Philippine Islands.

Ecology.—Disturbed primary forest and secondary forest, also reported from a logged-over dipterocarp forest and a coconut plantation; on limestone and laterite; sea level up to 700 m.

Phenology records.—Flowering: January to June; fruiting: February, August, September.

Preliminary conservation assessment.—Least Concern (**LC**). Last collected in 1997.

Distinguishing characteristics.—Monochotomous inflorescences; leaf blades usually 4–8 cm long, glabrous or at most puberulent; primary inflorescence internodes less that 2 cm long, branches ascending, and relatively closely placed, leaf-like bracts often present; corolla minutely puberulous.

Remarks.—This was one of the earliest species described from the Philippines. It was among the collections made in the archipelago by Thaddaeus Haenke, a Czech botanist, during the Malaspina Expedition in the latter part of the eighteenth century. The type preserved at PR is in very good condition and we are assuming that it was the specimen De Candolle had in hand when describing his new species. This is a very common climber in the Philippines and demonstrates a great deal of variability. One of the variations is towards very narrow-leafed individuals but we have not described such variation formally, mainly because a geographical pattern seems to be absent. We wish to point out that the following collections (cited above) are good examples of this variation: *Sulit PNH 21504* (Biliran Island), *Ramos BS 43230* (Bohol Island) and, *Ramos & Edaño BS 75546* (Catanduanes). Another collection we refer to here for lack of more information is *Sulit & Gojar PNH 2687*. This collection from the southern part of Luzon is somewhat distinct in the close placement of the nodes at the branch tips and the stipules, which appear to be thicker and blunter than those demonstrated by other collections cited for this taxon. The relationship between this taxon and *P. sarmentosa* Blume may be very close.

17. Psychotria cuernosensis Elmer, Leafl. Philipp. Bot. 2:521. 1908; Merr., Enum. Philipp. Fl. Pl. 3:554. 1923. (**Fig. 18**). TYPE: PHILIPPINES. NEGROS: Negros Oriental Prov.: Cuernos Mountains, Dumaguete, mossy forests at ca. 1500 m, May 1908, *Elmer 10076* (LECTOTYPE: L, designated here; ISOLECTOTYPES: A, BISH, BM, GH, K, MO, NY, P, US, W).

Psychotria mearnsii Merr., Philipp. J. Sci. 10:126. 1915. TYPE: PHILIPPINES. MINDANAO: Misamis Prov.: Mt. Malindung, along Malabug River at ca. 500 m, May 1906, *Mearns & Hutchinson FB 4770* (LECTOTYPE: US, designated here).

Climber or scandent shrub (tangled branches occur several meters from ground), exact height/length not recorded, with adventitious roots; young stems glabrous. Stipules valvate, ovate, 3–7 mm long, glabrous or pubescent; apex obtuse to acute with 1–2 aristate tips. Leaves: petioles 0.2–0.8 cm long, usually glabrous; leaf blades elliptic to broadly lanceolate or ovate, (1.6–)2.1–4.2(–6.1) × 0.7–2.3 cm, subcoriaceous to chartaceous, glabrous above, glabrous or infrequently pubescent below at least on veins; secondary veins (4–)5(–6) pairs; tertiary venation prominent to obscure; base acute to obtuse; apex acute to acuminate. Inflorescence/infructescence monochotomous (often appearing trichotomous due to a short principal axis), 1.5–6 cm long, glabrous to pubescent; principal axis (peduncle) 0.3–2.5 cm long, with 2–3 primary nodes and opposite branching at each node, each branch branched once more or terminating in a cymule (with < 70 flowers per inflorescence and often much fewer); bracts deltate to narrowly triangular, 1–1.5 mm long, margins distinctly serrate; bracteoles deltate to narrowly triangular, ca. 0.3 mm long, margins usually distinctly serrate. Flowers 4- or 5-merous, pedicels 0.9–2 mm long; calyx (incl. hypanthium) 1.7–2 mm long, glabrous; calyx lobes shallowly triangular, 0.4–0.9 mm long; corolla: color unknown; corolla tube 2–4(–6) mm long, external surface glabrous, densely villous at throat within; corolla lobes 1–2 mm long; stamens: filaments ca. 0.5 mm (long styled) or 1–2 mm long (short styled); anthers ca. 0.75 mm (long styled) or ca. 1 mm long (short styled); style 3–4 mm long (long styled) or 1–2 mm long (short styled), lobes ca. 0.7 mm (short styled) or 1.5 mm long (long styled). Fruit obovoid-ellipsoid, 6–7 mm long, ribbed, white, glabrous; calyx limb and lobes persistent; pedicels 1.5–2 mm long; pyrenes ribbed on outer surface; endosperm entire.

Other specimens examined: **Mindoro. Mindoro Oriental Prov.:** north coast, Subaan River headwaters, inland from San Teodoro, *Coode 5662* (K, L); Mt. Halcon, *Edaño PNH 3253* (PNH), *PNH 3306* (A), *PNH 3366* (PNH), *Merrill 6170* (NY), Merrill *6176* (NY, US), *Ridsdale 1714* (K). **Prov. unknown:** Mt. Ilong, *Edaño PNH 3350* (A, L, PNH); **Panay. Capiz Prov.:** Mt. Bulilao, *Martelino & Edaño BS 35739* (K, P); Mt. Madiaas, *Ramos & Edaño BS 30646* (NY). **Antique Prov.:** Culasi, Hangod Tubig, Mt. Madia-as, *Gaerlan & Fuentes PPI 4940* (BRIT, K); western side of Mt. Baloy, Iki, Valderama, *Garcia & Romero PPI 15837* (BRIT). **Antique Prov.:** without further data. *McGregor BS 32237* (NY). **Leyte. Without further data:** *Wenzel 762* (BM, MO). **Negros. Negros Oriental Prov.:** Mt. Malbug, *Edaño PNH 7121* (A, L, PNH); Cuernos de Negros, *Edaño PNH 7188* (A, PNH), *PNH 7193* (A, PNH); Matangwa River, *Edaño PNH 7218* (A, L, PNH). **Negros Occidental Prov.:** Mt. Canlaon, *Edaño PNH 21970* (K, L, PNH); Negros Island, Canalon National Park, Mt. Canlaon, *Stone et al. PPI 6205* (BRIT, K). **Mindanao. Surigao Prov.:** Mt. Kabatuan, Alegria, *Mendoza & Convocar PNH 10551* (A). **Agusan Prov.:** Mt. Malaampayan, *Mendoza & Convocar PNH 10818* (A, PNH). **Misamis Occidental Prov.:** Mt. Malimdang, *Gaerlan, Sagcal & Romero PPI 10797* (K); *Gaerlan, Sagcal & Romero PPI 10836* (BRIT, K). **Bukidnon Prov.:** Mt. Candoon, *Ramos & Edaño BS 38787* (K, P); Mt. Katanglad, *Sulit PNH 10075* (A, BM, K, L, PNH). **Davao Prov.:** Mt. Apo, *Clemens BS 15664* (NY, UC), *s.n.* (UC), *Elmer 11424* (A, BISH, BM, GH, K, L, MO, NY, W), *11503* (A, BISH, BM, GH, K, L, MO, NY, W); Kidapawan, Mt. Apo, *Gaerlan, Alvarez & Garcia PPI 2533* (K).

Distribution.—Luzon, Mindoro, Leyte, Panay, Negros, and particularly Mindanao.

Ecology.—Primary and secondary forest, including high altitude 'moss forest'; 500–2000 m.

Phenology records.—Flowering: January to August; fruiting: March to June.

Preliminary conservation assessment.—Least Concern (**LC**). Last collected in 1992.

Distinguishing characteristics.—Endosperm entire (non ruminate); relatively compact inflorescence with far fewer than 70 flowers; corolla tube 2–4 mm long.

Remarks.—The leaves of this species look superficially like those of *P. diffusa*, and many sheets of *P. cuernosensis* are identified as *P. diffusa*. However, the nature of the inflorescence is enough to distinguish *P. cuernosensis* from *P. diffusa*. Merrill (l.c.), compared his *P. mearnsii* with *P. elliptifolia* Elmer, from which it is easily separable. Had he compared *P. mearnsii* with *P. cuernosensis* he may not have described his new taxon.

Fig. 18. *Psychotria cuernosensis* (*Elmer 10076*) (L, lectotype).

18. Psychotria elliptifolia Elmer, Leafl. Philipp. Bot. 1:66. 1906; Merr., Enum. Philipp. Fl. Pl. 3:555. 1923. (**Fig. 19**). TYPE: PHILIPPINES. LUZON: Quezon Prov.: Mt. Banahão, wet mossy region at 1600 m, Apr 1906, *Elmer 7943* (LECTOTYPE: A, designated here; ISOLECTOTYPES: K, L, NY, US, W).

Climber or scandent shrub, height/length not recorded, with adventitious roots; young stems glabrous to pubescent. Stipules valvate, ovate-oblong, (1.2–)2–5(–8.7) mm long, sometimes distinctly chartaceous, glabrous; apex obtuse or rounded, usually with 1 aristate tip. Leaves: petioles 0.3–0.5 cm long, glabrous; leaf blades elliptic, oval-elliptic to oblanceolate, 2–4.1 × 1–1.7 cm, subcoriaceous, glabrous, or rarely with pubescence on midvein below; secondary veins 4–6 pairs; tertiary venation manifest to obsure, or ± invisible; base acute to cuneate; apex acute or apiculate. Inflorescence/infructescence trichotomous, 1.6–5 cm long, glabrous; principal axis 0.6–2 cm long, each axis with 2–3 primary nodes and opposite branching at each node, secondary branches branching 1 or 2 more times, ultimate branches terminating in cymules; bracts very broadly ovate, 0.5–0.7 mm long, margins distinctly fimbriate; bracteoles like the bracts, 0.2–0.4 mm long. Flowers 4- or 5-merous, pedicels 0.5–1 mm long; calyx (incl. hypanthium) 1–1.8 mm long, glabrous; calyx lobes ± obsolete, or triangular, ca. 0.5 mm long; corolla: white; corolla tube ca. 1 mm long, external surface glabrous, internal surface glabrous; corolla lobes ca. 1 mm long; stamens: filaments ca. 1 mm long; anthers ca. 0.5 mm long (?long styled); style 1–2 mm long, lobes ca. 0.5 mm long. Fruit ellipsoid-obovoid or ellipsoid, 3.1–4.3(–6) mm long, striated, white, glabrous; calyx limb and lobes persistent; pedicels 1.5–2.5 mm long; pyrenes ribbed on outer surface; endosperm weakly ruminate.

Other specimens examined: **Luzon. Abra Prov.:** without further data: *Ramos BS 7061* (P). **Mountain Prov.:** Bontoc, *Van Steenis 17945* (L, PNH); Mt. Pukis, *Ramos & Edaño BS 37820* (A, K, P, US); [Pauai (?)] Benguet to Mt. Data, *Clemens 16418* (UC); Mt. Data, *Stern & Rojo 2316* (PNH); Barangay Garling, Mt. Amuyao, Barlig, *Reynoso, Garcia & Sagcal PPI 14598* (BRIT, K). **Nueva Vizcaya Prov.:** Caraballo mountain, *Loher 13645* (UC), *s.n.* (UC 229265); Mt. Umingan, *Ramos & Edaño BS 26300* (A, K, NY, P, US). **Rizal Prov.:** Mt. Angilog, *Ramos BS 40782* (A, K P). **Quezon Prov.:** Lucban, *Elmer 8061* (A, K, L, MO, NY, W). **Laguna Prov.:** Mt. Banahao, *Ocampo BS 27994* (A, MO).

Distribution.—Restricted to central and north central Luzon.

Ecology.—Forest, including stunted 'mossy' forest; 1500–2000(–2700) m.

Phenology records.—Flowering: February to May, August, November; fruiting: August, September.

Preliminary conservation assessment.—Vulnerable (**VU B1ab(iii)**). Last collected in 1994.

Distinguishing characteristics.—Scandent shrub or climber; stipules relatively large, leaf-like, often persistent and drying orange-brown; inflorescence trichotomous, relatively contracted, and usually under 5 cm long; flowers few, never more than 70; corolla tubes ca. 1 mm long, and glabrous at throat.

Remarks.—This taxon is similar to *P. cuernosensis* in many ways, including relatively few-flowered and contracted inflorescences, and would be easy to combine with it. However, the stipules are conspicuously different, and the flowers are less than half the size than in *P. cuernosensis*. While studying representatives of these taxa we sought evidence to indicate that the difference in flower size was merely the result of heteromorphy, which is so common in *Psychotria* and other Rubiaceae. We did not, however, find conclusive evidence for this one way or the other, and the difference in the stipules convinced us that these taxa should remain distinct.

19. Psychotria lianoides Elmer, Leafl. Philipp. Bot. 3:1029. 1911; Merr., Enum. Philipp. Fl. Pl. 3:557. 1923. (**Fig. 20**). TYPE: PHILIPPINES. MINDANAO: Davao Prov.: Todaya (Mt. Apo), humid forests along Baruring River on trail to Talon at ca. 1300 m, Sep 1909, *Elmer 11645* (LECTOTYPE: P, designated here; ISOLECTOTYPES: A, BISH, BM, GH, K, L, MO, NY, US, W).

Climber or scandent shrub (forming dense masses over the tops of tree), height/length not recorded, adventitious roots not recorded; young stems glabrous. Stipules unknown. Leaves: petioles 0.8–1.4 cm long, glabrous; leaf blades ovate-elliptic, 5.5–9 × 3.3–6 cm, stiff-chartaceous to coriaceous (apparently somewhat folded along midrib in field, i.e. conduplicate), glabrous above, pubescent below on midrib; secondary veins 9–10 pairs; tertiary venation manifest; base truncate to subcordate; apex abruptly acuminate. Inflorescence/infructescence trichotomous, 8–21 cm long, puberulous to pubsecent, sometimes glabrous; principal axis 5.5–9.5 cm long, each axis with 3–4 primary nodes and opposite branching at each node, secondary branches branching 2–4 more times, ultimate

FIG. 19. *Psychotria elliptifolia* (*Elmer 7943*) (A, lectotype).

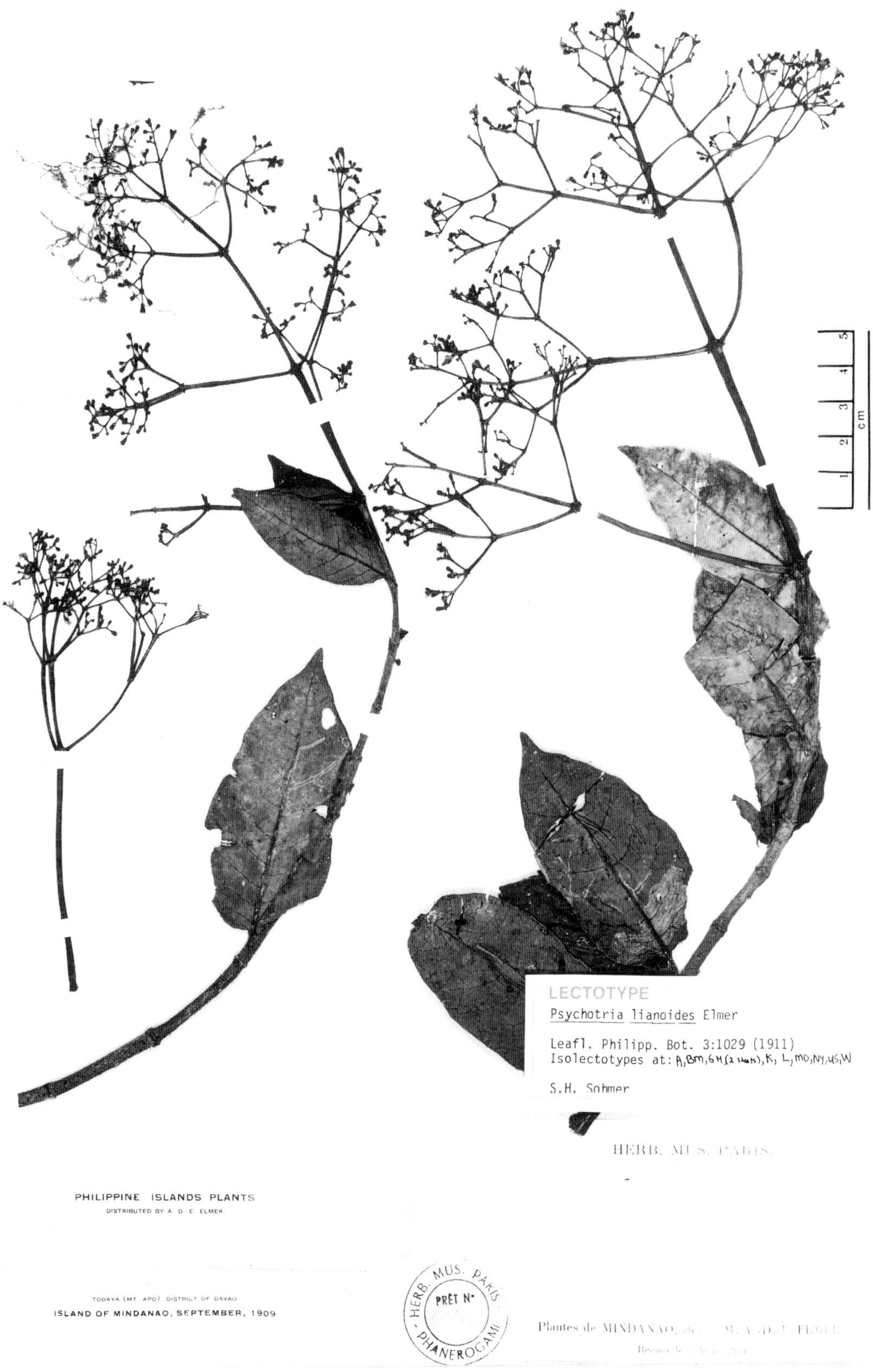

FIG. 20. *Psychotria lianoides* (*Elmer 11645*) (P, lectotype).

branches terminating in numerous cymules; bracts leaf-like, narrowly ovate-elliptic, ca. 3.5 cm long (but often caducous), margins entire; bracteoles triangular to narrowly triangular, ca. 0.7 mm long, margins serrate (?colleters). Flowers 4-merous, pedicels 2.2–3.5 mm long; calyx (incl. hypanthium) ca. 1.4 mm long, glabrous; calyx lobes ± obsolete; corolla: white, corolla tube ca. 2 mm long, external surface glabrous, ? internal surface glabrous; corolla lobes ca. 2 mm long; stamens: filaments 2–3 mm long (short styled); anthers ca. 0.5 mm long; style ca. 2 mm long (short styled), lobes ca. 0.5 mm long. Fruit ellipsoid, 3–4 mm long, ± smooth, white, glabrous; calyx limb and lobes persistent; pedicels ca. 3.5 mm long; pyrenes smooth on outer surface; endosperm unknown.

Other specimens examined: Known only from the type collection.

Distribution.—Known only from Mt. Apo, Mindanao.

Ecology.—Humid forest; 1200–1300 m.

Phenology records.—Flowering: September; fruiting: unknown.

Preliminary conservation assessment.—Critically Endangered (**CR B1ab(iii)**). Only collected once in 1909.

Distinguishing characteristics.—Strong climbing habit; leaves truncate to subcordate at base, and reported to be conduplicate in the field; inflorescence large, diffuse; flowers small, corolla tubes ca. 2 mm long.

Remarks.—The nature of the flower and inflorescence relates this taxon to the *P. diffusa*, but the leaves appear to indicate a relationship with *P. ramosissima*.

20. Psychotria ramosissima Elmer, Leafl. Philipp. Bot. 1:355. 1908; Merr., Enum. Philipp. Fl. Pl. 3:562. 1923. (**Fig. 21**). TYPE: PHILIPPINES. LUZON: Quezon Prov.: Lucban, dense shrubby forest of gulches at ca. 200 m, May 1907, *Elmer 9178* (LECTOTYPE: US, designated here; ISOLECTOTYPES: A, K, L, MO, W).

Climber or scandent shrub, height/length not recorded (but climbing relatively high onto the branches of tall trees), with adventitious roots; young stems glabrous. Stipules valvate, ovate-deltate, 3.5–6 mm long, glabrous; apex round or obtuse. Leaves: petioles (0.5–)1–2.1 cm long, glabrous; leaf blades oblong-elliptic to ovate-oblong, 6.8–14.2 × 2.4–6 cm, thick-chartaceous to subcoriaceous (often reported as folded abaxially along its midrib (conduplicate) in the field), glabrous above, pubescent on veins below; secondary veins (6–)8–10 pairs; tertiary venation manifest to obscure; base obtuse to acute; apex abruptly or short-acuminate. Inflorescence/infructescence trichotomous, (10.5–)15–30 cm long, glabrous, or puberulous to pubescent; principal axis 5–15 cm long, each axis with 3–4 primary nodes and opposite branching at each node, secondary branches branching 2–4 more times, ultimate branches terminating in large cymules, flowers numerous; bracts absent (?caducous); bracteoles triangular, ca. 0.6 mm long, pubescent. Flowers 4- or 5-merous, pedicels 2.7–5.2 mm long; calyx (incl. hypanthium) 0.4–0.6 mm long, glabrous; calyx lobes tooth-like; corolla: light cream to white; corolla tube 2–4 mm long, external surface glabrous, densely villous at throat within; corolla lobes 1–2 mm long; stamens: filaments 2–3 mm long (short styled); anthers ca. 0.5 mm long; style 1.5–2 mm long, lobes ca. 0.5 mm long. Fruit ellipsoid, 3–4.1 mm long, smooth, probably white, glabrous; calyx limb and lobes persistent; pedicels 2.7–5.5 mm long; pyrenes smooth on outer surface; endosperm ruminate.

Other specimens examined: **Luzon. Sorsogon Prov.:** Lake Bulusan, *Sulit & Gojar PNH 2633* (A, L, MO). **Samar. Northern Samar Prov.:** Catubig River, *Ramos BS 24313* (A, P), *BS 24498* (A, BISH, BM, GH, L, MO, NY, P, US). **Western Samar Prov.:** Wright, *Sulit PNH 6074* (A, L). **Leyte. Without further data:** *Wenzel 693* (A, BM, GH, MO). **Bohol. Bohol Prov.:** without further data: *Ramos BS 43062* (A, W).

Distribution.—The islands of Luzon, Leyte, Samar, and Bohol.

Ecology.—Forest; ca. 100–600 m.

Phenology records.—Flowering: February to May, July, August, October; fruiting: unknown.

Preliminary conservation assessment.—Vulnerable (**VU B1ab(iii)**). Last collected in 1947.

Distinguishing characteristics.—Robust climber producing tangles of branches at the tops of the vegetation as it climbs; large inflorescences; leaves folded or conduplicate when living, base acute to obtuse; flowers small, corolla tubes generally 1–2 mm long.

Remarks.—*Psychotria ramosissima* is very similar to *P. lianoides* Elmer, but has obtuse to acute leaf bases as compared to the truncate or subcordate bases of the latter.

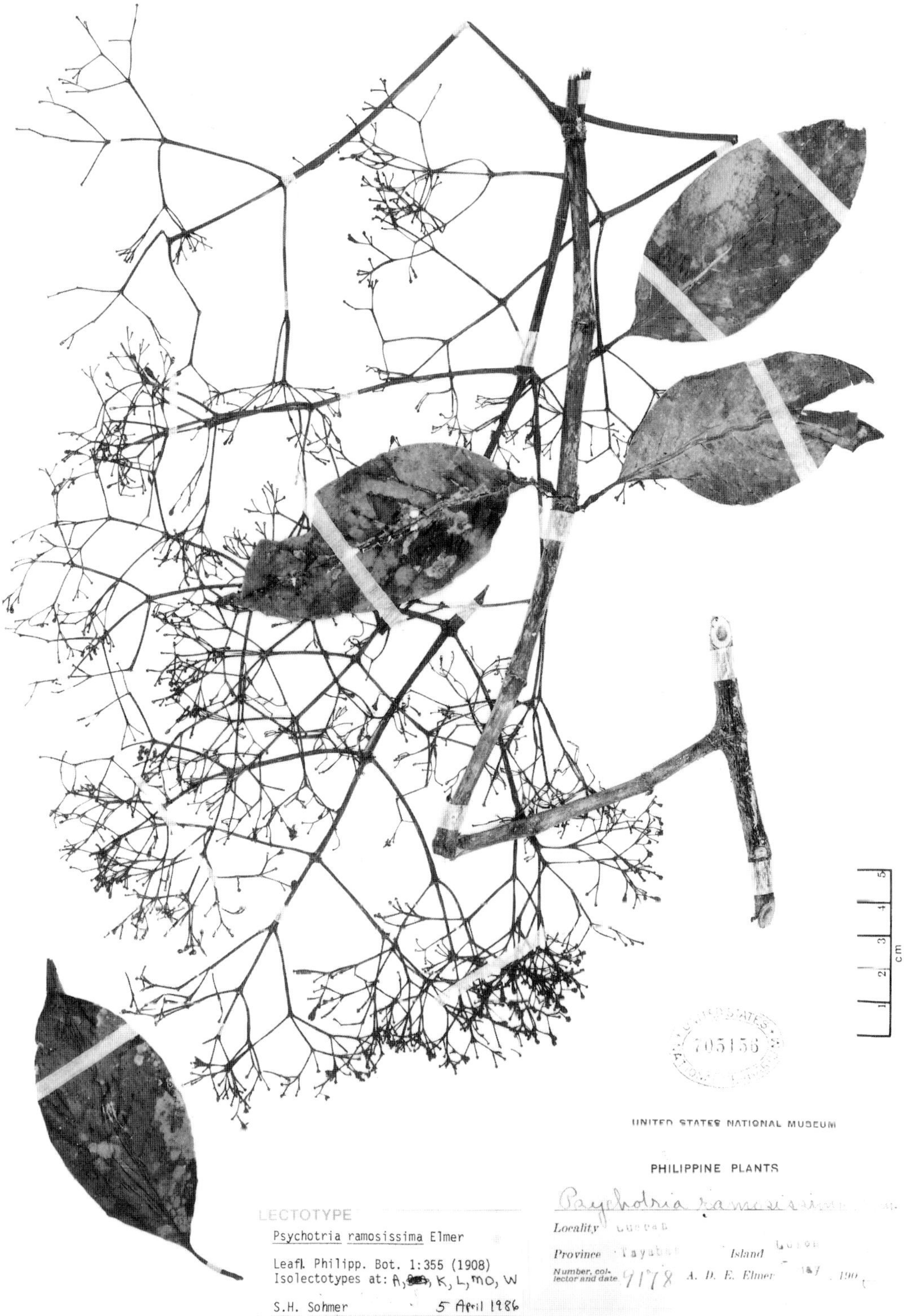

Fig. 21. *Psychotria ramosissima* (*Elmer 9178*) (US, lectotype).

21. Psychotria lanaensis (Merr.) Merr., Enum. Philipp. Fl. Pl. 3:557.1923. *Grumilea lanaensis* Merr., Philipp. J. Sci. 10:136. 1915. TYPE: PHILIPPINES. MINDANAO: Lanao Prov.: Lanao, Camp Keithley, in swamps at ca. 750 m, Jun 1907, *Clemens 924* (not seen).

Climber, height/length not recorded, with adventitious roots; young glabrous. Stipules poorly known, ?broadly ovate, ca. 3 mm wide; apex rounded. Leaves: petioles 0.5–0.7 cm long, glabrous; leaf blades elliptic to elliptic-obovate, 4.9–6.4 × 1.5–2.5 cm, chartaceous to subcoriaceous, glabrous; secondary veins 4–7 pairs; tertiary venation prominent; base acute to obtuse; apex acute or abruptly acuminate. Inflorescence/infructescence monochotomous, 3.4–6.6 cm long, glabrous or puberulous; principal axis (peduncle) 2.8–4.2 cm long, with 2–3 primary nodes and opposite branching at each node, secondary branches branching 2–3 more times, ultimate branches terminating in cymules; bracts leaf-like, narrowly elliptic, or elliptic-obovate, 1.5–3.5 cm long, margins entire; bracteoles narrowly triangular to deltate, 0.5–1.5 mm long. Flowers poorly known (immature), 5-merous, pedicels ca. 0.5 mm long; calyx (incl. hypanthium) 1.5–1.8 mm long, puberulous; calyx lobes ± deltate, ca. 0.5 mm long; corolla: color unknown; corolla tube ca. 2 mm long, external surface puberulous, villous at throat within; corolla lobes ca. 1.8 mm long; stamens: unknown; style unknown. Fruit ovoid, ca. 5 mm long, ± smooth, white, glabrous; calyx limb and lobes persistent; pedicels ca. 0.5 mm long; pyrenes smooth on outer surface; endosperm prominently ruminate.

Other specimens examined: **Luzon. Mountain Prov.:** Baynian, Banaue, *Conklin & Buwaya PNH 80449* (L). **Mindoro. Minodoro Occidental Prov.:** Mt. Calavite, *Ramos BS 39370* (A, K, L, P). **Sibuyan. Romblon Prov.:** Mt. Giting-giting, *Reynoso & Espirita PPI 118622* (K). **Mindanao. Surigao del Norte Prov.:** Dinagat Island, KROMINCO, Mt. Redondo, open pit mine, *Gaerlan, Sagcal & Fernando PPI 4661* (BRIT).

Distribution.—Luzon, Mindoro, Sibuyan, and Mindanao.

Ecology.—Forest (primary and disturbed), including swamp forest; 650–700 m.

Phenology records.—Flowering: April; fruiting: May, June, September.

Preliminary conservation assessment.—Near Threatened (**NT**). Last collected in 1991.

Distinguishing characteristics.—Inflorescence strongly trichotomous, bracts relatively long, narrow and often conspicuous; leaves glabrous with prominent tertiary venation; corolla tube ca. 2 mm long and puberulent on the external surface.

Remarks.—Other than inflorescence morphology, there is little to distinguish *P. lanaensis* from *P. ixoroides*, although Merrill (l.c.) states that its fruit are very different from those of the latter. Most *P. lanaensis* speciemens lack flowers and fruits, and we are uncertain as to what the specimens cited here from Mindoro and Luzon really represent. Therefore, until some representation of the type collection is found or collections from the type locality are made—should forests still exist there—the relationships and status of this taxon will remain uncertain. It should be noted, however, that Merrill (1915b:137) clearly says that the fruits of *P. lanaensis* are entirely different from *P. ixoroides*, as the pyrenes are stated to be smooth, and not at all ribbed.

22. Psychotria diffusa Merr., Philipp. J. Sci. 1 (Suppl. 1):134. 1906; Merr., Enum. Philipp. Fl. Pl. 3:554. 1923. (**Fig. 22**). TYPE: PHILIPPINES. LUZON: Bataan Prov.: Mt. Mariveles, Nov 1904, *Elmer 6811* (LECTOTYPE: NY, designated here; ISOLECTOTYPE: K).

Psychotria microphylla Elmer, Leafl. Philipp. Bot. 2:522. 1908. TYPE: PHILIPPINES. NEGROS: Negros Oriental Prov.: Cuernos Mountains, Dumaguete, moss-covered forests at 4500 m, Mar 1908, *Elmer 9505* (LECTOTYPE: L, designated here; ISOLECTOTYPES: A, BISH, BM, K, MO, NY, US, W).

Climber or scandent shrub (with climbing branches), height/length not recorded, with adventitious roots; young stems sparsely pubescent. Stipules probably valvate, ovate to narrowly ovate, 3–11 mm long, glabrous or pubescent; apex acute-acuminate or with 2 aristate points. Leaves: petioles 0.5–1.8 cm long, glabrous or pubescent; leaf blades ovate-elliptic, elliptic to obovate-elliptic, or broadly elliptic, (3.2–)3.5–7(–9) × 1.5–4.5 cm, chartaceous to subcoriaceous, glabrous above and below, or pubescent to puberulous on veins below, sometimes densely pubescent below; secondary veins 4–8 pairs; tertiary venation manifest to obscure; base acute to narrowly acute, truncate, round, or nearly cordate, rarely obtuse; apex abruptly acute or acuminate. Inflorescence/infructescence trichotomous, 5–19.5(–23) cm long, glabrous to densely pubescent; principal axis 3–8 cm long, each axis with 2–3 primary nodes and opposite branching at each node, secondary branches branching 2–4 more times,

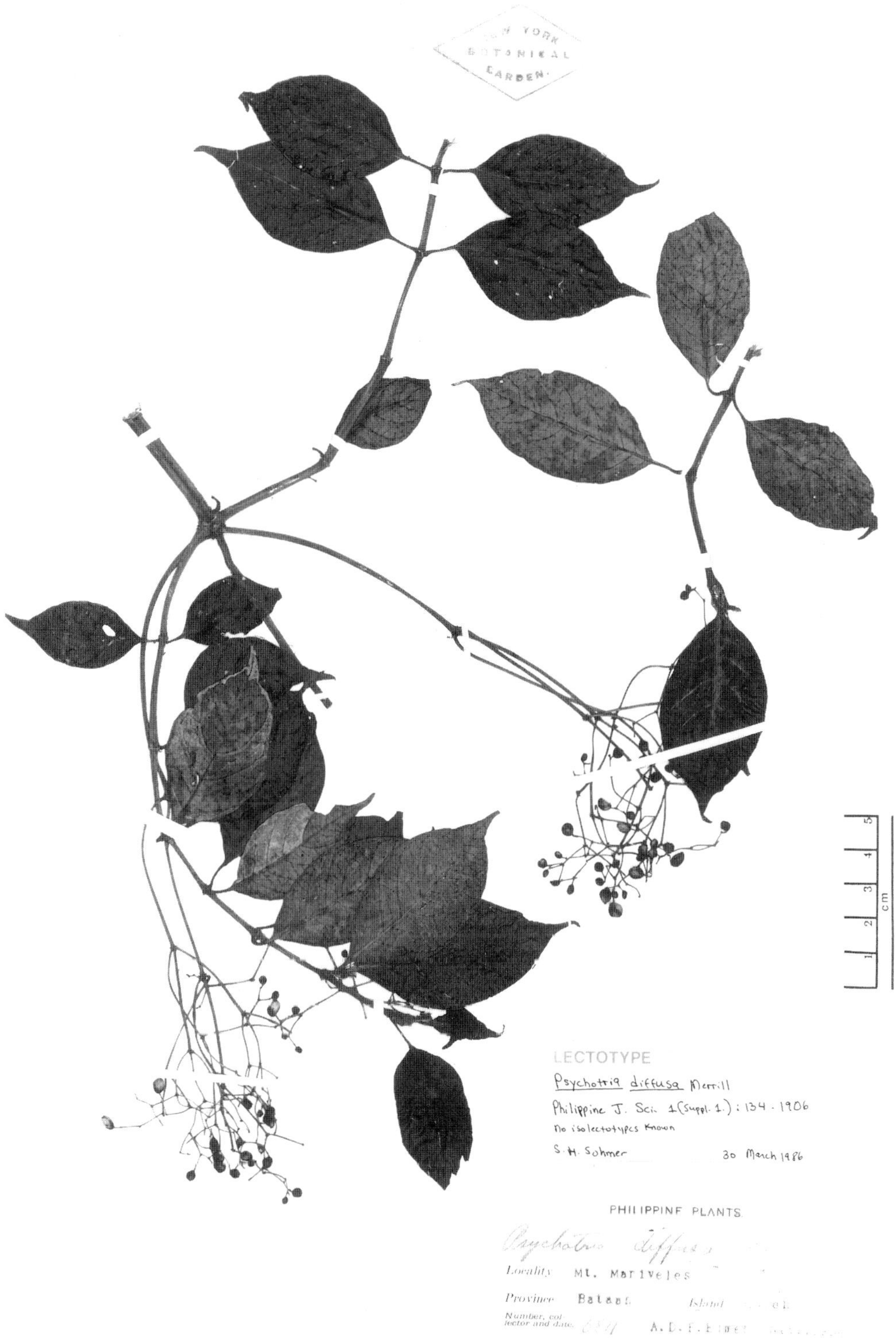

Fig. 22. *Psychotria diffusa* var. *diffusa* (*Elmer 6811*) (NY, lectotype).

ultimate branches terminating in widely spaced cymules (with > 30 flowers); bracts deltate to triangular, 0.8–1.5 mm long, sometimes absent (caducous), margins distinctly serrate; bracteoles deltate to triangular, 0.5–1.3 mm long, margins distinctly serrate. Flowers 4- or 5-merous; pedicels 1–3 mm long; calyx (incl. hypanthium) 0.8–1.1 mm long, glabrous; calyx lobes deltate to triangular, 0.3–0.4 mm long, rarely obscure; corolla: white or yellowish green; corolla tube 0.7–1.1 mm long, external surface glabrous, villous at throat within; corolla lobes ca. 0.4 mm long; stamens: filaments ca. 1 mm long; anthers 0.5–0.8 mm long (long styled); style 1–2 mm long, lobes ca. 0.5 mm long. Fruit ellipsoid or narrowly ellipsoid, 3–5 mm long, striated, white, glabrous or pubescent; calyx limb and lobes persistent; pedicels 1–3 mm long; pyrenes ± smooth on outer surface; endosperm ruminate.

Distribution.—Widely distributed throughout the Philippines.

Ecology.—Humid forest, including secondary forest; 350–2500 m

Phenology records.—Flowering: February to August; fruiting: March to November.

Preliminary conservation assessment.—Near Threatened (**NT**). Last collected in 1996.

Distinguishing characteristics.—Leaf blades acuminate, acute, obtuse, round or nearly cordate at base, often with pronounced abruptly acuminate apices; inflorescence trichotomous, diffuse, widely spreading, with numerous flowers; bracts and bracteoles with distinctly serrate margins; corolla tubes often less than 1 mm long; fruit small, narrowly ellipsoid, 3–5 mm long.

Remarks.—This is one of the most widely distributed Philippine *Psychotria* species and shows considerable morphological variability. There seems to have been adaptive radiation along certain lines such as leaf size, shape, and plant vestiture. Most of the variation in leaf size cannot be segregated satisfactorily, and recognition of one particular kind (a more pronounced acuminate apex, for example) would automatically lead to the recognition of several to many such variations with no possibility of consistently separating them on the characters chosen. At this time three varieties are recognized, which can be separated by the following key.

KEY TO THE VARIETIES OF *PSYCHOTRIA DIFFUSA*

1. Leaf base truncate or at least rounded, often slightly cordate ______ **22c.** var. **agusanensis**
1. Leaf base acute, narrowly acute, obtuse, rarely rounded.
   2. Inflorescence axes densely pubescent ______ **22b.** var. **cervina**
   2. Inflorescence axes glabrous or puberulent, or at most thinly pubescent ______ **22a.** var. **diffusa**

22a. Psychotria diffusa Merr. var. diffusa

Leaf base acute, acuminate or obtuse. Inflorescence axes glabrous or at most thinly pubescent.

Other specimens examined: **Luzon. Cagayan Prov.:** Baguio Cove, *Allen 382-81* (BISH). **Isabela Prov.:** Mt. Moises, *Clemens 17024* (UC); Sierra Madre Mountains, Sangcad, *Gutierrez PNH 78198* (K). **Mountain Prov.:** Mt. Polis, *Celestino PNH 7992* (A, PNH); Mabonglong, Banaue, *Conklin & Buwaya PNH 80449* (A, K, L); Bayninan, Banaue, *Conklin & Buwaya PNH 80627* (A, L, PNH). **Ifugao Prov.:** Mt. Data, *Ramos & Edaño BS 40188* (A, P, US); Lepanto, *Ramos BS 5959* (L, US); Semegar, *Barbon & Pedersen PPI 12806* (BRIT); Barangay Pula, Mt. Baladau, Banaue, *Reynoso, Garcia & Sagcal PPI 14516* (BRIT, K). **Benguet Prov.:** Mt. Lusong, *Ramos & Edaño BS 40447* (A, BM); Sablan, *Elmer 6258* (K, NY, P); Baguio, *Williams 1155* (K, NY); Mt. Tonglon, *Loher 1501* (K, US), *Ramos BS 5348* (BISH); Barangay Pula, Mt. Hagada, *Reynoso, Garcia & Sagcal PPI 14490* (BRIT, K); without further data: *Loher 1502* (K). **Nueva Vizcaya Prov.:** Mt. Alzapan, *Ramos & Edaño BS 45611* (UC), *BS 45707* (BM, NY, UC, W). **Rizal Prov.:** Mt. Irid, *Ramos & Edaño BS 48437* (BM, NY, P, UC, W); Montalban, *Loher 12054* (UC); Mt. Lumutan, *Ramos & Edaño BS 29735* (A); Rolay, *Loher 6439* (K); Angilog, *Loher 6382* (K), Oriud, *Loher 6411* (K). **Rizal Prov.:** without further data: *Loher 6439* (K). **Bataan Prov.:** Mt. Mariveles, Lamao River, *Whitford 152* (K, NY, P, US), *1208* (K, NY); Mt. Bataan, *Sohmer 12007* (BISH). **Laguna Prov.:** Paete, *Foxworthy BS 8989* (BISH); Mt. Maquiling, *Ramos BS 13648* (MO, S), *Salvoza FB 28590* (A). **Quezon Prov.:** Infanta, *Robinson BS 9444* (L); Philipp. National Botanic Garden, *Sohmer 12014* (BISH); Lucban, *Elmer 7953* (A, K, L), *9236* (A, K, L, MO, NY, US, W); Mt. Banahao, *Whitford 930* (K, NY); Polillo Island, *Salvoza FB 29654* (UC); Tagkawayan, *Barbon, Garcia & Fernando PPI 12189* (K). **Camarines Sur Prov.:** Mt. Potianay, *Edaño BS 75974* (A, NY); Mt. Isarog, Naga City, *Barbon, Romero & Fuentes PPI 8237* (BRIT, K); *ibid.*, *Barbon, Romero & Fuentes PPI 8012* (BRIT, K). **Albay Prov.:** Castilla, *Vidal 3066* (A, K). **Sorsogon Prov.:** Irosin (Mt. Bulusan), *Elmer 15507* (A, BM, K, MO, NY, P, S, UC, US). **Sorsogon Prov.:** without further data: *Ramos BS 23526* (A, BISH, BM, K, L, NY, P, PNH, US). **Catanduanes. Catanduanes Prov.:** without further data: *Ramos & Edaño BS 75249* (K, NY, UC). **Mindoro. Mindoro Occidental Prov.:** Mt. Calavite, *Ramos BS 39454* (A, BM, UC); Mt. Halcon, *Ridsdale 1756* (K, L). **Samar. Western Samar Prov.:** Mt. Sohoton, *Madulid et al. PNH 11784* (L). **Panay. Capiz Prov.:** Mt. Madiaas, *Ramos & Edaño BS 30659* (L, NY). **Leyte. Leyte Prov.:** Mt. Mamban, *Edaño PNH 12152* (A, BM, L, PNH, UC, US); Biliran Island, *Sulit PNH 21541* (K, L, PNH). **Leyte. Without further data:** *Wenzel 887* (A, BM, L, MO, US), *1502* (A, BM, GH, MO, NY). **Negros. Negros Oriental Prov.:** Cuernos Mountains, Dumaguete, *Elmer 9662* (A, BISH, BM, K, LM MO, NY, W), *10191* (A, BM, K, L, MO, NY, W). **Negros Occidental Prov.:** La Carlota City, Barangay Ara-al, Mt.

Canalon National Park, Sitio Guintubdan, *Reynoso, Garcia & Sagcal PPI 14246* (BRIT); Canlaon National Park, Mt. Canlaon, *Stone et al. PPI 6202* (BRIT, K). **Mindanao. Davao Prov.:** Mt. Galintan, *Ramos & Edaño BS 48908* (NY, S, UC). **Palawan. Palawan Prov.:** without further data: *Foxworthy BS 705* (US). **PHILIPPINES. Without further data:** *unknown collector US 629668* (L, PNH).

Distribution.—Widely distributed throughout the Philippines.

Ecology.—Humid forest, including secondary forest; 500–2500 m.

Phenology records.—Flowering: March to August; fruiting: March to November.

Preliminary conservation assessment.—Near Threatened (**NT**). Last collected in 1996.

Distinguishing characteristics.—Leaf base acute, acuminate or obtuse, with pronounced abruptly acuminate apices; inflorescence glabrous or at most thinly pubescent.

Remarks.—Variation in habit has made collectors of this species refer to it as a scandent shrub, scandent subshrub, or climber. We believe that taxa like *P. diffusa* are capable of coping with different environments and will be shrubby with a great profusion of branches in open areas, or will climb relatively tall trees to reach the light in shady areas.

22b. Psychotria diffusa Merr. var. **cervina** Elmer, Leafl. Philipp. Bot. 2:592. 1909; Merr., Enum. Philipp. Fl. Pl. 3:555. 1923. (**Fig. 24**). *Psychotria cervina* (Elmer) Elmer, Leafl. Philipp. Bot. 3:1022. 1911. TYPE: PHILIPPINES. NEGROS: Negros Oriental Prov.: Cuernos Mountains, Dumaguete, in forest of windswept ridge at ca. 1300 m, Apr 1908, *Elmer 9800* (LECTOTYPE: A, designated here; ISOLECTOTYPES: BM, K, L, MO, NY, US, W).

Leaf base acute, acuminate, or obtuse, never rounded or slightly cordate. Inflorescence axes densely pubescent.

Other specimens examined: None—known only from the type collection.

Distribution.—Negros.

Ecology.—Forest, including mossy forest; 1200–1300 m.

Phenology records.—Flowering: March, April; fruiting: unknown.

Preliminary conservation assessment.—Extinct (**EX**). Only collected once in 1908.

Distinguishing characteristics.—Inflorescence axes densely pubescent.

Remarks.—The only feature with which this taxon can be separated from the nominal variety is the densely pubescent inflorescence axes. We are convinced that should more material become available from the type locality, variation might be found to link this variety with the nominal variety. Elmer (l.c.) apparently used material from the same area to describe *P. microphylla*, here treated as a synonym of *P. diffusa* var. *diffusa*.

22c. Psychotria diffusa var. **agusanensis** (Elmer) Merr., Philipp. J. Sci. 10:123. 1915; Merr., Enum. Philipp. Fl. Pl. 3:555. 1923. (**Fig. 23**). *Psychotria agusanensis* Elmer, Leafl. Philipp. Bot. 5:1888. 1913. TYPE: PHILIPPINES. MINDANAO. Agusan Prov.: Cabadbaran (Mt. Urdaneta), Cawilanan Peak, wind-swept forested ridge at ca. 1300 m, Sept. 1912. *Elmer 13667* (LECTOTYPE L, designated here; ISOLECTOTYPES: A, BISH, BM, GH, K, MO, NY, P, UC, US, W).

Psychotria ovalis Elmer, Leafl. Philipp. Bot. 3:1023. 1911. TYPE: PHILIPPINES. MINDANAO: Davao Prov.: Todaya (Mt. Apo), dense ridge forest S of Mt. Calelan at ca. 1400 m, May 1909, *Elmer 10567* (LECTOTYPE: L, designated here; ISOLECTOTYPES: A, BISH, BM, GH, K, MO, NY, US, W).

Leaf base truncate or at least rounded, often somewhat cordate. Inflorescence axes glabrous to pubescent.

Other specimens examined: **Luzon. Camarines Norte Prov.:** Sipocot National Park, 368.4 km from Manila, *Sohmer 12239* (BISH). **Albay Prov.:** Barangay Sto. Cristo, Tiwi Side, Mt. Malinao, *Reynoso, Sagcal & Fernando PPI 21373* (K). **Mindanao. Agusan Prov.:** Tungao So., San Mateo Bo., *Mendoza PNH 41903* (L). **Bukidnon Prov.:** Mt. Katangland, *Sulit PNH 9846* (A, BM, L, PNH); Tangculan, *Ramos & Edaño BS 39099* (A, NY). **Davao Prov.:** Mt. Kampalili, *Edaño PNH 11570* (PNH).

Distribution.—Southern Luzon and Mindanao.

Ecology.—Forest; 350–1800 m.

Phenology records.—Flowering: February, March, June; fruiting: unknown.

Preliminary conservation assessment.—Near Threatened (**NT**). Last collected in 1995.

Distinguishing characteristics.—Leaf bases truncate or at least rounded, often somewhat cordate.

Remarks.—There is obviously a transition between *Psychotria diffusa* var. *agusanensis* and the other two varieties recognized here, both in terms of geography and morphology. Leyte appears to have populations that are transitional between this variety and the next. Some of the higher elevation

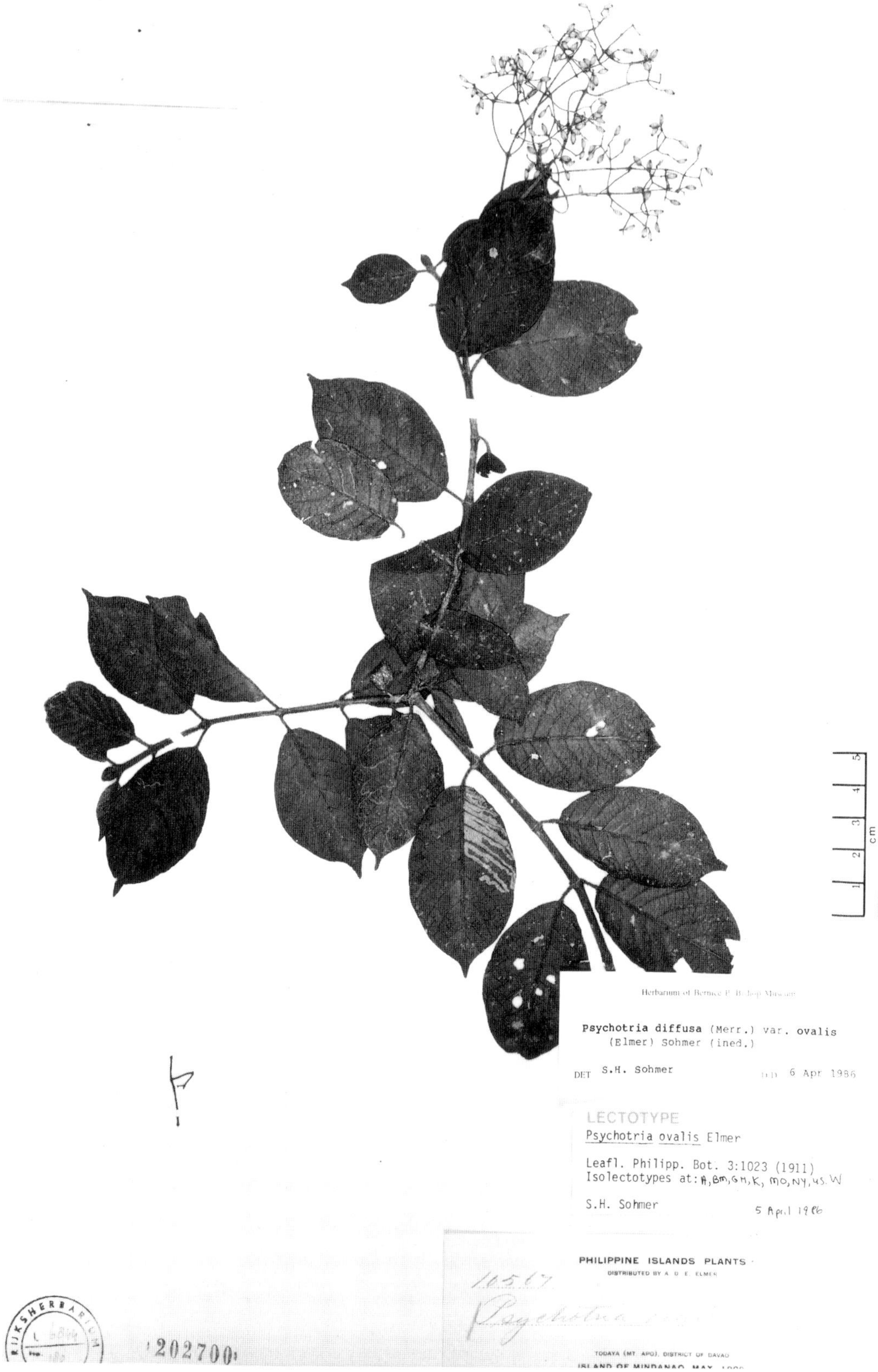

FIG. 23. *Psychotria diffusa* var. *agusanensis* (*Elmer 10567*) (L, lectotype of *Psychotria ovalis*).

FIG. 24. *Psychotria diffusa* var. *cervina* (*Elmer 9800*) (A, lectotype).

collections were made from individuals that were reported to have rugose leaves. The venation pattern on most of the individuals is generally more pronounced than in the nominal variety.

II.—Linearis Species Group

Shrubs or small trees. Stipules valvate or rarely imbricate; in one species (*P. linearis*) margins fused and forming a cylindrical tube. Leaf blades less or more than 3 cm wide; never more than 20 cm long and often less than 15 cm long; chartaceous to subcoriaceous. ***Inflorescence parts greatly reduced, axes and branches absent or more or less so. Flowers pedicellate.*** Fruits red, rarely purple; pyrene inner (adaxial) surface flat; seed producing a red color when placed in 70% ethanol, due to a soluble seed coat pigment (SCP positive), rarely SCP negative; endosperm ruminate or entire, usually dark. Comprising six species.

Note.—Text in ***bold italics*** denotes features useful for the placement of taxa within the respective species group.

KEY TO THE SPECIES OF THE LINEARIS SPECIES GROUP

1. Leaf blades linear or narrowly elliptic, < 1.5 cm wide at widest point, often < 0.8 cm wide, 4.5–11.2(–13) cm long ______ **23. P. linearis**
1. Leaf blades elliptic, ovate, obovate or oblanceolate, if < 2 cm wide at widest point, never > 5 cm long.
    2. Leaf base cordate, truncate, or round.
        3. Leaf blades > 8 cm long ______ **24. P. madulidii**
        3. Leaf blades < 6 cm long ______ **31. P. subcucullata***
    2. Leaf base narrowly acute, acute, obtuse or at most rounded, never truncate or cordate at base.
        4. Leaf blades 13.5–15.8 cm long ______ **25. P. cephalophora**
        4. Leaf blades < 11 cm long.
            5. Leaf blades 6–10.1 cm long, base usually round, obtuse to cuneate ______ **26. P. longipedicellata**
            5. Leaf blades 3.6–9.2 cm long, base acute or cuneate.
                6. Leaf blades 5.7–9.2 cm long ______ **27. P. cuneata**
                6. Leaf blades 3.6–6 cm long ______ **28. P. yagawensis**

*This taxon is not a member of this species group, and, it is included in this key because of variability in key characteristics.

23. Psychotria linearis Bartl. ex DC., Prodr. 4:522. 1830; Merr., Enum. Philipp. Fl. Pl. 3:557. 1923. (**Fig. 25**). TYPE: PHILIPPINES. LUZON: 'in ins. Manilla prope Sorzogon.', *Haenke s.n.* (HOLOTYPE: PR).

Amaracarpus longifolius Elmer, Leafl. Philipp. Bot. 1:1. 1906. TYPE: PHILIPPINES. LUZON: Tayabas Prov.: Atimonan, Mar 1905, *Merrill 4000* (HOLOTYPE: PNH; ISOTYPES: K, P).

Psychotria depauperata Merr., Philipp. J. Sci. 12:174. 1917. TYPE: PHILIPPINES. ALABAT: Quezon Prov.: back of Sangirin on dry ridges in virgin forest at ca. 100 m, 23 Dec 1916, *Merrill 10487* (LECTOTYPE: US, designated here; ISOLECTOTYPES: A, BM, K, L, NY, P, PNH).

Shrub, 0.5–1 m high; young stems pubescent. Stipules valvate, ovate to ± triangular, (1.8–)2–8 mm long, pubescent; apex acute to acuminate. Leaves: petioles 0.5–1.5 cm long, pubescent; leaf blades linear, narrowly elliptic to narrowly lanceolate or oblanceolate, 4.5–11.2(–13) × 0.9–1.5 cm, chartaceous to subcoriaceous, glabrous above, pubescent below at least on veins; secondary veins 6–12(–14) pairs; tertiary veins manifest; base acute to narrowly acute; apex acute or obtuse. Inflorescence/infructescence simple (much reduced), lacking axes and branches, composed of (1–)3–8 pedicellate flowers, sparsely pubescent; bracts and bracteoles absent. Flowers 4- or 5-merous, pedicels 3–7.5 mm long; calyx (incl. hypanthium) ca. 1.2 mm long, upper parts (limb and lobes) sparsely pubescent to pubescent; calyx lobes deltate, ca. 0.2 mm long, corolla: white; corolla tube length unknonwn, external surface pubescent, pubescent at throat within; corolla lobes unknown; stamens: unknown style ca. 0.6 mm long, lobes ca. 0.6 mm long (long and short styled). Fruit obovoid, 5–8 mm long, ribbed, red, glabrous; calyx limb and lobes persistent; pedicels 4–8.2 mm long; pyrenes sharply ridged on outer surface; endosperm entire.

Other specimens examined: **Luzon. Quezon Prov.:** Atimonan, *Merrill 4000* (NY); Manban, *Curran FB 10201* (K, US); Mt. Pular, *Ramos BS 19377* (K); Alabat Island, *Ramos & Edaño BS 48211* (A, BM, K, NY, P, SING, UC, US, W). **Quezon Prov.:** without further data: *Edaño BS 26954* (A, BM, K, P, US). **Nueva Ecija Prov.:** *Ramos & Edaño BS 26452* (A, UC, US). **Rizal Prov.:** Mt. Irid, *Ramos BS 41949* (A, L, W); Mt. Susong Dalaga, *Ramos & Edaño BS 29419* (A, K, P, US); Manila, *unknown collector 98* (W); San Isidro, *Ramos BS 335* (US). **Rizal Prov.:** without further data: *Loher 14828* (UC), *Ramos BS 13569* (BM, K, L, P), *Ahern's collector FB 3375* (K).

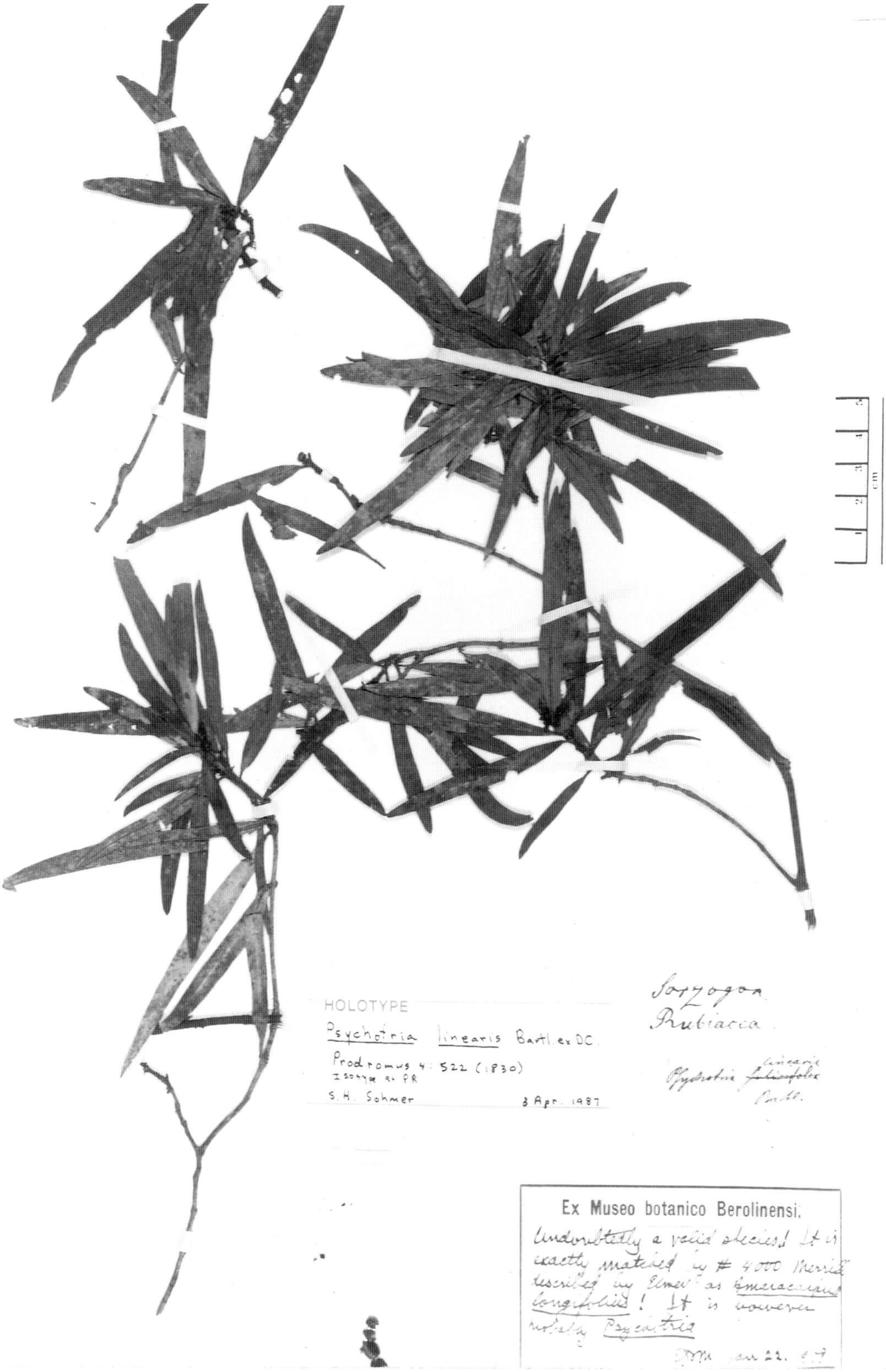

Fig. 25. *Psychotria linearis* (*Haenke s.n.*) (PR, holotype).

Camarines Prov.: without further data: *Simeon FB 28744* (A). **Samar. Western Samar Prov.:** Basey, *Fontanoza FB 31143* (NY); Mt. Sohoton, *Gutierrez et al. PNH 118039* (L).

Distribution.—Luzon, Alabat and Samar.

Ecology.—Relatively dry ridge forest; 100–500 m.

Phenology records.—Flowering: August, November; fruiting: January, March August to October, December.

Preliminary conservation assessment.—Vulnerable (**VU B1ab(iii)**). Last collected in 1970.

Distinguishing characteristics.—Leaf blades linear to narrowly oblanceolate; inflorescence sessile, flowers several to many and on relatively long and thin pedicels, floral parts pubescent; endosperm entire (non ruminate).

Remarks.—After studying the holotype of *P. linearis* at PR (by S.H. Sohmer) it was obvious that *P. depauperata* was synonymous with the former. There appear to be two morphological tendencies in leaf shape in *P. linearis*, with one extreme resulting in relatively long linear leaf blades and the other resulting in relatively shorter, elliptic leaf blades. The former is known as *P. linearis* and the latter what was once referred to as *P. depauperata*; the type of *P. linearis* is actually more like the *P. depauperata* variant. In all other ways they are very similar and intermediate forms link the leaf shape extremes.

24. Psychotria madulidii Sohmer & A.P. Davis, sp. nov. (**Fig. 26**). TYPE: PHILIPPINES. LUZON: Cagayan Prov.: Mt. Cagua, Oct–Nov 1929, *Edaño BS 78408* (HOLOTYPE: A).

Frutex est, ramulis pubescentibus, stipulis 10 mm longis valvatis in apice biaristatis interdum pubescentilus, laminis 4.5 × 8.4–6 × 13.5 cm latis et longis chartaceis oblanceolatis vel obovatis vel subpanduratis apice acuto subacuto acuminatove basi cordata vel truncata vel rotundata supra glabris, in nervis sparse pubescentibus, petiolis 1.1–3.6 cm longis pubescentibus, inflorescetia sessili, pedicellis 3–10 mm longis, fructibus 6–7 mm longis obovoideis laevibus cum pilis dispersis licet rubris, pyrenis in dorso laevibus, endospermo valde runinato.

Shrub or small tree, height not recorded; young stems pubescent. Stipules valvate, lanceolate to narrowly triangular-lanceolate, ca. 10 mm long, pubescent or glabrous; apex with 2 aristate tips. Leaves: petioles 1.1–3.6 cm long, pubescent; leaf blades oblanceolate to obovate to nearly pandurate, 8.4–13.5 × 4.5–6 cm, chartaceous, glabrous above, sparsely pubescent below on veins; secondary veins 8–12 pairs; tertiary veins manifest to rather prominent; base cordate or at least truncate to rounded; apex acute to abruptly acute or acuminate. Inflorescence/infructescence simple (much reduced), lacking axes and branches, composed of 3–5 pedicellate flowers, sparsely pubescent to pubescent; bracts and bracteoles absent. Flowers unknown, presumed pedicellate. Fruit obovoid, 6–7 mm long, smooth, probably red, sparsely pubescent; calyx limb and lobes persistent but not prominent; pedicels 3–10 mm long; pyrenes smooth on outer surface; endosperm strongly ruminate.

PARATYPES: **Luzon. Cagayan Prov.:** Mt. Cagua Volcano, *Edaño BS 78400* (A), *BS 78401* (A), *BS 78427* (A). **Palawan. Palawan Prov.:** Mt. Mantalingajan, *Edaño BS 77556* (A).

Distribution.—Cagayan province on Luzon and perhaps Mt. Mantalingajan on Palawan.

Ecology.—Unknown.

Phenology records.—Flowering: unknown; fruiting: October, November.

Preliminary conservation assessment.—Vulnerable (**VU B1ab(iii)**). Last collected in 1929.

Distinguishing characteristics.—Inflorescence sessile, fruit clearly pedicellate; flowers presumed pedicellate; leaf blades truncate to cordate; twigs pubescent.

Remarks.—This species can be confused with *P. bataanensis* and *P. elliptilimba* because of their truncate or cordate leaves but the inflorescence in *P. madulidii* lacks axes and branches. The species has been named for Dr. Domingo Madulid, a friend and colleague (of S.H. Sohmer) who expedited much of the fieldwork for this revision. We are certain *Edaño BS 77556* belongs here, even though its stipules are broader than normal and densely pubescent.

25. Psychotria cephalophora Merr., Philipp. J. Sci. 3:438. 1908; Merr., Enum. Philipp. Fl. Pl. 3:554. 1923. (**Fig. 27**). *Grumilea cephalophora* (Merr.) Merr., Philipp. J. Sci. 10:133. 1915. TYPE: PHILIPPINES. CAMIGUIN: Batanes Prov.: in forests, 27 Jun 1907, *Fenix BS 4048* (LECTOTYPE: US, designated here).

Small tree or shrub, height not recorded; young stems glabrous. Stipules valvate, ± narrowly triangular, ca. 6 mm long, glabrous; apex acuminate. Leaves: petioles 1.5–1.8 cm long, glabrous; leaf blades ob-

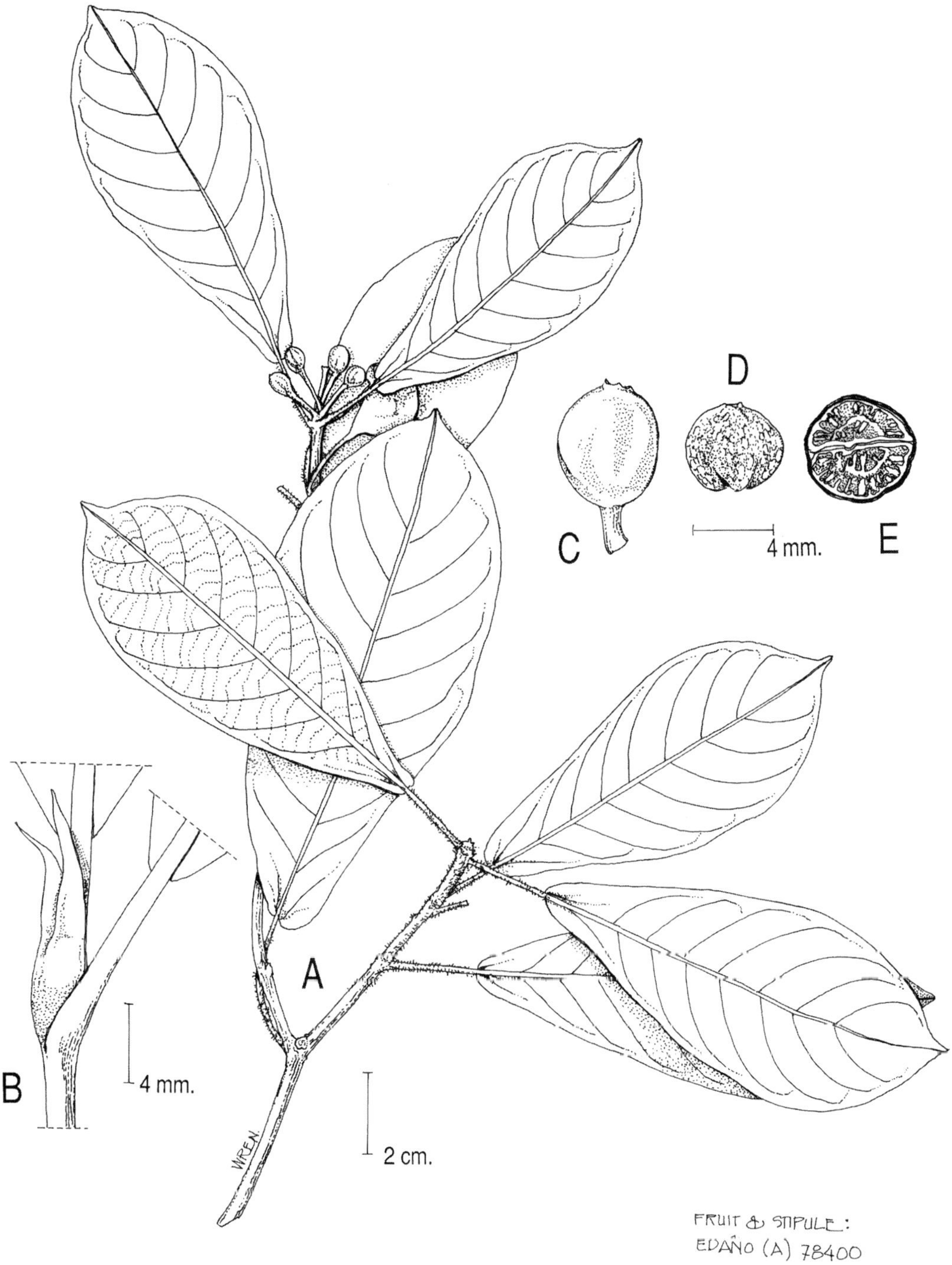

Fig. 26. ***Psychotria madulidii*. A, habit of fruiting branch (*Edaño BS 78408*) (A, holotype); B, stipule; C, fruit; D, pyrene; E, fruit in cross section (*Edaño, BS 78400*) (A).**

long-elliptic to obovate, 13.5–15.8 × 5.8–6.5 cm, chartaceous, glabrous above, pubescent below along margin of midrib; secondary veins 13 or 14 pairs; tertiary veins manifest to obscure; base narrowly acute; apex abruptly acuminate to acuminate. Inflorescence/Infructescence simple (much reduced), lacking axes and branches, composed of < 5 pedicellate flowers/fruits, glabrous; bracts and bracteoles unknown (?absent). Flowers unknown, presumed pedicellate. Fruit obovoid, ca. 6 mm long, smooth,

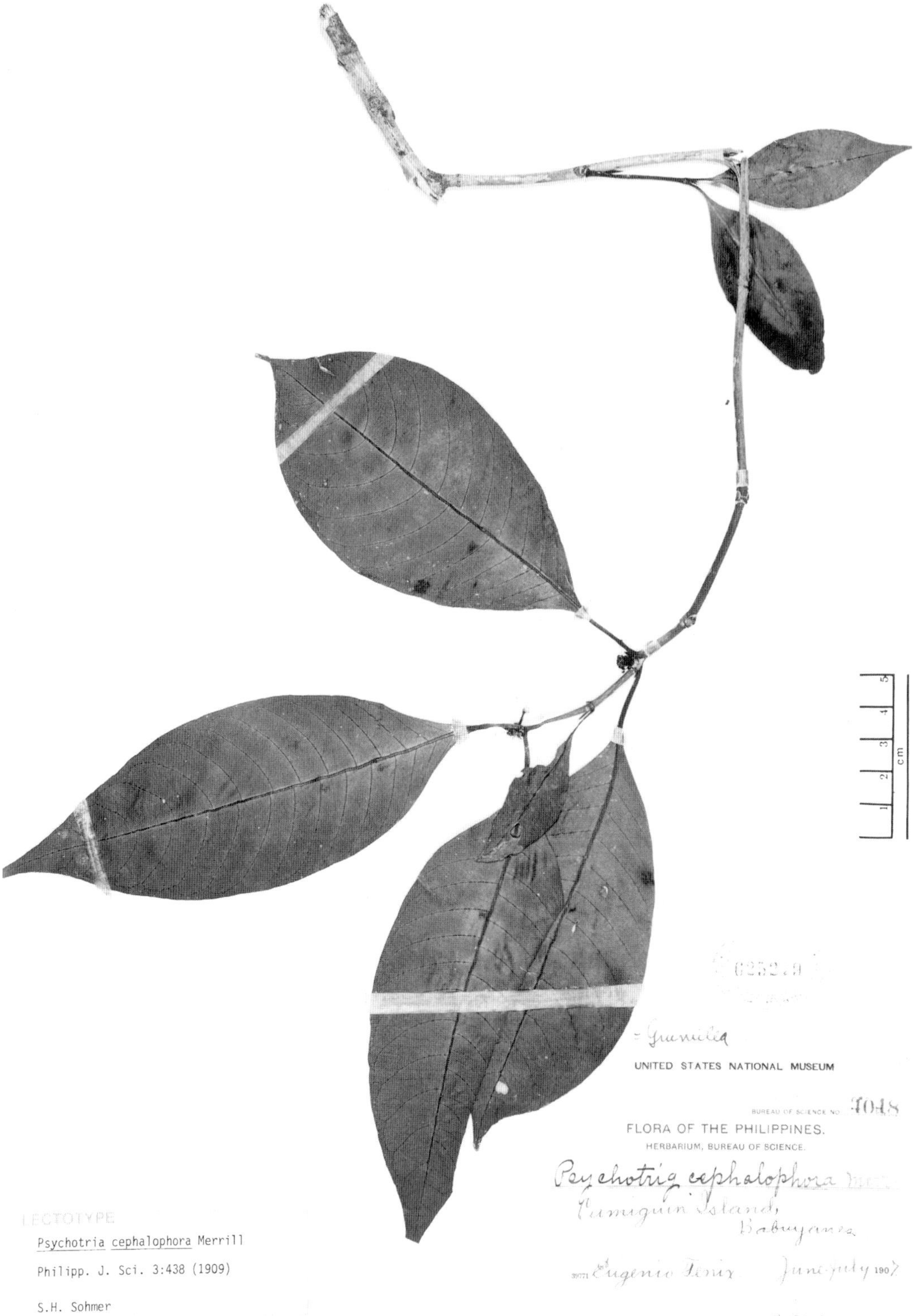

FIG. 27. *Psychotria cephalophora* (*Fenix BS 4048*) (US, lectotype).

probably red, glabrous; calyx limb and lobes not persistent; pedicels ca. 5 mm long; pyrenes smooth on outer surface; endosperm ruminate.

Other specimens examined: **Catanduanes. Catanduanes Prov.:** without precise locality: *Ramos & Edaño BS 75114* (NY, UC) *vel aff.* **Panay. Capiz Prov.:** Libacão, *Martelino & Edaño BS 35307* (NY). **Mindanao. Agusan Prov.:** Butuan, Tungao So., San Mateo Bo., Antipolo Line, 30 km, *Mendoza PNH 42178* (PNH).

Distribution.—Catanduanes, Panay and Mindanao.

Ecology.—Forest; at low to medium elevation.

Phenology records.—Flowering: unknown; fruiting: June.

Preliminary conservation assessment.—Vulnerable (**VU B1ab(iii)**). Last collected in 1961.

Distinguishing characteristics.—Leaf blades more than 10 cm long; twigs glabrous; inflorescence sessile; fruit pedicellate; endosperm ruminate.

Remarks.—If this is a 'good' species, then its glabrous nature and ruminate endosperm distinguish it from other members of its group. Specimens identified as *P. cephalophora* from Taiwan (e.g. *Chang 8721*, K) are similar to the the Philippine species but are not conspecific.

26. Psychotria longipedicellata Elmer, Leafl. Philipp. Bot. 1:22. 1906; Merr., Enum. Philipp. Fl. Pl. 3:557. 1923. (**Fig. 28**). *Grumilea longipedicellata* (Elmer) Merr., Philipp. J. Sci. 10:133. 1915. TYPE: PHILIPPINES. LUZON: Quezon Prov.: Lagumanoc, Sep 1904, *Whitford 907* (LECTOTYPE: K, designated here; ISOLECTOTYPES: NY, P, US).

Shrub 2–3 m high; young stems pubescent to glabrous. Stipules valvate, ovate to ovate-triangular, 5–10 mm long, pubescent; apex round or acute. Leaves: petioles 1–1.7 cm long, pubescent; leaf blades rather clustered at apex of branches, usually obovate to elliptic, 6–10.1 × 3.3–3.9 cm, chartaceous to subcoriaceous, glabrous above, pubescent below along margins of midrib; secondary veins (8–)10–13(–14) pairs; tertiary veins manifest to rather prominent; base rounded to obtuse, or cuneate; apex obtuse or round, often with a short and rather blunt acuminate tip. Inflorescence/infructescence simple (much reduced), lacking axes and branches, composed of (5–)10–15 pedicellate flowers, puberulous to glabrous; bracts unknown (?absent); bracteoles triangular, ca. 0.2 mm long. Flowers 5-merous, pedicels ca. 5 mm long; calyx (incl. hypanthium) ca. 1 mm long, glabrous; calyx lobes obscure; corolla: color unknown; corolla tube ca. 2 mm long, external surface glabrous, villous or pilose at throat within; corolla lobes ca. 2mm long; stamens: filaments ca. 1 mm long (short styled); anthers ca. 0.5 mm long; style ca. 2 mm long (short styled), lobes ca. 0.5 mm long. Fruit obovoid, 6–7 mm long, ± smooth to weakly ribbed, red or purple, glabrous; calyx limb and lobes ± persistent; pedicels 6–15 mm long; pyrenes weakly ribbed on outer surface; endosperm ruminate on inner surface only.

Other specimens examined: **Luzon. Quezon Prov.:** Mt. Dingalan, *Ramos & Edaño BS 26591* (A, P, US); Lagumanoc, *Merrill 3361* (BM, K, NY, P, US). BURIAS. **Masbate. Masbate Prov.:** without further data: *Clark FB 1727* (BM, NY).

Distribution.—Found in Quezon province, Luzon, and on Burias Island.

Ecology.—Forest; low altitude.

Phenology records.—Flowering: June; fruiting: September, November.

Preliminary conservation assessment.—Critically Endangered (**CR B1ab(iii)**), or Extinct (**EX**). Last collected in 1910.

Distinguishing characteristics.—Leaf base often rounded; petioles relatively long for this species group; inflorescence sessile, flowers and fruit pedicellate; endosperm ruminate.

27. Psychotria cuneata Elmer, Leafl. Philipp. Bot. 9:3261. 1934. (**Fig. 29**). TYPE: PHILIPPINES. LUZON. Sorsogon Prov.: Mt. Bulusan, Irosin, woody ravine at medium altitude, Sep 1916, *Elmer 17170* (LECTOTYPE: UC, designated here; ISOLECTOTYPES: A, BISH, BM, GH, K, L, NY, MO, P, PNH, US).

Shrub, height not recorded; young stems pubescent. Stipules valvate, traingular to ovate-triangular, 3–7 mm long, glabrous; apex obtuse or with an abruptly acuminate tip. Leaves: petioles 0.7–1.3 cm long, pubescent; leaf blades obovate to oblanceolate-elliptic, 5.7–9.2 × 2.3–3.5 cm, chartaceous, drying very dark brown above and yellowish brown below, glabrous above, pubescent along midvein below; secondary veins 6–9 pairs; tertiary veins manifest; base cuneate to acute; apex abruptly acuminate. Inflorescence/infructescence simple (much reduced), lacking axes and branches, composed of 3–7 pedicellate flowers, very sparsely pubescent; bracts present, 2 or 3 pairs (reported by Elmer, 1934); bracteoles unknown. Flowers poorly known, recorded as pedicellate; calyx (incl. hypanthium) ca. 1

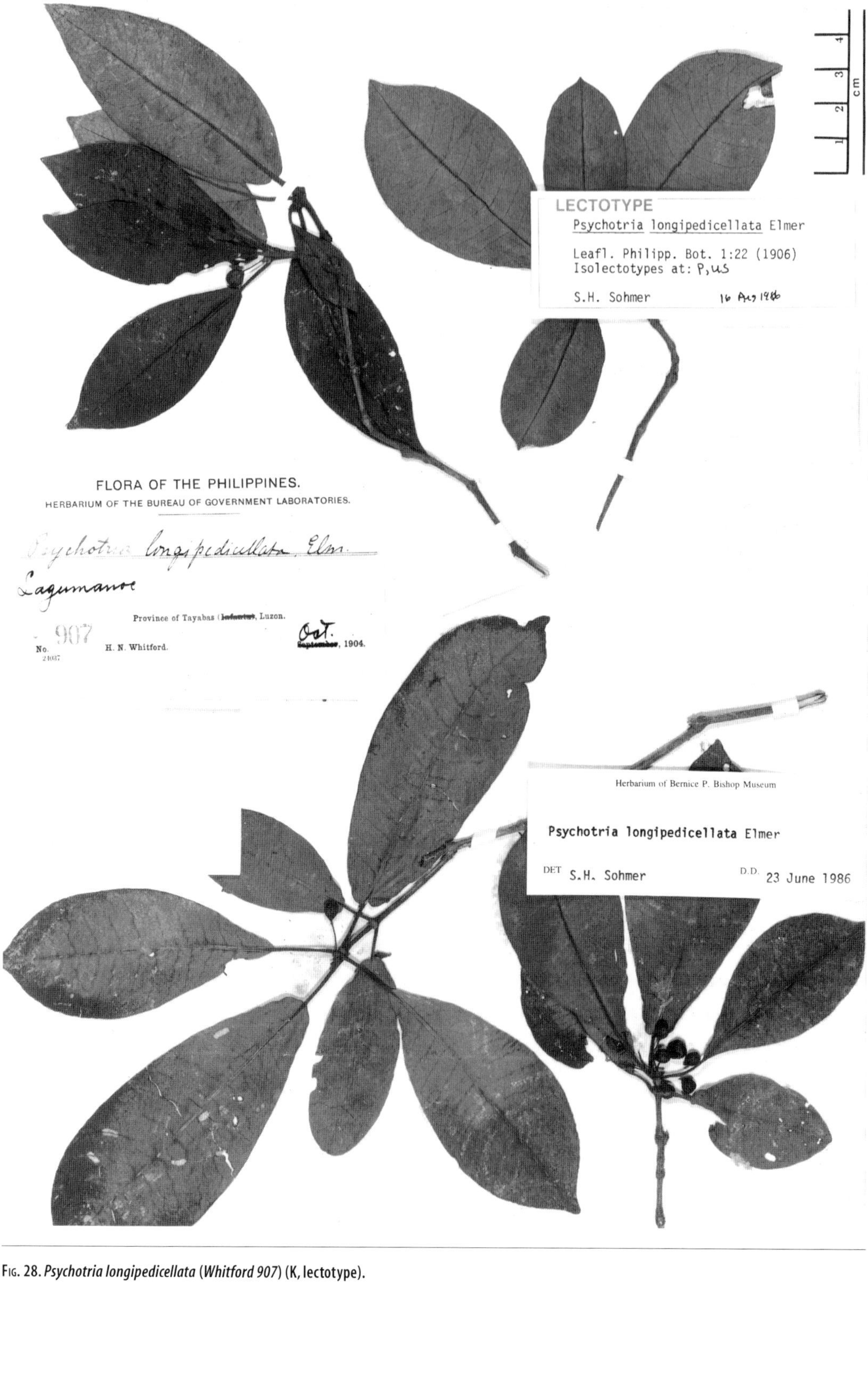

FIG. 28. *Psychotria longipedicellata* (*Whitford 907*) (K, lectotype).

Fig. 29. *Psychotria cuneata* (*Elmer 17170*) (UC, lectotype).

mm long, glabrous to sparsely pubescent; calyx lobes ± deltate, ca. 0.4 mm long, or ± obsolete; corolla: color unknown; corolla tube ca. 1 mm long, external surface glabrous, pubescent at throat within; corolla lobes not recorded; stamens: filaments unknown; anthers ca. 0.5 mm long. style unknown. Fruit obovoid, 7–11 mm long, ridged or ribbed, probably red, glabrous; calyx limb and lobes persistent; pedicels 0.5–1 cm long; pyrenes ridged or ribbed on outer surface; endosperm probably entire.

Other specimens examined: **Tablas. Romblon Prov.:** Mt. Pan-ayong, So. Balikawang, Barangay Marigondon Sur, San Andres, *Reynoso & Fuentes PPI 24616* (BRIT, K).

Distribution.—Known only from the type collected on Mt. Bulusan, Luzon, and one specimen collected on Tablas Island.

Ecology.—Forest, including disturbed forest; ravines and ridges; one record made on limestone; medium altitudes (e.g. ca. 450 m).

Phenology records.—Flowering: unknown; fruiting: August, September.

Preliminary conservation assessment.—Endangered (**EN B1ab(iii)**). Last collected in 1996.

Distinguishing characteristics.—Inflorescences much reduced, lacking axes and branches, composed of 3–7 pedicellate flowers/fruits; young stems pubescent; leaf base acute to cuneate.

Remarks.—It has been placed in this group because the two specimens bearing attached fruit have pedicellate flowers and greatly reduced inflorescences.

28. Psychotria yagawensis Sohmer & A.P. Davis, sp. nov. (**Fig. 30**). Type: PHILIPPINES. MINDORO: Mt. Yagaw, southeast slope at 330 m, 17 Jul 1953, *Conklin PNH 18687* (HOLOTYPE: A; ISOTYPES: L, PNH).

Frutices sunt stipulis 2–5 mm longis imbricatis glabris vel puberulis apice obtuso, laminis 1.8 × 3.6–2.6 × 6 cm latis et longis ellipticis apice et basi acuto vel cuneato chartaceis supra glabris infra saltem in costa pubescentibus nervis lateralilus 9–13 in dimidio quoque in sicco subviridibus-subrubis-badiis, petiolis 0.9–1.9 cm longis sparse pubescentibus, inflorescentia sessili, florilus sessilibus vel congregatis, in fructu pedicellis 0.3–0.9 cm longis, fructibus 5 mm longis globosis vel obovoideis glabris costatis, endospermo non ruminato.

Shrub, height not recorded; young stems glabrous. Stipules imbricate, triangular to narrowly triangular, 2–5 mm long, glabrous or puberulent; apex obtuse to acute. Leaves: petioles 0.9–1.9 cm long, glabrous to sparsely pubescent; leaf blades elliptic, 3.6–6 × 1.8–2.6 cm, chartaceous, drying greenish reddish brown, glabrous above, puberulous below mainly along midrib and secondary veins; secondary veins 9–13 pairs; tertiary veins manifest to prominent; base acute to cuneate; apex acute. Inflorescence/infructescence simple (much reduced), lacking axes and branches, with several flowers/fruits at the shoot apex, or with a very short main axis (peduncle) bearing a small umbel of (2–)3 flowers, glabrous; principal axis (peduncle), if present, ca. 0.3 cm long; bracts and bracteoles absent. Flowers unknown. Fruit globose to obovoid, ca. 5 mm long, ribbed, color unknown, glabrous; calyx limb and lobes persistent but obscure; pedicels 3–9 mm long; pyrenes ribbed or ridged on outer surface; endosperm entire.

PARATYPES: None—known only from the type collection.

Distribution.—Known only from Mt. Yagaw on Mindoro.

Ecology.—Forest; ca. 330 m.

Phenology records.—Flowering: unknown; fruiting: July.

Preliminary conservation assessment.—Critically Endangered (**CR B1ab(iii)**) or possibly Extinct (**EX**). Last collected in 1953.

Distinguishing characteristics.—Inflorescences sessile, flowers pedicellate; leaf blades small, 3.6–6 cm long, elliptic, base acute to cuneate; endosperm entire (non ruminate).

Remarks.—This species is very similar to *P. cuneata* and can be distinguished from it only by smaller leaves. Perhaps if future collections are made, the variation in leaf blade size will be shown to be greater than what is demonstrated by the single collection from Mindoro. If the leaf size begins to intergrade with *P. cuneata*, it may then be possible to combine these taxa.

III.—Tayabensis Species Group

Shrubs or small trees. Stipules valvate or rarely calyptrate. Leaf blades less or more than 3 cm wide; never more than 25 cm long; chartaceous to subcoriaceous. ***Inflorescence parts greatly reduced, axes***

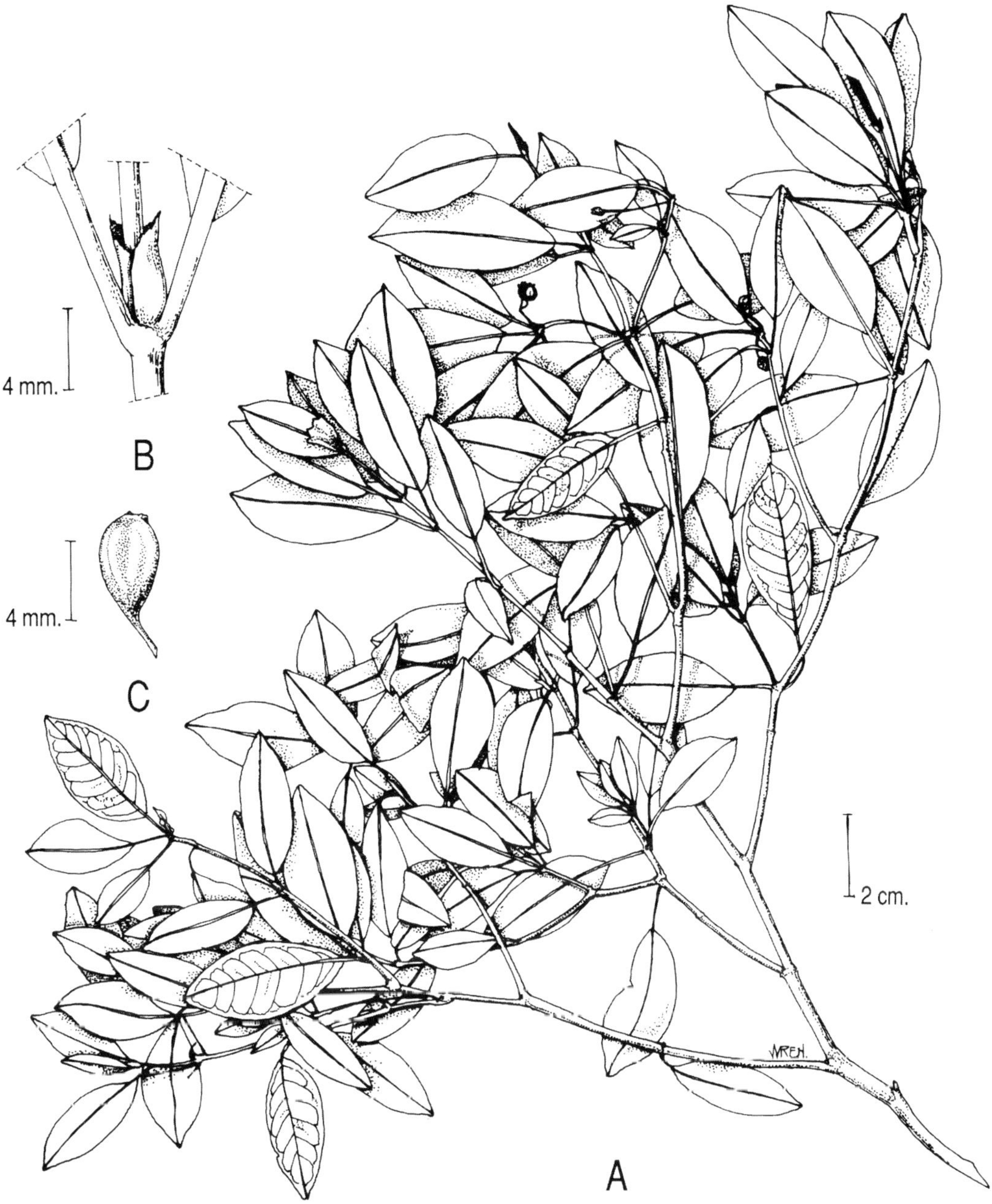

FIG. 30. *Psychotria yagawensis* (*Conklin PNH 18687*) (A, holotype). **A,** habit of fruiting and vegetative branches; **B,** stipule (*Conklin PNH 18687*) (PNH, isotype); **C,** fruit (*Conklin PNH 18687*) (L, isotype).

and branches absent or more or less so. Flowers sessile or rarely pedicellate. Fruits usually red, sometimes yellow, purple or black; pyrene inner (adaxial) surface flat; seed producing a red color when placed in 70% ethanol, due to a soluble seed coat pigment (SCP positive), rarely SCP negative; endosperm ruminate, usually dark. Comprising eight species.

Note.—Text in ***bold italics*** denotes features useful for the placement of taxa within the respective species group.

KEY TO THE SPECIES OF THE TAYABENSIS SPECIES GROUP

1. Leaf blades pubescent above; young stems and petioles pubescent ____ **29. P. papillata**
1. Leaf blades glabrous above; young stems and petioles pubescent or glabrous.
   2. Fruits heart-shaped in longitudinal section (or outline), apex with 2 pronounced lobes ____ **30. P. yenii**
   2. Fruits globose, turbinate, obovoid, or somewhat diplococcous in logitudinal section (or outline); apex not lobed.
      3. Leaf base base obtuse to subcordate ____ **31. P. subcucullata**
      3. Leaf base acute to obtuse, sometimes attenuate.
         4. Leaf blades 2.2–7.1 cm long, usually < 5 cm long.
            5. Plants usually with 1 fruit per inflorescence ____ **32. P. amaracarpoides**
            5. Plants with > 3 fruits per inflorescence.
               6. Leaf blades pubescent below along veins, grayish when dry (herbarium specimens) ____ **33. P. urdanetensis**
               6. Leaf blades sparsely pubescent below, at least on veins, brownish when dry (herbarium specimens) ____ **34. P. nagapatensis**
         4. Leaf blades 4.6–22 cm long; usually > 7 cm long.
            7. Stipules entire at apex; leaf blades broadly elliptic to broadly obovate; fruit 7–11 mm long, usually with pronounced ribbing when dry ____ **35. P. subsessiliflora**
            7. Stipules cleft at apex; leaf blades narrowly elliptic to obovate or oblanceolate; fruit (3–) 4–10 mm long, usually with minute wrinkling when dry ____ **36. P. tayabensis**

29. Psychotria papillata (Merr.) Merr., Enum. Philipp. Fl. Pl. 3:560. 1923. (**Fig. 31**). *Grumilea papillata* Merr., Philipp. J. Sci. 10:138. 1915. TYPE: PHILIPPINES. SAMAR. Prov. unknown, Cauayan Valley, Apr 1914, *Ramos BS 1701* (LECTOTYPE: MO, designated here; ISOLECTOTYPES: A, BM, GH, L, NY, P, SING).

Shrub, 1–3 m high; young stems (and all young parts) prominently pubescent (hairs rusty brown when dry, probably white *in vivo*). Stipules appearing calyptrate (tightly rolled), ± linear, 6–20 mm long, pubescent; apex acute. Leaves: petioles 0.7–2.2 cm long, pubescent; leaf blades elliptic to elliptic-obovate, 5–12.6 × 1.8–4.5 cm, chartaceous, pubescent on both surfaces; secondary veins 7–13 pairs; tertiary venation manifest to obscure; base acute to obtuse; apex acute to acuminate. Inflorescence/infructescence simple (much reduced), lacking axes and branches, composed of 1–2(–3+) very shortly pedicellate or sessile flowers, pubescent; bracts and bracteoles unknown (?absent). Flowers unknown. Fruit obovoid to globose, 6–9 mm long, smooth, black or dark red, pubescent; calyx limb and lobes persistent but not prominent; pedicels 0.5–2 mm long, or absent; pyrenes smooth on outer surface; endosperm ruminate.

Other specimens examined: **Catanduanes. Catanduanes Prov.:** Mt. Abucay, *Ramos & Edaño BS 75625* (NY, SING, UC). **Catanduanes Prov.:** without further data: *Ramos BS 30517* (A, K, P). **Samar. Western Samar Prov.:** Loquilocon, *McGregor BS 43850* (K, NY, UC); *Wright, Sulit & Conese PNH 6205* (A, L, PNH).

Distribution.—Samar and Catanduanes.

Ecology.—Forest; below 1000 m.

Phenology records.—Flowering: unknown; fruiting: April to June, November, December.

Preliminary conservation assessment.—Vulnerable (**VU B1ab(iii)**). Last collected in 1948.

Distinguishing characteristics.—Inflorescence sessile; leaf blades hairy above. The shape and size of the hairs are similar to those of *P. pilosella*.

Remarks.—Only two species of Philippine *Psychotria* possess leaf blades with hairs on the upper surface, *P. papillata* and *P. pilosella*. The nature of the leaf investiture and sessile inflorescence distinguish *P. papillata* from all other Philippine *Psychotria*.

30. Psychotria yenii Sohmer & A.P. Davis, sp. nov. (**Fig. 32**). TYPE: PHILIPPINES. MINDANAO: South Cotabato Prov.: Tasaday Forest, dipterocarp forest at 1220 m, Aug or Dec 1972, *Yen 173* (HOLOTYPE: BISH).

Arbores vel frutices sunt, laminis chartaceis glabris late elliptis 6 × 16–7 × 18 cm latis et longis nervis lateralibus 8 in dimidio quoque apice acuto basi acuta vel obtusa, petiolis 7–8 mm longis, inflorescentia sessili, fructibus sessilibus glabris in sicco corrugatis cordatis 5–8 mm longis sessilibus vel subsessilibus, pyrenis in dorso laevibus, endospermio ruminato.

Tree or shrub, height unknown; young stems glabrous. Stipules unknown. Leaves: petioles 0.7–0.8 cm long, glabrous; leaf blades broadly elliptic, 16–18 × 6–7 cm, chartaceous, glabrous; secondary veins 7–9 pairs; tertiary venation weak; base acute to obtuse; apex acute. Inflorescence/infructescence simple (much reduced), lacking axes and branches, composed of ca. 11 pedicellate flowers/fruits, glabrous;

FIG. 31. *Psychotria papillata* (*Ramos BS 1701*) (MO, lectotype for *Grumilea papillata*).

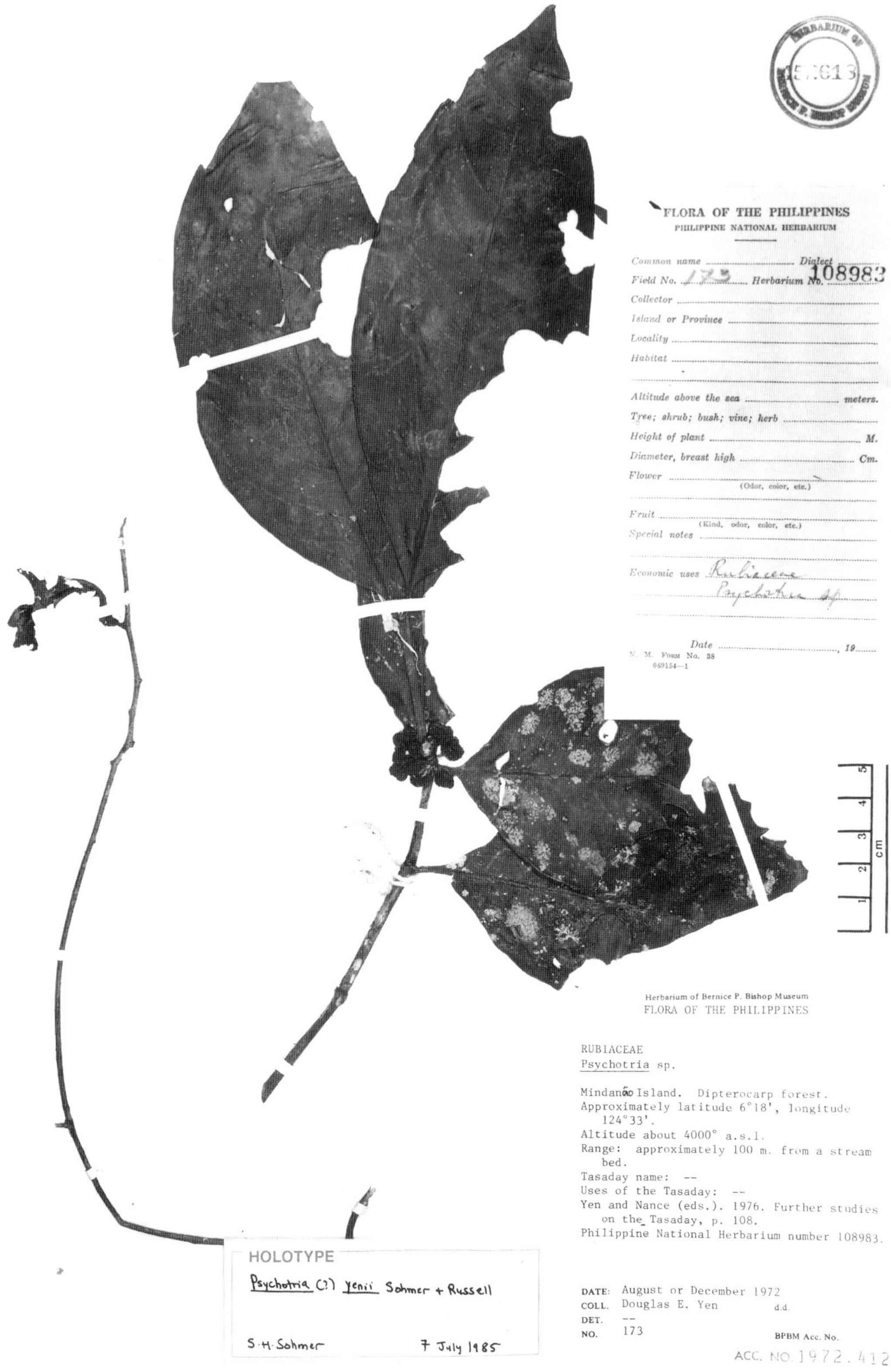

FIG. 32. *Psychotria yenii* (*Yen 173*) (BISH, holotype).

bracts and bracteoles absent. Flowers unknown. Fruit obovoid, ± heart-shaped (apex with 2 pronounced lobes), 5–8 mm long, wrinkled when dry, colour unknown, glabrous; calyx limb and lobes not persistent; pedicels 1–5 mm long, or ± or absent; pyrenes smooth on outer surface; endosperm ruminate.

PARATYPES: None—known only from the type collection.

Distribution.—This taxon is from the Tasaday area of Mindanao.

Ecology.—Dipterocarp forest; ca. 1220 m.

Phenology records.—Flowering: unknown; fruiting: August or December.

Preliminary conservation assessment.—Critically Endangered (**CR B1ab(iii)**). Only collected once in 1972.

Distinguishing characteristics.—Sessile inflorescences, fruit heart-shaped in longitudinal section.

Remarks.—The very unusual fruit shape distinguishes this taxon from all other *Psychotria* species in the Philippines.

31. Psychotria subcucullata Merr., Philipp. J. Sci. 29:492. 1926. (**Fig. 33**). TYPE: PHILIPPINES: BOHOL: Bohol Prov.: Kalingohan, rocky slopes in forest from 300–700 m, 20 Aug 1923, *Ramos BS 43144* (LECTOTYPE: UC, designated here; ISOLECTOTYPES: A, K, P, US).

Shrub, ca. 1 m high; young stems glabrous to pubescent; Stipules valvate, narrowly ovate, 4–6 mm long, glabrous but margin ciliate; apex acute. Leaves: petioles 0.3–1.3 cm long, pubescent; leaf blades narrowly oblong to lanceolate to oblanceolate, 3–5.4 × 0.9–1.6 cm, subcoriaceous, glabrous above, pubescent along midrib below; secondary veins 9–15 pairs; tertiary venation manifest to prominent; base obtuse to subcordate, apex acute to very abruptly acuminate. Inflorescence/infructescence simple (much reduced), lacking axes and branches, composed of 1–5 shortly pedicellate to sessile flowers, sparsely pubescent to pubescent; bracts and bracteoles absent. Flowers unknown. Fruit obovoid, 4–5 mm long, ribbed, reported to be yellow, probably turning red or black, sparsely pubescent to pubescent; calyx limb and lobes persistent; pedicels 0.5–1.2 mm long, or or absent; pyrenes ribbed on outer surface; endosperm weakly ruminate.

Other specimens examined: **Bohol. Bohol Prov.:** Bilar, *Ramos BS 43083* (A, BM, K, P, SING, UC, W).

Distribution.—Known only from Bohol.

Ecology.—Forest, on rocky slopes; 300–700 m.

Phenology records.—Flowering: unknown; fruiting: August to October.

Preliminary conservation assessment.—Critically Endangered (**CR B1ab(iii)**), or possibly Extinct (**EX**). Last collected in 1923.

Distinguishing characteristics.—Leaf blades small, 3–5.4 cm long, base often nearly cordate; inflorescence much reduced, lacking axes and branches, composed of 1–5 shortly pedicellate to sessile flowers.

Remarks.—This taxon bears some resemblance to the Linearis Species Group but as it possesses valvate, and not tubular, stipules, it belongs here. Due to the resemblance, it also has been keyed with the Linearis group.

32. Psychotria amaracarpoides (Merr.) Merr., Enum. Philipp. Fl. Pl. 3:552. 1923. (**Fig. 34**). *Grumilea amaracarpoides* Merr., Philipp. J. Sci. 10:134. 1915. TYPE: PHILIPPINES. LUZON: Zambales Prov.: Mt. Tapulao, Nov–Dec 1907, *Curran & Merritt FB 8120* (LECTOTYPE: US, designated here; ISOLECTOTYPE: NY).

Shrub, 2–3 m high, young stems pubescent (hairs prominent, appressed). Stipules valvate, ovate or ovate-cylindrical, 3–7 mm long, pubescent; apex cleft, lobes nearly aristate. Leaves: petioles 0.4–0.8(–1.2) cm long, pubescent; leaf blades narrowly elliptic, 2.2–5 × 1–1.8 cm, chartaceous, glabrous above, pubescent below at least or midrib; secondary veins 5–8 pairs; tertiary venation manifest to prominent; base acute to obtuse; apex acute to obtuse. Inflorescence/infructescence simple (much reduced), lacking axes and branches, composed of 1–2(?–3+) sessile flowers, sparsely pubescent to glabrous; bracts and bracteoles absent. Flowers unknown. Fruit obovoid, 5–9 mm long, smooth to very slightly ribbed, red, glabrous; calyx limb and lobes persistent; pedicels absent; pyrenes slightly ribbed on outer surface; endosperm ruminate.

Fig. 33. ***Psychotria subcucullata*** **(*Ramos BS 43144*) (UC, lectotype).**

Other specimens examined: **Luzon. Abra Prov.:** Mt. Posooy [Mt. Poosoy (?)], *Ramos BS 26996* (A, US). **Mountain Prov.:** Bontoc subprov., Mt. Masapilid, *Ramos & Edaño BS 37948* (BM, GH, L, UC). **Nueva Vizcaya Prov.:** Caraballo Mountains, *Loher s.n.* (K,UC). **Rizal Prov.:** Mt. Tokduanbanoy, *Ramos & Edaño BS 48592* (NY,UC). **Bataan Prov.:** Mt. Mariveles, *Elmer 6975* (K,NY), Mt. Mariveles, Lamao River, *Merrill 3194* (K,NY), *Whitford 1193* (K,NY, US), *Meyer FB 2837* (K,NY); without further data, *Whitford s.n.* (K). **Mindoro. Mindoro Occidental Prov.:** Paluan, *Ramos BS 39634* (A, US); Mindoro, without further data, *Merritt FB 8638* (US).

Distribution.—Central and northern Luzon and Mindoro.

Ecology.—Forest; 1000–1600 m.

Fig. 34. *Psychotria amaracarpoides* (*Curran & Merritt FB 8120*) (US, lectotype of *Grumilea amaracarpoides*).

Phenology records.—Flowering: unknown; fruiting: January to March, June, October to December.

Preliminary conservation assessment.—Vulnerable (**VU B1ab(iii)**). Last collected in 1926.

Distinguishing characteristics.—Leaves small, 2.2–5 cm long; inflorescence simple (much reduced), lacking axes and branches, normally composed of 1–2 sessile flowers; fruit sessile and often solitary.

Remarks.—In leaf size and shape this species bears some resemblance to the genus *Amaracarpus* Blume.

33. Psychotria urdanetensis Elmer, Leafl. Philipp. Bot. 5:1893. 1913; Merr., Enum. Philipp. Fl. Pl. 3:563. 1923. (**Fig. 35**). TYPE: PHILIPPINES. MINDANAO: Agusan del Norte Prov.: Mt. Urdaneta (Cabadbaran), humid forest near Lake Danao at ca. 1600 m, Sep 1912, *Elmer 13703* (LECTOTYPE: UC; ISOLECTOTYPES: A, BISH, BM, GH, K, L, MO, NY, P, US, W).

Shrub, 1–2 m high, young stems pubescent. Stipules valvate, ovate, 9–13 mm, pubescent; apex acute. Leaves: petioles 0.4–1.1 cm long, pubescent; leaf blades narrowly elliptic, 3.5–7.1 × 0.8–1.6 cm, chartaceous, drying grayish in color, glabrous above, pubescent below along veins; secondary veins 6–8 pairs; tertiary venation manifest to obscure; base acute or obtuse; apex acute to acuminate. Inflorescence/infructescence simple (much reduced), lacking axes and branches, composed of 1–4 sessile flowers, sparsely pubescent to glabrous; bracts and bracteoles absent. Flowers unknown. Fruit obovoid, 6–8 mm long, smooth, color unknown, glabrous to sparsely pubescent; calyx limb and lobes persistent but not prominent; pedicels absent; pyrenes smooth or very weakly ribbed on outer surface; endosperm ruminate.

Other specimens examined: Known only from the type collection.

Distribution.—Mt. Urdaneta, Mindanao.

Ecology.—Humid forest, including mossy forest; 1500–1600 m.

Phenology records.—Flowering: unknown; fruiting: September.

Preliminary conservation assessment.—Critically Endangered (**CR B1ab(iii)**), or possibly Extinct (**EX**). Last collected in 1912.

Distinguishing characteristics.—Inflorescence (much reduced), lacking axes and branches, composed of 1–4 sessile flowers; leaf blades 3.5–7.1 cm long, glabrous above, drying grayish.

Remarks.—The size and shape of the leaves make this taxon morphologically very close to *P. nagapatensis*. There is also a very strong resemblance to members of the Linearis Species Group.

34. Psychotria nagapatensis Merr., Philipp. J. Sci. 17:454. 1921; Merr., Enum. Philipp. Fl. Pl. 3:559. 1923. (**Fig. 36**). TYPE: PHILIPPINES. LUZON: Ilocos Norte Prov.: Mt. Nagapatan, in forests on dry slopes at 700 m, 8 Aug 1918, *Ramos BS 33193* (LECTOTYPE: A, designated here; ISOLECTOTYPES: BM, K, P, US).

Shrub, ca. 1 m high, young stems pubescent. Stipules valvate, narrowly ovate or cylindrical, 4–9 mm long, pubescent; apex acute or cleft. Leaves: petioles 0.5–1.2 cm long, sparsely to densely pubescent; leaf blades obovate to oblanceolate, 3–6 × 1.2–2 cm, chartaceous, drying brownish in color, glabrous above, pubescent below, at least on veins; secondary veins 5–12(–15) pairs; tertiary venation manifest to obscure; base acute to obtuse; apex narrowly acute, or very abruptly acuminate. Inflorescence/infructescence simple (much reduced), lacking axes and branches, composed of 2–10 shortly pedicellate or sessile flowers, sparsely pubescent to glabrous; bracts and bracteoles absent. Flowers ± unknown (very immature). Fruit globose to obovoid, 4–7 mm long, smooth, red, glabrous; calyx limb and lobes persistent, quite obvious; pedicels 0.4–1 mm long, or absent; pyrenes smooth on outer surface; endosperm ruminate.

Other specimens examined: **Panay. Antique Prov.:** western side of Mt. Baloy, Iki, Valderama, *Garcia & Romero PPI 15811* (BRIT, K). **Antique Prov.:** without further data: *McGregor BS 32386* (NY). **Capiz Prov.:** Mt. Salibongbong, *Martelino & Edaño BS 35576* (A, K). **Bohol. Bohol Prov.:** Logarita, Bilar, *Reynoso, Alvarez & Fuentes PPI 7784* (BRIT, K).

Distribution.—Found on Luzon, Panay, and Bohol.

Ecology.—Forest, includimg forest on dry slopes; 700–1600 m.

Phenology records.—Flowering: unknown; fruiting: April, March, August.

Preliminary conservation assessment.—Near Threatened (**NT**). Last collected in 1995.

Distinguishing characteristics.—Inflorescence much reduced, lacking axes and branches, composed of 2–10 shortly pedicellate or sessile flowers; leaf blades often obovate, 3–6 cm long, drying brownish; fruits small, smooth (unribbed), several to many per infructescnce.

Fig. 35. *Psychotria urdanetensis* (*Elmer 13703*) (UC, lectotype).

Fig. 36. *Psychotria nagapatensis* (*Ramos BS 33193*) (A, lectotype).

Remarks.—This taxon is morphologically very similar to *P. urdanetensis* Elmer from Mindanao. These two taxa could well be combined should additional evidence bridge the gap between them involving leaf blade vestiture and their color when dry.

35. Psychotria subsessiliflora Elmer, Leafl. Philipp. Bot. 1:24.1906; Merr., Enum. Philipp. Fl. Pl. 3:563. 1923. (**Fig. 37**). *Grumilea subsessiliflora* (Elmer) Merr., Philipp. J. Sci. 10:131. 1915. TYPE: PHILIPPINES. LUZON: Tayabas (Quezon) Prov.: Atimonan, Aug–Sep 1904, *Whitford 722* (LECTOTYPE: P, designated here; ISOLECTOTYPE: US).

Psychotria edanoi Merr., Enum. Philipp. Fl. Pl. 3:555. 1923, as 'edanoii.' *Grumilea brachybotrys* Merr., Philipp. J. Sci. 12:172. 1917. Non *Psychotria brachybotrys* Ridl., J. Fed. Malay States Mus. 4:38. 1909. TYPE: PHILIPPINES. LUZON: Nueva Ecija Prov.: Mt. Umingan, Aug 1916, *Ramos & Edaño BS 26256* (HOLOTYPE: PNH; ISOTYPE: K).

Small tree or shrub, 1–6 m high, young stems glabrous or sometimes pubescent. Stipules valvate, appearing twisted before opening, narrowly triangular, 5–13 mm long, glabrous or pubescent, margins sometimes ciliate; apex acuminate. Leaves: petioles 0.6–2.6 cm long, glabrous or pubescent; leaf blades broadly elliptic to broadly obovate, 7–14.8(–17) × 2.8–6(–7.9) cm, chartaceous to subcoriaceous, glabrous or pubescent below; secondary veins 7–14 pairs; tertiary venation manifest to obscure; base obtuse to acute or infrequently attenuate; apex abruptly acuminate to acuminate, or acute. Inflorescence/infructescence simple (much reduced), lacking axes and branches, or with a short main axis, or rarely trichotomous with two small branches from the base of the main axis, with 1 or 2 clusters of sessile flowers, glabrous to pubescent; bracts and bracteoles absent. Flowers 4- or 5-merous, pedicels absent; calyx (incl. hypanthium) ca. 1.5 mm long, glabrous; calyx lobes obscure; corolla: color unknown; corolla tube 2–2.5 mm (long styled), 2–3 mm (short styled), external surface glabrous, villous at throat within; corolla lobes 1–1.5 mm long (long styled), 1–2 mm long (short styled); stamens: filaments ca. 0.3 mm long (long styled), ca. 2 mm long short styled); anthers 0.7–1 mm long; style ca. 2 mm long (long styled), ca. 1 mm long (short styled), lobes ca. 0.7 mm long. Fruit globose to obovoid, 7–11 mm long, ribbed or ± smooth, often distinctly wrinked when dry, black or red at maturity, glabrous; calyx limb and lobes persistent but not obvious; pedicels absent; pyrenes ribbed to distinctly ribbed on outer surface; endosperm ruminate.

Distribution.—Throughout Luzon, Mindoro and Negros.

Phenology records.—Flowering: February to May, September; fruiting: all year round.

Ecology.—Forest; below 250 m.

Preliminary conservation assessment.—Near Threatened (**NT**). Last collected in 1996.

Distinguishing characteristics.—Usually glabrous throughout, young shoots always glabrous; leaf blades broadly elliptic or obovate; inflorescences much reduced, lacking axes and branches, or with a short main axis, or rarely trichotomous with two small branches from the base of the main axis, with 1 or 2 clusters of sessile flowers; fruit large, shiny, black, drying with prominent ribs.

Remarks.—This species is fairly common wherever appropriate forest remains. An individual shrub in full fruit is extremely ornamental. The shiny, black fruits are very prominent and probably taken by a number of native birds. Most individuals do not have inflorescence axes but some appear to possess them. It becomes difficult in some instances to separate some individuals from *P. pauciflora* Bartl. ex DC. in the Subalpina Species Group, particularly if such individuals also have small branches at the base of the axis/axes. There are two varieties that may be separated by the following key.

KEY TO THE VARIETIES OF *PSYCHOTRIA SUBSESSILIFLORA*

1. Leaf blades usually glabrous, or at most with sparse hairs ____ **35a.** var. **subsessiliflora**
1. Leaf blades usually pubescent below ____ **35b.** var. **carinata**

35a. Psychotria subsessiliflora var. subsessiliflora

Small trees or shrubs, 1–6 m tall. Stipules 5–13 mm long. Leaf blades usually glabrous below; petioles glabrous, 0.6–2.6 cm long; secondary veins 7–12(–14) per side.

Other specimens examined: **Luzon. Cagayan Prov.:** Tamdagan River, *Edaño BS 79519* (A), *BS 79524* (A), *BS 79537* (A); Baguio Cove, *Allen 84-81* (BISH); Mt. Caboloan, *Edaño BS 78606* (A), *Edaño BS 78620* (A); Mt. Babatngin, *Edaño BS 79388* (A), *BS 79418* (A), *BS 79435* (A), *BS 79444* (A), *BS 79446* (A); Pagikpik, *Edaño BS 79488* (A). **Kalinga-Apayao Prov.:** without further data: *Fenix BS 28058* (A, MO, US). **Ilocos Norte Prov.:** *Ramos BS 32960* (SING). **Isabela Prov.:** Mt. Moises, *Clemens 16845* (UC); Palanan Municipality, Sierra Madre Mountain Range, Barangay San Isidro, hills N. of Sitio Diago, *Co 3428* (K). **Quezon Prov.:** Mt. Cadig, *Yates BS 25516* (A, US), *BS 25545* (UC); Lucban, *Elmer 9200* (A, L); Quezon National Forest Park, Pinaglabanan Trail, *Sohmer & Reynoso 12078* (BISH), *12079* (BISH); Llawinan River, Philipp. National Bot. Garden, *Sohmer 12021* (BISH); Polillo Island, Anibawan, *Castro, PNH 6508* (L), *vel aff.*; Alabat Island, without further data, *Ramos & Edaño BS 48075* (BM, NY, UC, US). **Bataan Prov.:** Bataan Mountain, *Sohmer 11989* (BISH), *Sohmer 12024* (BISH). **Laguna Prov.:** San Antonio, *Ramos BS 15077* (K, P). **Camarines Norte Prov.:** Paracale,

FIG. 37. *Psychotria subsessiliflora* (*Whitford 722*) (P, lectotype).

Ramos & Edaño BS 33612 (BM, P); Bicol National Park, *Croat 53037* (MO); Mt. Kadig, *Edaño PNH 40163* (L, PNH), *PNH 40195* (BM, CANB, K, L); Basud, Tuaca, Mt. Nilisan, *Reynoso, Romero & Fuentes PPI 1217* (BRIT, K). **Camarines Sur Prov.:** Mt. Potianay, *Edaño BS 75916* (NY). **Albay Prov.:** Mayon Volcano, *Mendoza PNH 18275* (L, PNH). **Sorsogon Prov.:** Mt. Bulusan, south slope, *Edaño & Gutierrez PNH 38482* (A, BISH, BM, K, L), *PNH 38485* (L, PNH), *PNH 38495* (A, PNH), *Sulit PNH 2748* (A, MO); Mt. Bulusan National Park, Barangay Inlagadian, Casiguran, *Garcia & Fernando PPI 22790* (BRIT, K); *Garcia & Fernando PPI 22809* (BRIT); SW of Bulusan National Park, Barangay Cogon, Irosin, *Garcia & Fernando PPI 25018* (BRIT, K); Irosin, *Elmer 14977* (A, BISH, BM, GH, K, L, MO, NY, US), *15256* (A, BM, GH, K, L, MO, NY, P, S, UC), *16805* (A, BISH, BM, K, L, MO, NY, P, S, UC). **Sorsogon Prov.:** without further data: *Ramos BS 23386* (A, NY, P). **Catanduanes. Cataduanes Prov.:** San Miguel, Barangay Sulung, Mt. Saan, *Barbon, Garcia & Alvarez PPI 2393* (K). **Catanduanes Prov.:** without further data: *Ramos BS 30522* (L, P). **Mindoro. Mindoro Oriental Prov.:** Mt. Halcon, *Ramos & Edaño BS 40565* (A, K); Busay, Naujan, *Celestino & Castro PNH 1913* (A, L, MO); Laylay, 10 km from Swedish Match Hilleshog's tree farm, *Sohmer 12292* (BISH), *12293* (BISH), *12295* (BISH). **Samar. Western Samar Prov.:** Mt. Sarauag, *Edaño PNH 15257* (A, CANB, K, L). **Leyte. Leyte Prov.:** Biliran Island, Mt. Suiro, *Sulit PNH 21495* (L). **Negros. Negros Occidental Prov.:** Victoria Mountains, *Denoga FB 30601* (UC), *FB 30602* (UC). **Mindanao. Camiguin Prov.:** Camiguin Island, So. Sinablayan, Barangay Botay (boundry of Sagay) Guinsiliban, *Gaerlan, Fernando & Majaducon PPI 19962* (K). **Zamboanga del Norte Prov.:** Tulalu, *Frake PNH 38192* (L).

Distribution.—Luzon, Leyte, Negros, and Mindoro.

Ecology.—Forest, including sub-mossy secondary forest, and plantations; 35–900 m.

Phenology records.—Flowering: February to May, September; fruiting: all year round.

Preliminary conservation assessment.—Near Threatened (**NT**). Last collected in 1996.

35b. Psychotria subsessiliflora var. **carinata** (Quisumb. & Merr.) Sohmer & A.P. Davis, comb. et stat. nov. (**Fig. 38**). *Psychotria carinata* Quisumb. & Merr., Philipp. J. Sci. 37:205.1928. TYPE: PHILIPPINES. LUZON: Quezon Prov.: Casiguran in damp forest, 28 May 1925, *Ramos & Edaño BS 45278* (LECTOTYPE: US, designated here; ISOLECTOTYPES: A, BM, NY, US).

Shrubs, ca. 1 m tall. Stipules 6–12 mm long. Leaf blades usually hairy below; petioles usually pubescent, 0.7–2.3 cm long; secondary veins 8–14 per side.

Other specimens examined: **Luzon. Cagayan Prov.:** Gonzaga, *Edaño BS 78205* (A), *BS 78207* (A), *BS 78215* (A), *BS 78236* (A), *BS 78265* (A), *BS 78275* (A), *BS 78304* (A); Mt. Cagua, *Edaño BS 78362* (A), *BS 78406* (A); Mt. Balatongan, *Edaño BS 78530* (A). **Isabela Prov.:** Sierra Madre Mountains, San Mariano, *Gutierrez PNH 78165* (A, L). **Quezon Prov.:** Casiguran, *Ramos & Edaño BS 45325* (A, K, NY, UC), *BS 45518* (K, NY, UC); Mt. Dingalan, *Ramos & Edaño BS 26623* (A, K). **Camarines Norte Prov.:** Mt. Kadig, *Edaño PNH 40163* (L, PNH), *PNH 40195* (BM, CANB, K, L); **Camarines Sur Prov.:** Sarapan, *Edaño BS 76155* (NY), *BS 76161* (NY). **Catanduanes. Catanduanes Prov.:** Dakulangpatag, *Ramos & Edaño BS 75570* (NY). **Catanduanes Prov.:** without further data: *Ramos & Edaño BS 75439* (UC).

Distribution.—Central and northern Luzon, and Catanduanes Island.

Ecology.—Forest; low elevation.

Phenology records.—Flowering: unknown; fruiting: May to September.

Preliminary conservation assessment.—Near Threatened (**NT**). Last collected in 1961.

36. Psychotria tayabensis Elmer, Leafl. Philipp. Bot. 1:354. 1908; Merr., Enum. Philipp. Fl. Pl. 3:563. 1923. (**Fig. 39**). *Grumilea tayabensis* (Elmer) Merr., Philipp. J. Sci. 10:133. 1915 TYPE: PHILIPPINES. LUZON: Tayabas (Quezon) Prov.: Lucban, May 1907, *Elmer 7581* (LECTOTYPE: US, designated here; ISOLECTOTYPES: A, K, L, MO, NY, W).

Small tree or shrub, 1–2(–6) m high, young stems glabrous or pubescent. Stipules valvate but twisted in bud, ± triangular to narrowly triangular, 5–15 mm long, glabrous or pubescent; apex cleft. Leaves: petioles 0.5–3(–3.5) cm long, pubescent or glabrous; leaf blades narrowly elliptic to obovate or oblanceolate, 9.2–21.5 × (1.2–)1.6–5.4 cm, chartaceous to subcoriaceous, pubescent or glabrous below; secondary veins 7–20 pairs; tertiary venation prominent to obscure; base acute to attenuate; apex abruptly acuminate to acuminate or acute. Inflorescence/infructescence simple (much reduced), lacking axes and branches, composed of 1–18 sessile, or infrequently pedicellate, flowers, glabrous to pubescent; bracts deltate to triangular, ca. 5 mm long, sometimes absent; bracteoles ?absent. Flowers 4- or 5-merous, pedicels absent; calyx (incl. hypanthium) 1.5–3 mm long, glabrous or pubescent; calyx lobes ± deltate, 0.75–1 mm long; corolla: white; corolla tube 2–3 mm long, external surface glabrous, villous at throat within; corolla lobes 2–2.5 mm long (long styled); stamens: filaments ca. 1 mm long; anthes ca. 0.8 mm long; style ca. 5 mm long (long styled), lobes ca. 1 mm long. Fruit obovoid to globose, (3–)4–10 mm long, smooth, red or red-purple, glabrous; calyx limb and lobes persistent but not prominent; pedicels absent; pyrenes smooth on outer surface; endosperm ruminate.

Distribution.—Throughout Luzon, Mindoro, Leyte, Negros, Alabat, Catanduanes, Panay, and Samar.

FIG. 38. *Psychotria subsessiliflora* var. *carinata* (*Ramos & Edaño BS 45278*) (US, lectotype of *Psychotria carinata*).

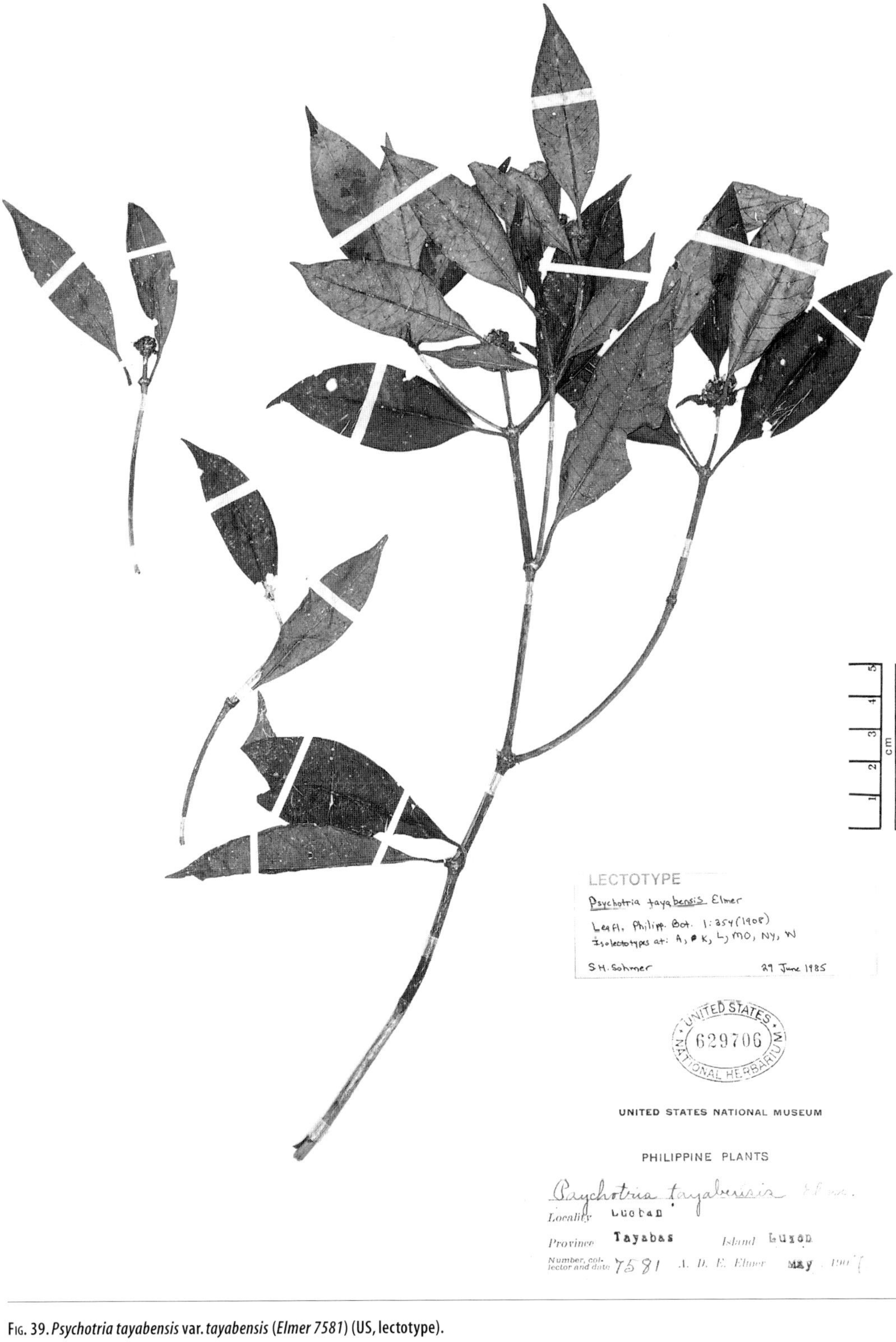

FIG. 39. *Psychotria tayabensis* var. *tayabensis* (*Elmer 7581*) (US, lectotype).

Ecology.—Forest; sea level to 1300 m.

Phenology records.—Flowering: February, May, September; fruiting: March to August, October, November.

Preliminary conservation assessment.—Near Threatened (**NT**). Last collected in1996.

Distinguishing characteristics.—Leaf blades 4.6–21.5 cm long; stipule apex cleft; inflorescence simple (much reduced), lacking axes and branches, composed of 1–18 sessile, or infrequently pedicellate, flowers; fruits several to many, wrinkled when dry.

Remarks.—This species is relatively common, extremely variable, and consists of a number of more-or-less discrete varieties that can be separated by the following key.

KEY TO THE VARIETIES OF *PSYCHOTRIA TAYABENSIS*

1. Leaf blades usually <10 cm long.
   2. Stipules, petioles, lower leaf blade surface and other vegetative organs usually pubescent ______ **36a.** var. **tayabensis**
   2. All parts glabrous ______ **36b.** var **castanea**
1. Leaf blades usually >10 cm long.
   3. Petioles usually pubescent; leaf blades usually with tertiary venation prominent below, forming a network visible with a10× lens ______ **36d.** var. **fasciculiflora**
   3. Petioles usually glabrous; leaf blades usually without prominent network of tertiary veins.
      4. Leaf blades < 3 cm wide ______ **36e.** var. **pinagteponanensis**
      4. Leaf blades > 3 cm wide ______ **36c.** var. **euphlebia**

36a. Psychotria tayabensis var. tayabensis

Small tree or shrub, 1–4 m tall, young stems glabrous to pubescent. Stipules 5–13 mm long, glabrous to pubescent. Leaves: petioles 0.5–2.2 cm, glabrous to pubescent; leaf blades narrowly elliptic to narrowly obovate, 4.6–14.4 × 1.6–5.3 cm, pubescent below; secondary veins 7–14 per side; base acute or attenuate; apex abruptly acuminate or acute; tertiary venation obscure to manifest. Inflorescence with up to 12 flowers. Flowers 5-merous. Fruit globose or obovoid, 5–10 mm long, sessile or pedicellate.

Other specimens examined: **Luzon. Cagayan Prov.:** Claveria, *McLean & Catalan 141* (UC); Baguio Cove, *Allen 53-81* (BISH), *337-81* (BISH); Mt. Caboloan, *Edaño BS 78585* (A). **Kalinga-Apayao Prov.:** Mt. Magnas, *Edaño PNH 19770* (L). **Ilocos Norte Prov.:** Mt. Quebrada, *Edaño PNH 17831* (L). **Ifugao Prov.:** Sumigar, Mt. Nabito, *Barbon & Pedersen et al. PPI 12894* (K). **Quezon Prov.:** Lalawinan River, *Sohmer 12013* (BISH); Lucban, *Elmer 7934* (A, K, L, MO, NY, W), *7952* (A, K, L); Tagkawayan, *Hermida 6* (US). **Rizal Prov.:** Montalban, *Loher 12125* (UC); Mt. Lumutan, *Ramos & Edaño BS 29625* (L, P); San Isidro, *Ramos BS 12108* (BM, K, L, P). **Laguna Prov.:** Mt. Banahao, *Duyag BS 76685* (NY), *Curran & Merritt FB 7870* (K); Majayjay, *Rivera & Duyag BS 49972* (UC). **Laguna Prov.:** without further data: *Robinson BS 6070* (US). **Camarines Sur Prov.:** Botol River, *Edaño BS 76297* (NY), *BS 76306* (K, NY). **Sorsogon Prov.:** Lake Agangay, *Sinclair & Edaño 9633* (K, US); Mt. Bulusan, Talatala Ridge, vicinity of Aguingai Valley, *Sulit & Gojar PNH 2683* (A); Irosin, *Elmer 15466* (A, BISH, BM, K, MO, NY, US). **Panay. Capiz Prov.:** Libaco, *Martelino & Edaño BS 35521* (K). **Catanduanes. Catanduanes Prov.:** without further data: *Ramos & Edaño BS 75193* (UC). **Mindoro. Without further data:** *Merritt FB 6883* (A, K). **Leyte. Without further data:** *Wenzel 1620* (A, BM, GH, NY). **Bohol. Bohol Prov.:** Demyao, *Ramos BS 42679* (A, UC); Kalingohan, *Ramos BS 43360* (A, UC, US, W). **Negros. Negros Occidental Prov.:** Canalon National Park, Mt. Canalon, Wasay trail, *Stone et al. PPI 6278* (K). **Negros Oriental Prov.:** Lake Balinsasayao, Tanjay, *Edaño PNH 6756* (A). **PHILIPPINES. Without further data:** *Loher 6446* (K).

Distribution.—Luzon, Leyte, Mindoro, Panay, Bohol, Catanduanes and Negros.

Ecology.—Drier forest; from sea level to 1300 m.

Phenology records.—Flowering: February, May, September; fruiting: March to August, October, November.

Preliminary conservation assessment.—Near Threatened (**NT**). Last collected in 1994.

Distinguishing characteristics.—Leaves small; young stems pubescent, petioles, leaf blades and inflorescence axes pubescent.

Remarks.—This is either the most common of the varieties of *P. tayabensis* or the easiest to collect.

36b. Psychotria tayabensis var. **castanea** (Merr.) Sohmer & A.P. Davis, comb. et stat. nov. (**Fig. 40**). *Psychotria castanea* Merr., Philipp. J. Sci. 17:446. 1921; Merr., Enum. Philipp. Fl. Pl. 3:554. 1923. TYPE: PHILIPPINES: LUZON: Ilocos Norte Prov.: Mt. Nagapatan, Aug 1918, *Ramos BS 33212* (LECTOTYPE: A, designated here; ISOLECTOTYPES: BM, K, P, US).

Shrub, height unknown, glabrous throughout. Stipules 5–9 mm long, glabrous. Leaves: petioles 6–1.7 cm long, glabrous; leaf blades narrowly elliptic to oblanceolate, 5–9.4 × 1.2–1.8 cm, generally glabrous

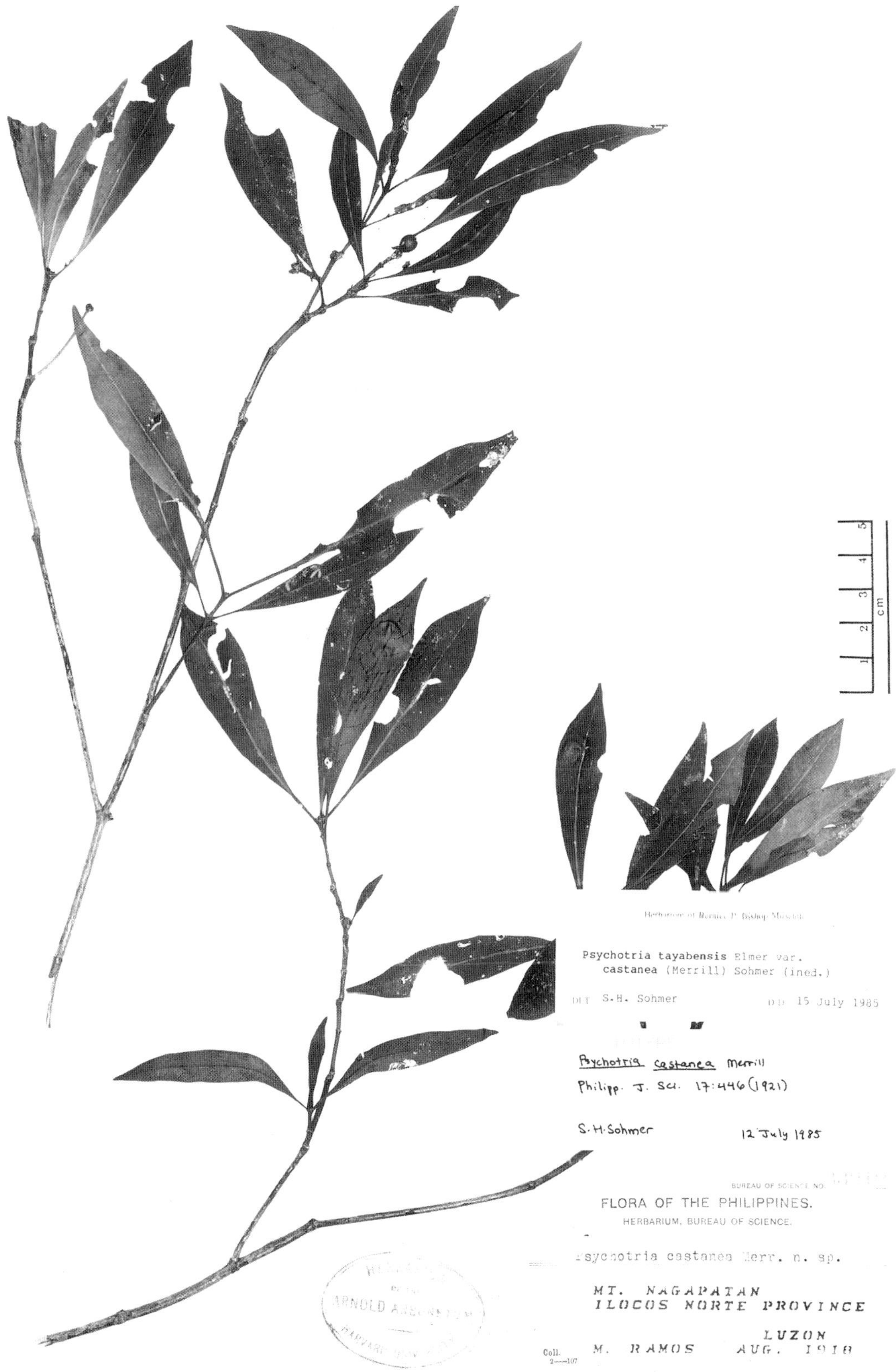

Fig. 40. *Psychotria tayabensis* var. *castanea* (*Ramos BS 33212*) (A, lectotype of *Psychotria castanea*).

above and below; secondary veins 8–15 per side; base attenuate; apex acuminate to acute; tertiary venation obscure to manifest. Inflorescence unknown. Flowers unknown. Fruit obovoid, 4–8 mm long, sessile.

Other specimens examined: **Luzon. Ilocos Norte Prov.:** Bangui to Claveria, *Ramos BS 33061* (A, BM, K, P); Mt. Quebrada, *Edaño PNH 17886* (PNH); Mt. Semenublan, east of Nueva Era, *Iwatsuki et al. P-258* (L).

Distribution.—Only in Ilocos Norte Prov., Luzon.

Ecology.—Drier forest, on slopes; ca. 700 m.

Phenology records.—Flowering: unknown; fruiting: August.

Preliminary conservation assessment.—Endangered (**EN**). Last collected in1975.

Distinguishing characteristics.—Plant glabrous; leaf blades 5.1–9.4 cm long.

Remarks.—Only one recent collection has been made other than the collections originally cited by Merrill (1921).

36c. Psychotria tayabensis var. **euphlebia** (Merr.) Sohmer & A.P. Davis, comb. et stat. nov. (**Fig. 43**). *Psychotria euphlebia* Merr., Philipp. J. Sci. 8:55. 1913; Merr., Enum. Philipp. Fl. Pl. 3:555. 1923. *Grumilea euphlebia* (Merr.) Merr., Philipp. J. Sci. 10:133. 1915. TYPE: PHILIPPINES. LUZON: Cagayan Prov.: Abulug River, collector unknown *BS 14582* (not seen).

Psychotria tricarpa Merr., Philipp. J. Sci. 17:460. 1921. TYPE: PHILIPPINES. PANAY: Capiz Prov.: Jamindan, Apr–May 1918, *Ramos & Edaño BS 31224* (LECTOTYPE: A, designated here; ISOLECTOTYPES: K, US).

Shrub, ca. 1 m tall, young stems sparsely pubescent. Stipules 5–10 mm long, pubescent at base. Leaves: petioles 1–3.2 cm long, pubescent; leaf blades elliptic to subobovate, 8–17.6 × 3.6—5 cm, pubescent below; secondary veins 10–19 per side; base acute to obtuse; apex abruptly acuminate or acute; tertiary venation obscure to manifest. Inflorescence with 10–18 flowers. Flowers 4–5-merous. Fruit globose, 5-9 mm long, sessile.

Other specimens examined: **Luzon. Cagayan Prov.:** Mt. Caboloan, *Edaño BS 78595* (A), *BS 78613* (A), *BS 78619* (A); Claveria, Barangay Kilkiling, Taggat Watershead Area, *Garcia, Fuentes & Romero PPI 18167* (BRIT, K). **Cagayan Prov.:** without further data: *Velasco FB 26976* (K, P). **Kalinga-Apayao Prov.:** Mt. Magnas, *Edaño PNH 19780* (K, L). **Isabela Prov.:** Mt. Moises, *Clemens 16846* (UC), *16847* (UC), *17021* (UC); Sierra Madre Mountains, Sangcad, Palanan, *Gutierrez PNH 78268* (A, K, L). **Quezon Prov.:** Mt. Cadig, *Yates BS 25513* (A, K, L, US); Alabat Island, without further data, *Ramos & Edaño BS 48228* (A, NY, P, SING, UC). **Camarines Sur Prov.:** Sarapan, *Edaño BS 76164* (K, NY); Mt. Bagacay, *Ramos & Edaño BS 33831* (L).

Distribution.—Alabat and Panay Islands, and on Luzon (Cagayan, Isabela, Quezon, and Camarines provinces).

Ecology.—Forest, including mixed forest with dipterocarps; 400–1000 m.

Phenology records.—Flowering: December; fruiting: April to July.

Preliminary conservation assessment.—Near Threatened (**NT**). Last collected in 1995.

Distinguishing characteristics.—Leaf blades large, often drying orange-brown, tertiary venation usually not prominent, petioles often pubescent.

Remarks.—Based on the description, location, and specimens cited by Merrill (1923:555) but not available for this study we have decided to recognize the taxon as herein circumscribed. The placement of *P. tricarpa* here is also somewhat dubious, but this seems to be the best option for it within our present concept of *P. tayabensis*, with which species it definitely belongs. Except for the variable tertiary venation patterns and often pubescent petioles, there is not much overall difference between this variety and *P. tayabensis* var. *fasiculiflora*.

36d. Psychotria tayabensis var. **fasciculiflora** (Merr.) Sohmer & A.P. Davis, comb. et stat. nov. (**Fig. 41**). *Grumilea fasciculiflora* Merr., Philipp. J. Sci. 10:135. 1915. *Psychotria fasciculiflora* (Merr.) Merr., Enum. Philipp. Fl. Pl. 3:555. 1923. TYPE: PHILIPPINES. SAMAR: Yabong, on forested slopes, 15 Mar 1914, *Ramos BS 17519* (LECTOTYPE: BM, designated here; ISOLECTOTYPE: K).

Shrub or tree, 1–6 m tall, young stems glabrous. Stipules 7–15 mm long, pubescent at base and along apex. Leaves: petioles 0.7–3 cm long, glabrous; leaf blades narrowly elliptic to lanceolate, 9–21.5 × 2.3–5.4 cm, glabrous or barely pubescent along veins beneath; secondary veins 10–18 per side; base obtuse or cuneate; apex acuminate; tertiary venation prominent; Inflorescence probably with ca. 10 flowers. Flowers 4–5-merous. Fruit obovoid to globose, 6–8 mm long, mostly sessile.

Other specimens examined: **Samar. Northern Samar Prov.:** Catubig River, *Edaño BS 24804* (US), *Ramos BS 24315* (A, BM, K, P, US). **Western Samar Prov.:** Mt. Sarauag, *Edaño PNH 15255* (A, L), *PNH 15297* (A, K, L). **Samar. Prov. unknown:** Mt. Pulog, *Edaño PNH 15382* (PNH), *PNH 15434* (A, K, L), *PNH 15439* (A, K, L).

Fig. 41. *Psychotria tayabensis* var. *fasciculiflora* (*Ramos BS 17519*) (BM, lectotype of *Psychotria fasciculiflora*).

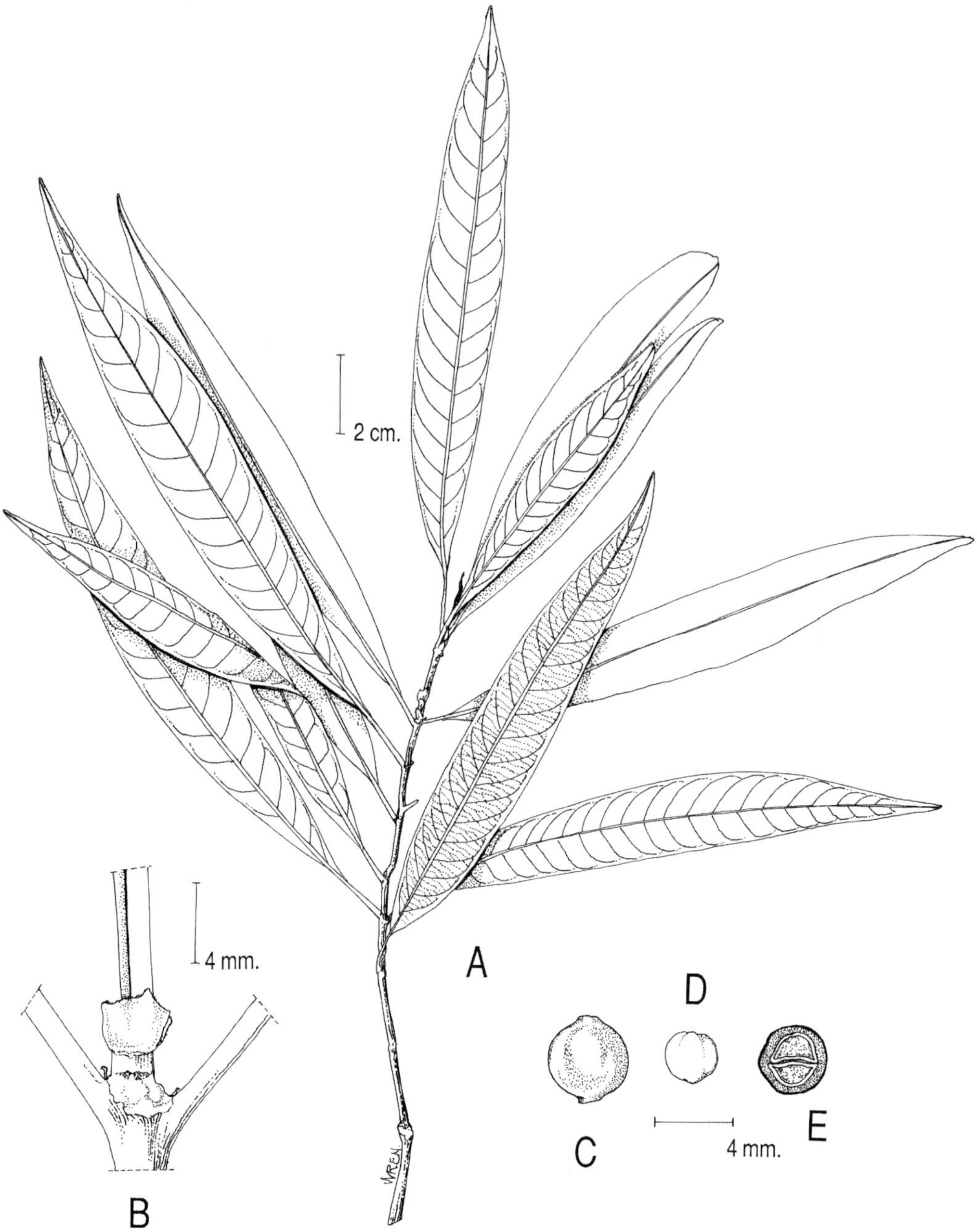

FIG. 42. ***Psychotria tayabensis*** **var.** ***pinagteponanensis*** **(*Edaño BS 78343*) (A, holotype). A, habit of vegetative branch; B, stipule; C, fruit; D, pyrene; E, fruit in cross section.**

Distribution.—Samar.

Ecology.—Forest; 100–200 m.

Phenology records.—Flowering: unknown; fruiting: March.

Preliminary conservation assessment.—Endangered (**EN**). Last collected in 1951.

Distinguishing characteristics.—Leaf blades robust, elliptic to lanceolate, drying orange-brown, petioles glabrous, tertiary venation pattern prominent.

Remarks.—This taxon most closely resembles *P. tayabensis* var. *pinagteponanensis* from which it is distinguished at this time (pending more collections of the latter) by its wider leaf blades. It is

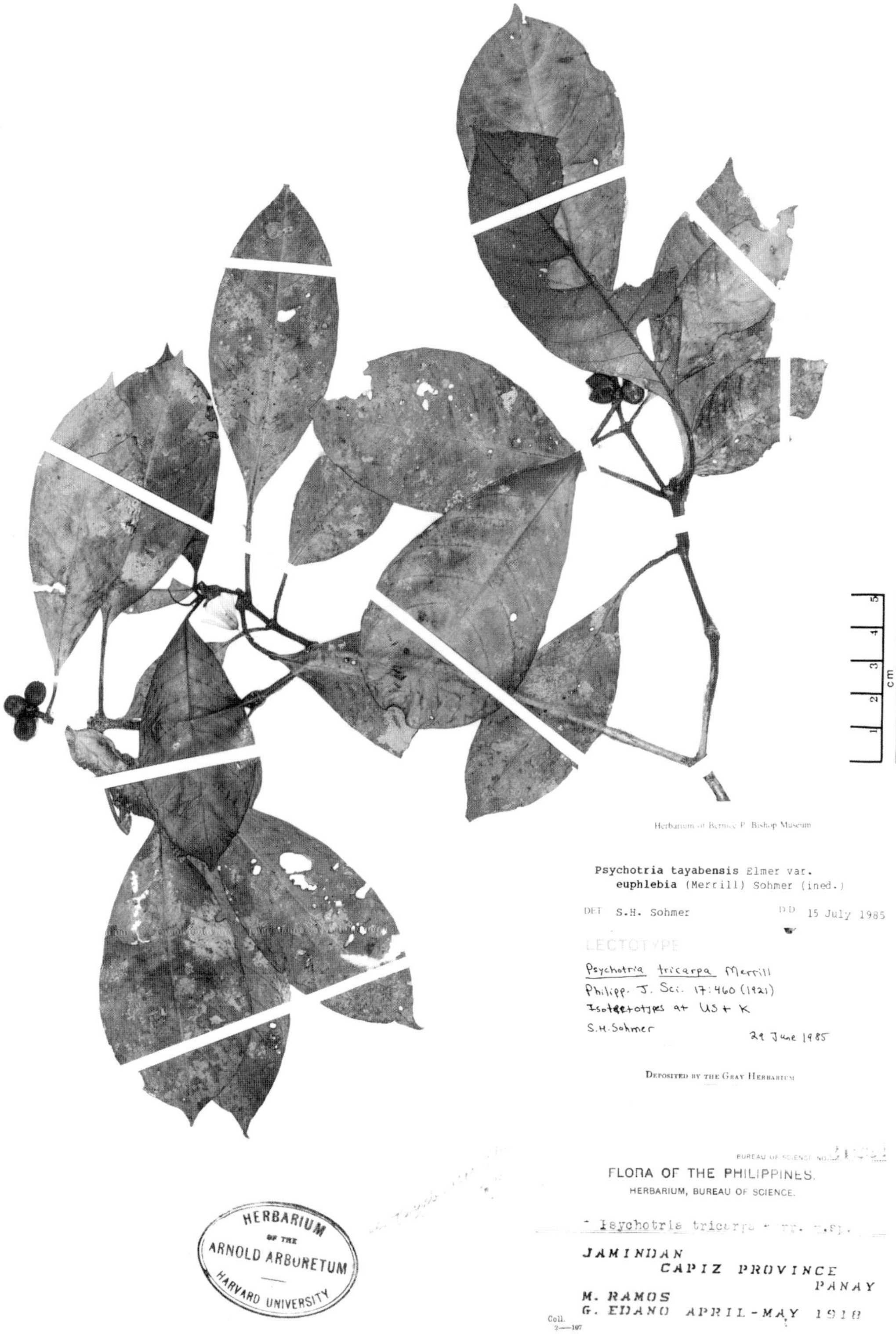

Fig. 43. *Psychotria tayabensis* var. *euphlebia* (*Ramos & Edaño BS 31224*) [A, lectotype of (?) *Psychotria tricarpa*].

separated from most of the other varieties by its much larger and robust leaves except for *P. tayabensis* var. *euphlebia*, from which it is distinguished by its often prominent tertiary venation and petioles usually glabrous.

36e. Psychotria tayabensis var. **pinagteponanensis** Sohmer & A.P. Davis, var. nov. (**Fig. 42**). TYPE: PHILIPPINES. LUZON: Cagayan Prov.: Pinagteponan River, Oct 1929, *Edaño BS 78343* (HOLOTYPE: A).

Licet fruticem est, stipulis infra tubum glabris formanlibus partilus alteris incognitis, laminis coriaceis vel semicoriaceis glabris sed infra in midnerva pubescentibus anguste ellipticis 1.2 × 4.2–2.3 × 15.7 cm latis et longis nervis lateralibus 12–20 in dimidio quoque apice acuminato basi acuta vel acuminata in sicco badiis, fructibus globosis 3–4 mm longis, pyrenis in dorso laevibus, endospermio ruminato.

Shrub or treelet, height unknown, glabrous throughout. Stipules ca. 6 mm long, glabrous. Leaves: petioles 0.8–1.1 cm long, glabrous; leaf blades narrowly elliptic, 9.2–15.7 ×1.2–2.3 cm, glabrous or pubescent along midrib below; secondary veins 12–20 per side; base acute or narrowly acute; apex acuminate; tertiary venation obscure to manifest. Inflorescence unknown. Flowers unknown. Fruit globose, 3–4 mm long, pedicellate.

PARATYPES: **Luzon. Cagayan Prov.:** Pinagteponan River, *Edaño BS 78635* (A); Barangay San Juan, Praxedes, *Garcia, Fuentes & Romero PPI 18464* (K). **Camarines Sur Prov.:** Agosais, *Edaño BS 76128* (NY).

Distribution.—Luzon.

Ecology.—Forest, including mixed forest with dipterocarps; ca. 500 m.

Phenology records.—Flowering: unknown; fruiting: October.

Preliminary conservation assessment.—Vulnerable (**VU**). Last collected in 1995.

Distinguishing characteristics.—Leaves long, narrow, 1.2–2.3 cm wide; fruit relatively small, 3–4 mm long.

Remarks.—This taxon most closely resembles *P. tayabensis* var. *fasciculiflora* but lacks the often prominent tertiary venation of the latter and has narrower leaf blades.

IV.—Pilosella Species Group

Shrubs, small trees or rarely suffrutescent. Stipules valvate. Leaf blades less or more than 3 cm wide; never more than 30 cm long; chartaceous to subcoriaceous, sometimes coriaceous. Inflorescences with axes and branches, trichotomous or monochotomous (pedunculate), ***with clusters of sessile flowers/fruits terminating the principal axes. Flowers sessile, rarely pedicellate.*** Fruits usually red, sometimes yellow, purple or black; pyrene inner (adaxial) surface flat; seed producing a red color when placed in 70% ethanol, due to a soluble seed coat pigment (SCP positive), rarely SCP negative; endosperm ruminate, rarely entire, usually dark. Comprising 13 species.

Note.—Text in ***bold italics*** denotes features useful for the placement of taxa within the respective species group.

KEY TO THE SPECIES OF THE PILOSELLA SPECIES GROUP

1. Leaf blades strongly scabrous beneath, very much like rough sandpaper ____ **37. P. scaberula**
1. Leaf blades not scabrous beneath, or very weakly so, never like rough sandpaper.
   2. Leaf base usually subauriculate, or at least round or obtuse.
      3. Leaf blades pubescent below at least along midrib ____ **38. P. bataanensis**
      3. Leaf blades glabrous below ____ **39. P. elliptilimba**
   2. Leaf base cuneate, acute, narrowly acute, attenuate, never subauriculate or truncate, rarely rounded.
      4. Plants generally glabrous or sparsely pubescent; leaf blades glabrous; inflorescence axis/axes glabrous or pubescent.
         5. Plants < 0.5 m tall; petioles < 0.7 cm long ____ **40. P. pygmaea**
         5. Plants 1–7 m tall; petioles usually > 1 cm long.
            6. Leaf blades 12.5–19 cm long; fruits 11–14 mm long ____ **41. P. arborescens**
            6. Leaf blades < 16 cm long, usually < 12 cm long; fruits, where known, 7–11 mm long.
               7. Leaf blades drying grayish green or yellowish green; fruits without prominent calyx tube; infloresence main axis usually with 2 nodes at apex ____ **42. P. wenzelii**
               7. Leaf blades drying dark brown or dark reddish brown; fruits, where known, with prominent, persistent calyx tube (2–4 mm long); inflorescence main axis usually with 1 node at apex.
                  8. Leaf apex round or obtuse to sometimes abruptly short-acute; fruits with prominent, persistent calyx tube 2–4 mm long ____ **43. P. castroi**

8. Leaf apex acute or acuminate to abruptly acuminate; fruits with rather prominent, persistent calyx tube ≥ 2 mm long ____ **44. P. surigaoensis**

4. Plants pubescent, often densely so; leaf blades pubescent at least on lower surface; inflorescence axis/axes pubescent.

9. Inflorescence axes > 3 mm thick; leaf blades 17–25.5 cm long; calyx pubescent ____ **45. P. sulitii**

9. Inflorescence axes usually < 3 mm thick; leaf blades usually < 17 cm long; calyx glabrous or pubescent.

10. Plants prostrate, creeping, suffrutescent; stipules pubescent, cleft with the two lobes drawn into aristate tips ____ **46. P. pulgarensis**

10. Plants not prostate, trees or shrubs; stipules pubescent or glabrous; stipules entire or cleft with two short acute lobes.

11. Petioles usually glabrous; leaf blades narrowly elliptic to lanceolate, glabrous above, glabrous or sparsely pubescent below; leaf apex usually narrowly acuminate ____ **47. P. lancilimba**

11. Petioles usually pubescent; leaf blades generally obovate or oblanceolate, infrequently elliptic or lanceolate, usually pubescent above, always pubescent below; leaf apex acuminate, acute, obtuse, or sometimes round.

12. Leaf blades elliptic, obovate to narrowly oblanceolate, chartaceous to subcoriaceous, usually with hairs above as well as below; leaf base usually base usually narrowly acute, or sometimes attenuate; leaf apex acuminate to acute ____ **48. P. pilosella**

12. Leaf blades usually ovate-elliptic to obovate, subcoriaceous, usually glabrous above or sometimes scabrid; leaf base acute, infrequently narrowly acute; leaf apex acute to nearly round ____ **49. P. crispipila**

37. Psychotria scaberula Merr., Philipp. J. Sci. 17:459. 1921; Merr., Enum. Philipp. Fl. Pl. 3:563. 1923. (**Fig. 44**). TYPE: PHILIPPINES. MINDANAO: Surigao del Norte Prov; Dinagat Island, 11 May 1919, *Ramos & Pascasio BS 35159* (LECTOTYPE: US, designated here; ISOLECTOTYPE: K).

Shrub or small tree, height unknown, young stems glabrous. Stipules valvate, ovate, 1–2.5 cm long, puberulous; apex obtuse to acute. Leaves: petioles 2–4.5 cm long, glabrous; leaf blades broadly obovate to broadly obovate-elliptic, 10.5–22 × 5.6–10 cm, chartaceous to subcoriaceous, strongly scabrous below; secondary veins 13–19 pairs; tertiary venation manifest; base attenuate; apex round or obtuse with an acute or acuminate tip. Inflorescence/infructescence trichotomous, 2.5–4.8 cm long, glabrous; principal axis 2–4.3 cm long, generally unbranched, each axis terminating in a cluster of sessile flowers/fruit; bracts and bracteoles absent. Flowers 4- or 5-merous, pedicels absent; calyx (incl. hypanthium) 4–6 mm long (limb 2–4 mm long), glabrous; calyx lobes deltate, 1.5–2 mm long; corolla and other flower parts unknown. Fruit globose to obovoid, ca. 10 mm long, ribbed, color unknown, glabrous; calyx limb and lobes persistent, conspicuous (4–5 mm long); pedicels absent; pyrenes ribbed on outer surface; endosperm ruminate.

Other specimens examined: **Mindanao. Surigao del Norte Prov.:** Dinagat Island, without further data, *Ramos & Convocar BS 83912* (A).

Distribution.—Known only from Dinagat Island.

Ecology.—Forest; low elevation.

Phenology records.—Flowering: unknown; fruiting: May.

Preliminary conservation assessment.—Critically Endangered (**CR B1ab(iii)**) or possibly Extinct (**EX**). Last collected in 1931.

Distinguishing characteristics.—Inflorescences trichotomous; leaves relatively large and strongly scabrous below.

Remarks. The scabrous nature of the leaves distinguishes this taxon not only from the other members of its specific group, but from all other species of Philippine *Psychotria*.

38. Psychotria bataanensis Elmer, Leafl. Philipp. Bot. 1:23. 1906; Elmer, Merr., Philipp. J. Sci. 1 (Suppl. 1): 134. 1906; Merr., Enum. Philipp. Fl. Pl. 3:553. 1923. (**Fig. 45**). TYPE: PHILIPPINES. LUZON. Bataan Prov.: Mt. Mariveles, Nov 1904, *Elmer 6980* (LECTOTYPE: K, designated here; ISOLECTOTYPE: NY).

Psychotria alvarezii Merr., Philipp. J. Sci. 8:56. 1913. *Grumilea alvarezii* (Merr.) Merr., Philipp. J. Sci. 10:132. 1915. TYPE: PHILIPPINES. LUZON: Nueva Ecija Prov.: Lucban, Dec 1910, *Alvarez FB 22189* (not seen, presumed lost or destroyed).

Psychotria rizalensis Merr., Philipp. J. Sci. 8:56. 1913. TYPE: PHILIPPINES. LUZON. Rizal Prov.: Montalban, Jul 1905, *Loher 6345* (LECTOTYPE: K, designated here).

Shrub, 1–2 m high, young stems pubescent to puberulous. Stipules valvate to imbricate, triangular, 7–10 mm long, glabrous to puberulous; apex acute. Leaves: petioles 0.9–2.5 cm long, pubescent to

Fig. 44. *Psychotria scaberula* (*Ramos & Pascasio BS 35159*) (US, lectotype).

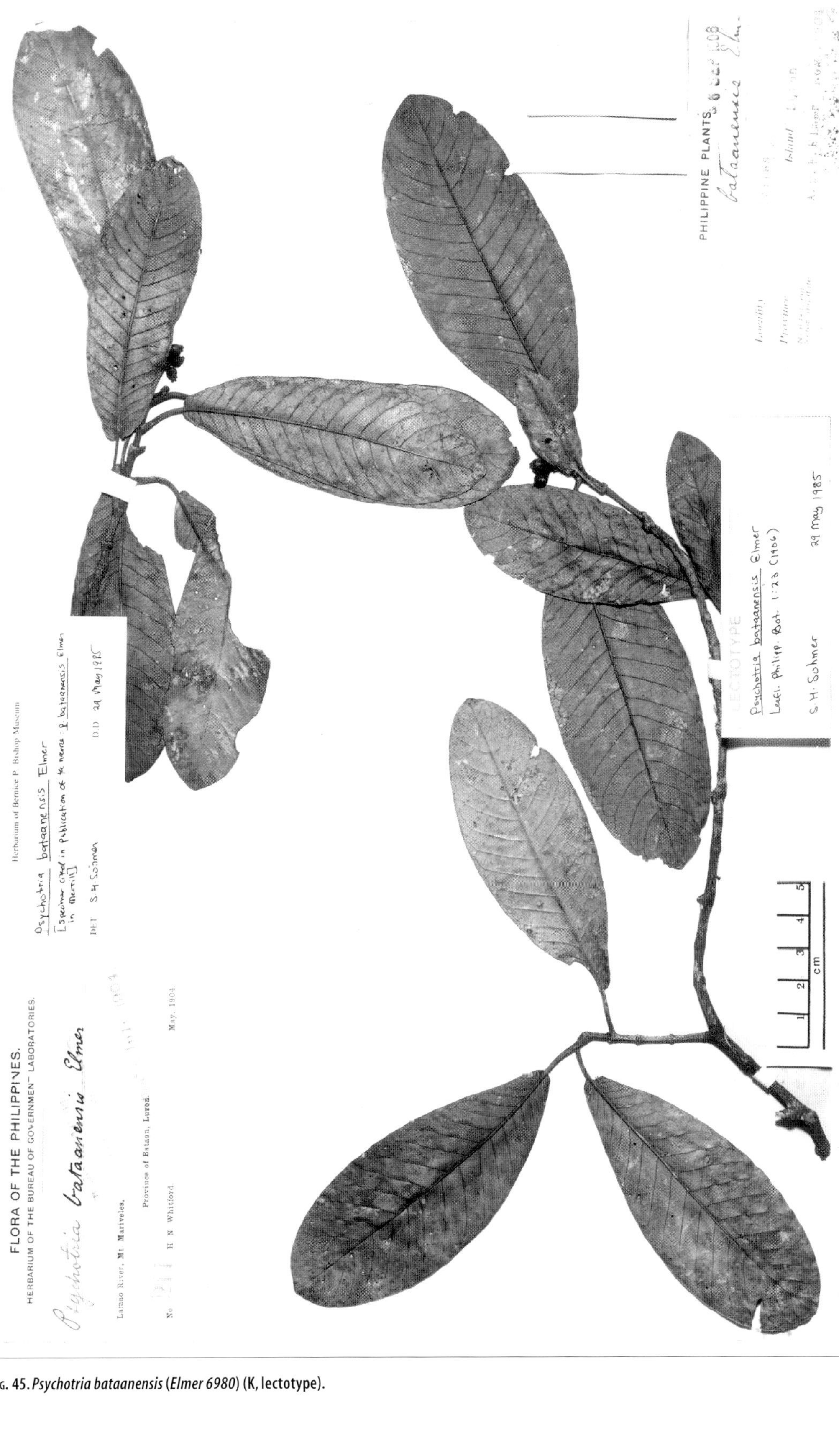

FIG. 45. *Psychotria bataanensis* (*Elmer 6980*) (K, lectotype).

puberulous; leaf blades elliptic to oblong, or elliptic-obovate, 6.4–13.5(–18.2) × 2.6–5.8(–8.9) cm, chartaceous to coriaceous, often pubescent below at least along midrib; secondary veins 11–18(–23) pairs; tertiary venation ± prominent; base round to obtuse, sometimes subauriculate; apex round to obtuse. Inflorescence/infructescence trichotomous, 1.4–4.8 cm long, glabrus to pubescent; principal axis 0.3–2.5 cm long, generally unbranched, each axis terminating in a cluster of sessile flowers/fruit; bracts and bracteoles absent. Flowers 4- or 5-merous, pedicels absent; calyx (incl. hypanthium) ca. 1 mm long, mainly glabrous but limb and lobes puberulous to pubescent; calyx lobes, shallowly triangular to deltate, 0.7–1.2 mm long; corolla: white; corolla tube 2–3 mm long, external surface glabrous, villous at throat within; corolla lobes 1–2 mm long (short styled); stamens: unknown; style ca. 1 mm long, lobes ca. 0.5 mm long (short styled). Fruit globose-turbinate, to ± obovoid, 6–8 mm long, ribbed, red, glabrous or sparsely pubescent at apex; calyx limb and lobes persistent; pedicels absent; pyrenes ribbed on outer surface; endosperm ruminate.

Other specimens examined: **Luzon. Nueva Ecija Prov.:** Mt. Umingan, *Ramos & Edaño BS 26323* (A, K, NY, US), *BS 26503* (A, BM, P). **Rizal Prov.:** Mt. Lumutan, *Ramos & Edaño BS 29794* (A, L, P). **Bataan Prov.:** Mt. Mariveles, *Merrill 3765* (BM, K, NY, P), *ibid.*, *Loher 3073* (K); Lamao River, *Borden FB 2078* (NY, US), *Ahern's collector FB 1508* (BM, K, NY), *Merrill 3180* (BM, K, NY, P), *Whitford 214* (K, NY, US), *Williams 411* (A, GH, K, NY, US), *613* (NY); Lamao Forest Reserve, *Foxworthy BS 1599* (US); Mt. Kuyapo, *Edaño PNH 4018* (A, PNH, SING), *PNH 4152* (A, L, PNH).

Distribution.—In the Nueva Ecija, Rizal, and Bataan provinces of Luzon.

Ecology.—Montane forest on ridges; 700–1600 m.

Phenology records.—Flowering: January, October; fruiting: January, May, July to December.

Preliminary conservation assessment.—Endangered (**EN B1ab(iii)**). Last collected in 1947.

Distinguishing characteristics.—Leaf blades rounded at both ends, base often subauriculate, pubescent below; inflorescences trichotomous with clusters of sessile flowers.

Remarks.—A taxon whose leaf-shape separates it from the other taxa of its group but it is otherwise similar to *P. crispipila*.

This species was published twice in the same year, by Elmer (1906) and Merrill (1906a). In this revision we assume that Elmer's description of *P. bataanensis* was first and that it has priority. If this is not the case this species should be cited as *P. bataanensis* Elmer ex Merr.

39. Psychotria elliptilimba Merr., Philipp. J. Sci. 17:448. 1921; Merr., Enum. Philipp. Fl. Pl. 3:555. 1923. (**Fig. 46**). TYPE: PHILIPPINES. CATANDUANES. Catanduanes Prov.: Santa Domingo River on low elevation forested slopes, 5 Dec 1917, *Ramos BS 30534* (LECTOTYPE: US, designated here; ISOLECTOTYPES: A, K, P).

Shrub, ca. 1 m high, young stems glabrous. Stipules valvate, ovate, 4–5 mm long, glabrous; apex acuminate. Leaves: petioles 1.2–1.9 cm long, glabrous; leaf blades oblong to elliptic, 7–9.6 × 2.5–4.2 cm, chartaceous to subcoriaceous, glabrous; secondary veins 8–9 pairs; tertiary venation obscure; base round or obtuse, sometimes subauriculate; apex rounded to broadly acute. Inflorescence/infructescence monochotomous, 5.4–6.5 cm long, glabrous; principal axis (peduncle) 0.9–13 cm long, with 1 primary node supporting 3 branches, each branch terminating in a cluster of sessile flowers/fruit; bracts leaf-like, ± elliptic, ca. 0.6 cm long, sometimes absent (?caducous); bracteoles present but poorly know (mostly caducous). Flowers unknown. Fruit turbinate to subglobose, 9–10.2 mm long, smooth to ribbed, color unknown, glabrous; calyx limb and lobes not persistent but disc often prominent; pedicels absent; pyrenes weakly ribbed on outer surface; endosperm ruminate.

Other specimens examined: Known only from the type collection.

Distribution.—Known only from Santa Domingo River, Catanduanes.

Ecology.—Forest; low elevation.

Phenology records.—Flowering: unknown; fruiting: December.

Preliminary conservation assessment.—Extinct (**EX**). Only collected once in 1917.

Distinguishing characteristics.—Plants glabrous; leaf blades: lateral veins 8–9 per side, base round or obtuse; flowers sessile.

Remarks.—This taxon is still known only from the type collection and is not well understood. The glabrous condition as well as the smaller number of lateral veins distinguish it from *P. bataanensis* Elmer.

Fig. 46. *Psychotria elliptilimba* (*Ramos, BS 30534*) (US, lectotype).

40. Psychotria pygmaea Merr., Philipp. J. Sci. 17:458. 1921; Merr., Enum. Philipp. Fl. Pl. 3:561. 1923. (**Fig. 47**). TYPE: PHILIPPINES. LUZON. Rizal Prov.: Mt. Susong-Dalaga in forests, 14 Aug 1917, *Ramos & Edaño BS 29271* (LECTOTYPE: US, designated here).

Shrub, ca. 0.5 m high, young stems glabrous. Stipules valvate, ovate, 5–10 mm long, glabrous; apex acuminate. Leaves: petioles 0.2–0.7 cm long, glabrous; leaf blades obovate to obovate-elliptic, 4.5–7.5 × 1.5–3 cm; chartaceous to subcoriaceous, glabrous; secondary veins 5–9 pairs; tertiary venation obscure; base acute; apex acute to obtuse. Inflorescence/infructescence trichotomous, compact, 1.2–2.1 cm long, glabrous; principal axis 0.5–0.8 cm long, each axis with 1 primary node, each node terminating in a cluster of sessile flowers/fruit; bract and bracteoles unknown. Flowers unknown. Fruit globose to ellipsoid, 6–8 mm long, ribbed or ridged, dark red or purple, glabrous; calyx limb and lobes persistent, rather obvious; pedicels absent; pyrenes prominently ridged on outer surface; endosperm entire.

Other specimens examined: Known only from the type collection.

Distribution.—Mt. Susong-Dalaga, Luzon.

Ecology.—Forest; altitude unknown.

Phenology records.—Flowering: unknown; fruiting: August.

Preliminary conservation assessment.—Critically Endangered (**CR B1ab(iii)**), or possibly Extinct (**EX**). Only collected once in 1917.

Distinguishing characteristics.—Small woody plant, less than 0.5 m tall; petioles less than 0.7 cm long; inflorescences compact, trichotomous with clusters of sessile flowers or fruit.

Remarks.—This taxon is still known only from the type collection with only a single specimen at US. It is unique among Philippine *Psychotria* by its short habit (the lectotype appears to be a whole plant only some 25 cm from roots to inflorescence) and compact, trichotomous inflorescence with clusters of sessile flowers or fruit.

41. Psychotria arborescens Elmer, Leafl. Philipp. Bot. 3:1030. 1911; Merr., Enum. Philipp. Fl. Pl. 3:552. 1923. (**Fig. 48**). *Grumilea arborescens* (Elmer) Merr in Philipp. J. Sci. 10:130. 1915. TYPE: PHILIPPINES. MINDANAO. Cotabato Prov.: Mt. Apo, Aug 1909, *Elmer 11263* (LECTOTYPE: US, designated here; ISOLECTOTYPES: A, BISH, BM, GH, K, L, MO, NY, W).

Small tree, ca. 7 m high, completely glabrous. Stipules valvate, ovate, ca. 10 mm long, glabrous; apex rounded. Leaves: petioles 1.7–3.5 cm long, glabrous; leaf blades obovate to oblanceolate-elliptic, or ± elliptic, 12.5–19 × 5–6.4 cm, subcoriaceous to coriaceous, glabrous; secondary veins 8–11 pairs; tertiary venation manifest; base attentuate; apex round or obtuse with an abruptly acute to acuminate tip. Inflorescence/infructescence monochotomous, 8.4–11.2 cm long, glabrous; principal axis (peduncle) 2.6–7.2 cm long, with 1 primary node, supporting 3–5 branches, second order branches terminating in clusters of sessile flowers/fruit; bracts ± triangular, 1.5–1.9 mm long; bracteoles deltate to triangular, ca. 0.5 mm long. Flowers unknown. Fruit turbinate to subglobose, 11–14 mm long, smooth, probably red, glabrous; calyx limb and lobes persitent but inconspicuous; pedicels absent; pyrenes very weakly ribbed on outer surface; endosperm ruminate.

Other specimens examined: **Leyte. Leyte Prov.:** Biluran [Biliran] Island, Mt. Suiro, *Sulit PNH 21713* (L). **Southern Leyte Prov.:** Cabalian, *Ramos BS 41582* (A, L, P, W). **Leyte. Prov. unknown:** Mt. Abucayan, *Edaño BS 41798* (A, K, L, P, SING, US, W).

Distribution.—Biliran, Leyte, and Mindanao Islands.

Ecology.—Forest; ca. 1000–1200 m.

Phenology records.—Flowering: unknown; fruiting: February, August, December.

Preliminary conservation assessment.—Vulnerable (**VU B1ab(iii)**). Last collected in 1954.

Distinguishing characteristics.—Small trees, completely glabrous; leaf blades relatively large, 12.5–19 cm long; inflorescence trichotomous with clusters of sessile flowers, axes flattened, 6–9.3 cm long.

42. Psychotria wenzelii (Merr.) Merr., Enum. Philipp. Fl. Pl. 3:564. 1923. (**Fig. 49**). *Grumilea wenzelii* Merr., Philipp. J. Sci. 10:140. 1915. TYPE: PHILIPPINES. LEYTE. Leyte Prov.: Buenavista near Jaro, forests, ca. 500 m, 1 Oct 1914, *Wenzel 1142* (LECTOTYPE: MO, designated here; ISOLECTOTYPES: A, BM).

Shrub, ca. 2 m high, glabrous throughout. Stipules imbricate, ± triangular, 5–10 mm long, glabrous; apex acute to acuminate. Leaves: petioles 0.6–1.5 cm long, glabrous; leaf blades elliptic to obovate,

FIG. 47. *Psychotria pygmaea* (*Ramos & Edaño BS 29271*) (US, lectotype).

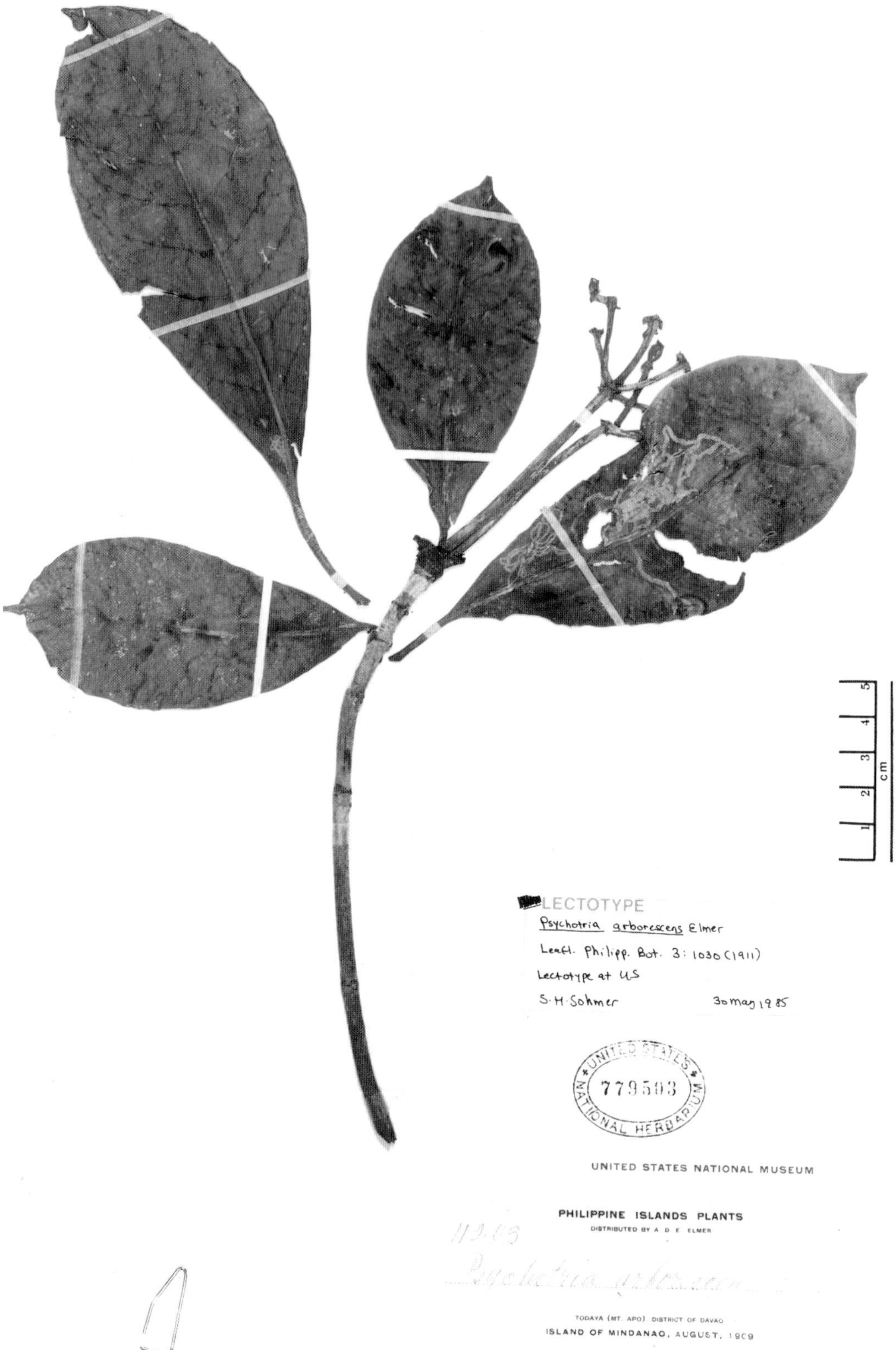

Fig. 48. *Psychotria arborescens* (*Elmer 11263*) (US, lectotype).

Fig. 49. *Psychotria wenzelii* (*Wenzel 1142*) (MO, lectotype of *Grumilea wenzelii*).

6.5–11 × 1.8–3.7 cm, chartaceous to subcoriaceous, drying yellow-green to gray-green above, and usually gray-green below, glabrous; secondary veins 7–11 pairs; tertiary venation manifest to quite prominent; base acute; apex abruptly acute or acuminate. Inflorescence/infructescence trichotomous, 1.7–4.8 cm long, glabrous; principal axis 0.6–1.1 cm long, each axis with 2 primary nodes on main axes, supporting branches terminating in 1–3 flowers/fruits; bracts and bracteols unknown (?absent). Flowers 4- or 5-merous, pedicels absent or ca. 0.3 mm long; calyx (incl. hypanthium) ca. 2 mm long, glabrous; calyx lobes irregular, ± triangular, ca. 0.7 mm long; corolla: white; corolla tube 2–2.5 mm long (long styled), external surface glabrous, villous at throat within; corolla lobes 2.5–2.7 mm long; stamens: filaments ca. 1 mm long (long styled); anthers ca. 0.5 mm long; style ca. 3 mm long (long styled), thickened below the lobes, nearly clavate, lobes ca. 1 mm long. Fruit globose, 7–8 mm long, smooth to weakly ribbed, usually very wrinkled when dry, probably red, glabrous; calyx limb and lobes absent or persistent but not conspicuous; pedicels 0.3–0.4 or ± absent; pyrenes ribbed on outer surface; endosperm ruminate.

Other specimens examined: Known only from the type collection.

Distribution.—Known only near Jaro on Leyte.

Ecology.—Forest; 500–700 m.

Phenology records.—Flowering: unknown; fruiting: October.

Preliminary conservation assessment.—Critically Endangered (**CR B1ab(iii)**) or possibly Extinct (**EX**). Only collected once in 1914.

Distinguishing characteristics.—Leaf blades drying grayish green or yellowish green; Inflorescences trichotomous, with flowers/fruit (sub)sessile at end of the main axes, sometimes apparently with two nodes at apex with several sessile flowers at each node.

Remarks.—This is a part of a species complex that includes *P. subsessiliflora* and, possibly, *P. pauciflora*. *Psychotria wenzelii* shares with the former the characteristic of having flowers and fruit sessile or subsessile. Unlike *P. subsessiliflora*, however, the inflorescence of *P. wenzelli* is clearly trichotomous. In overall morphology, *P. wenzelii* resembles *P. pauciflora*, but the latter taxon never has sessile or (sub)sessile flowers and fruit. Merrill (1923:564) cites two other collections of Wenzel besides the lectotype designated here: *Wenzel 622* and *Wenzel 882*. The former is *P. loheri* and the latter *P. paloensis*.

43. Psychotria castroi Merr. & Quisumb. ex Sohmer & A.P. Davis, sp. nov. (**Fig. 50**). TYPE: PHILIPPINES. SAMAR. Eastern Samar Prov.: Borongon, Pinamgasan in primary forest at 150 m, 23 Jun 1948, *Castro & Anonuevo PNH 5855* (HOLOTYPE: L; ISOTYPE: K).

Arbores parvae vel frutices 2–7 m alti glabri sunt, stipulis valvatis ovatis ca. 1 cm longis apice integro, laminis coriaceis glabris obovatis vel oblanceolatis 4.8 × 10–6.2 × 15 cm latis et longis nervis lateralibus 9–10 in dimidio quoque apice brevi acuto, petiolis glabris 1.7–3 cm longis, inflorescentia monochotoma glabra, pedunculo 0.8–2 cm longo in basi vel in lateri ramoso, ramulis cum floribus sessilibus vel fructibus terminatis, fructibus ut dicetur luteis vel versimiliter rubrescentibus fusiformi-turbinatis 0.7–1 cm longis et cum tubo calycis 2–4 cm longo persistenti coronatis, pyrenis in dorso costatis, endospermio ruminato.

Small tree or shrub, 2–7 m high, completely glabrous. Stipules valvate, ovate, ca. 10 mm long, glabrous; apex acute. Leaves: petioles 1.7–3 cm long, glabrous; leaf blades obovate to oblanceolate, 10–15 × 4.8–6.2 cm, coriaceous, drying dark brown or dark reddish brown, glabrous; secondary veins 9–10 pairs; tertiary venation manifest to obscure; base cuneate to acute; apex round or obtuse and sometimes with an abrupt acute tip. Inflorescence/infructescence monochotomous, 5.5–7.5 cm long, glabrous; principal axis (peduncle) 0.8–2 cm long, with 1 primary node, supporting 3 branches, each branch terminated by a cluster of sessile flowers/fruit. bracts leaf-like, ca. 3.7 cm long, or very depressed ovate, ca. 0.6 cm long, sometimes absent (caducous); bracteoles unknown. Flowers poorly known (calyx only); calyx (incl. hypanthium) 3–4 mm long, glabrous; calyx lobes deltate, 0.5–0.7 mm long. Fruit subglobose to turbinate, 7–10 mm long, ribbed, reported yellow but probably red at maturity, glabrous; calyx limb and lobes persistent, conspicuous, 2–4 mm long; pedicels absent; pyrenes weakly ribbed on outer surface; endosperm ruminate.

PARATYPE: **Samar. Prov unknown.** Mt. Apay, *Gutierrez et al. PNH 117439* (L).

Distribution.—Samar.

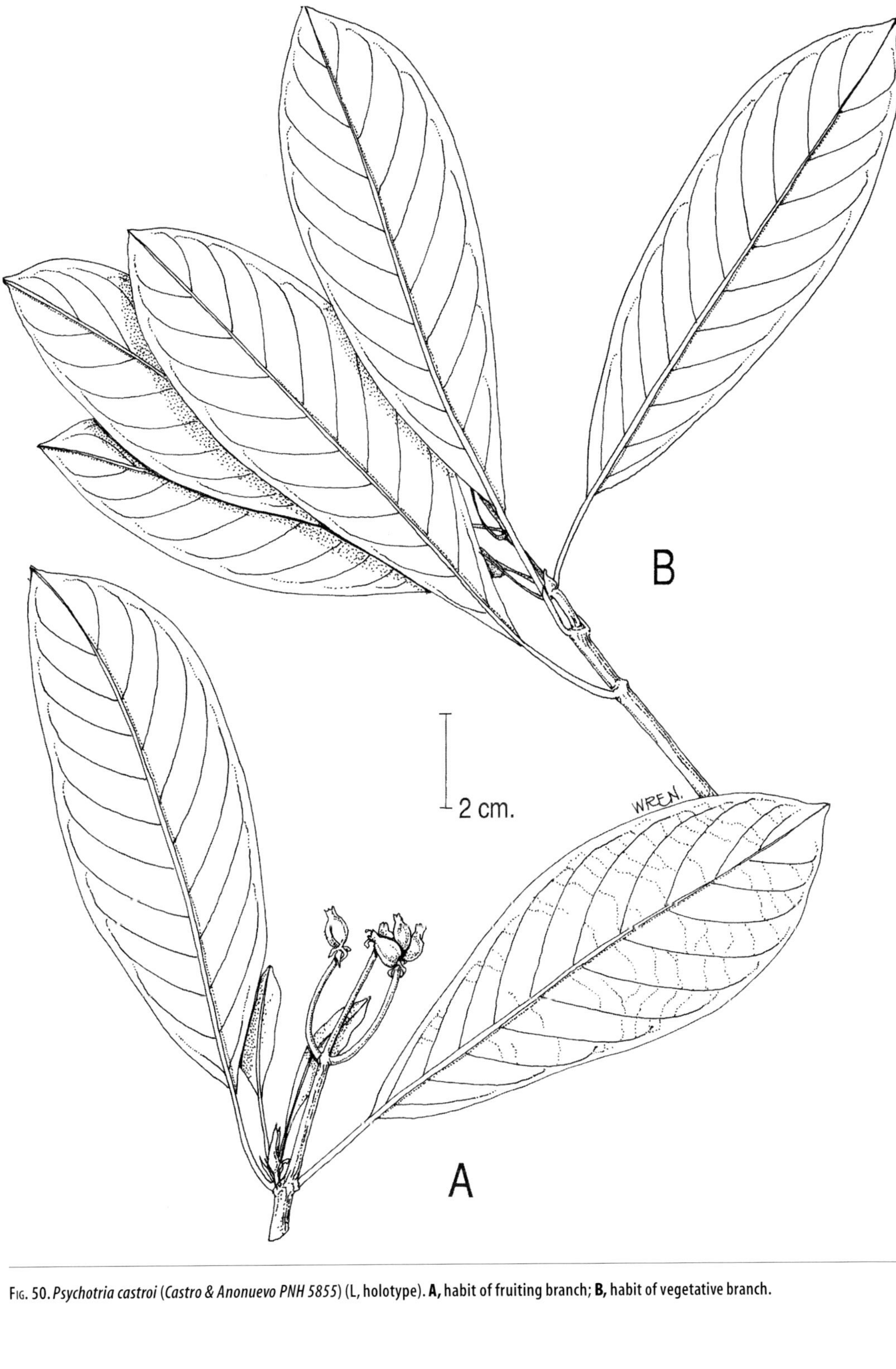

FIG. 50. *Psychotria castroi* (*Castro & Anonuevo PNH 5855*) (L, holotype). **A,** habit of fruiting branch; **B,** habit of vegetative branch.

Ecology.—Moist forest (only recorded from primary forest); low elevations (ca. 150 m).

Phenology records.—Flowering: unknown; fruiting: June.

Preliminary conservation assessment.—Critically Endangered (**CR B1ab(iii)**). Last collected in 1969.

Distinguishing characteristics.—Small tree or shrub, completely glabrous; leaf blades 10–15 cm long, apex normally round or obtuse; inflorescences trichotomous with main axes 1.5–4 cm long; flowers sessile; calyx limb conspicuous, persistent.

Remarks.—The relatively short, narrow inflorescence axes distinguish this taxon from *P. arborescens*. Interestingly, the other taxon that shares the conspicuous, persistent calyx tube limb of *P. castroi* is *P. scaberula* Merr., also found in Mindanao. These two species have a similar dry leaf color and overall inflorescence appearance. However, the leaves of *P. castroi* are much smaller than those of *P. scaberula*, and *P. castroi* is completely glabrous.

44. Psychotria surigaoensis Sohmer & A.P. Davis, sp. nov. (**Fig. 51**). TYPE: PHILIPPINES. MINDANAO. Surigao del Norte Prov.: Apr 1919, *Ramos & Pascasio BS 34581* (HOLOTYPE: BM; ISOTYPES: K, PNH).

Fruitices vel arbores, stipulis valvatis ovatis 5.7–9.7 mm longis apice integro, laminis chartaceis vel semicoriaceis glabris anguste ellipticis vel anguste oblanceolatis 1.5 × 5–3.5 × 15 cm latis et longis nervis lateralibus 7–12(–14) in dimidio quoque apice acuminato vel acuto basi attenuata vel acuminata, petiolis 0.5–3 cm longis, inflorescentia in basi monochotoma, rammis floribus sessilibus fructibus que terminantibus et usque 4 cm longis, floribus 4–5-meris dimorphicis, calyce et hypanthio 3–4 mm longis tubo brevi et lobis obscurus, corrolis albis tubo 2–3 mm longo lobis 2–3 mm longis.

Shrub or small tree, height unknown, young stems glabrous to puberulous or rarely pubescent (only very youngest part of stem). Stipules valvate, ovate, 5.7–9.7 mm long, glabrous to puberulous; apex acute. Leaves: petioles 0.5–3 cm long, glabrous to puberulous, rarely ± pubescent; leaf blades narrowly elliptic to narrowly oblanceolate, 5–15 × 1.5–3.5 cm, chartaceous to subcoriaceous, mostly glabrous but sometimes sparsely puberulous on the midrib and veins below, rarely slightly scabrid below; secondary veins 7–12(–14) pairs; tertiary venation obscure to invisible; base attenuate to narrowly acute; apex acuminate to acute, often abruptly acuminate. Inflorescence/infructescence monochotomous, 1.5–2.8 cm long, puberulous to pubescent; principal axis (peduncle) 0.3–0.9 cm long, with 1 primary node, supporting 3 branches, each branch terminated by a cluster of sessile flowers/fruit; bracts depressed ovate, ca. 1 mm long, puberulous; bracteoles numerous, ± ovate to depressed ovate, ca. 2 mm long. Flowers 4- or 5-merous, pedicels absent; calyx (incl. hypanthium) 3–4 mm long, glabrous; calyx lobes obscure, or ± deltate to triangular, 0.2–0.6 mm long; corolla: white: corolla tube 2–3 mm long, external surface glabrous, internal surface unknown; corolla lobes 2–3 mm long; stamens: filaments 1–2 mm long; anthers ca. 1 mm long; style unknown. Fruit subglobose to turbinate, 8–11 mm long, ribbed, bright orange, possibly red at maturity, glabrous; calyx limb and lobes persistent, rather prominent, 1.5–2 mm long; pedicels absent; pyrenes ribbed-ridged on outer surface; endosperm ruminate.

PARATYPES: **Mindanao. Surigao del Norte Prov.:** Dinagat Island, without further data, *Ramos & Convocar BS 84101* (SING); Cambinliw, Santiago, Loreto (Dinagat Island), *Gaerlan, Sagcal & Fernando PPI 4551* (BRIT, K), ibidem, *Gaerlan, Sagcal & Fernando PPI 4583* (BRIT, K). **Surigao del Norte/Sur Prov.:** without further data: *Ramos & Pascasio BS 34601* (K), *BS 35234* (NY).

Distribution.—Surigao del Norte Province of Mindanao (including Dinagat Island).

Ecology.—Forest, including dipterocarp forest; often found near rivers; ca. 50 m.

Phenology records.—Flowering: April; fruiting: August to September.

Preliminary conservation assessment.—Endangered (**EN B1ab(iii)**). Last collected in1991.

Distinguishing characteristics.—Plant completely glabrous or sparsely pubescent, the leaves rarely slightly scabrid below; inflorescence trichotomous with clusters of sessile flowers or fruit; leaf blades narrowly elliptic-oblanceolate, apex acute, or acuminate to abruptly acuminate.

45. Psychotria sulitii Merr. & Quisumb. ex Sohmer & A.P. Davis, sp. nov. (**Fig. 52**). TYPE: PHILIPPINES. SAMAR. Western Samar Prov.: Mt. Calbiga, Wright, in mid-montane forest at ca. 300 m, 23 May 1948, *Sulit PNH 6408* (HOLOTYPE: A; ISOTYPES: K, L).

Arbores parvae 6–7 m altae sunt, laminis coriaceis varie rigide chartaceis supra secus midnervum et infra pubescentibus obovatis vel late lanceolatis 7.5 × 17.2–8.5 × 25.5 cm latis et longis nervis lateralilus 17–20 in dimidio quoque nervis tertiis transversis subprominentibus apice rotundato sed puncto obtuse mucronato basi acuta vel cuneata, petiolis pubescentibus 2.5–3.8 cm longis,

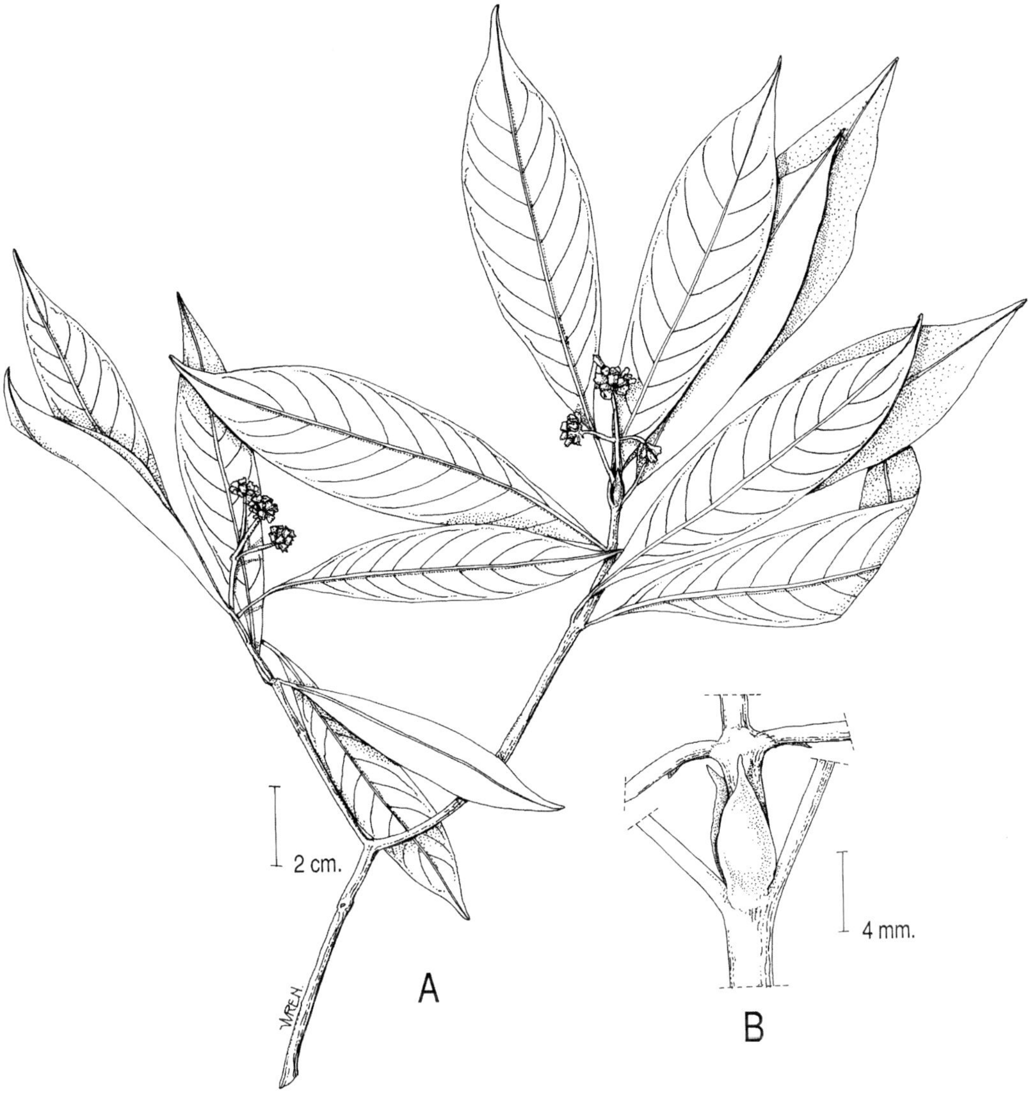

Fig. 51. ***Psychotria surigaoensis*** **(*Ramos & Pascasio BS 34581*) (BM, holotype). A, habit of fruiting branch; B, stipule.**

inflorescentia versimiliter trichotoma, 3 pedunculis 3–4 mm crassis bracteis ca. 2 cm longis floribus sessilibus multis congregatis 5-meris, hypanthio et tubo calycis valde pubescentibus 4–6.2 mm longis lobis 1.4–2 mm longis.

Small tree, 6–7 m high, young stems prominently pubescent; Stipules unknown. Leaves: petioles 2.5–3.8 cm long, pubescent; leaf blades obovate to broadly lanceolate, 17.2–25.5 × 7.5–8.5 cm, stiffly chartaceous to subcoriaceous, pubescent above, at least along midrib, and below; secondary veins 17–20 pairs; tertiary venation manifest, with somewhat prominent tertiary cross-venation; base acute to cuneate; apex round with a blunt mucronate tip. Inflorescence/infructescence trichotomous, ca. 4.2 cm long, prominently pubescent; principal axis, 1.7–2 cm long, rather thick (3–4 mm). each axis with 2 primary nodes, and opposite branching at each node, secondary branches terminating in clusters of sessile flowers/fruit; bracts ± ovate, ca. 2 cm long; bracteoles ovate, ca. 2.4 mm long. Flowers 5-merous, pedicels absent; calyx (incl. hypanthium) 4–6.2 mm long, prominently pubescent; calyx lobes triangular, 1.4–2 mm long; corolla and other flower parts unknown. Fruit red, otherwise unknown.

Paratypes: None—known only from the type collection.

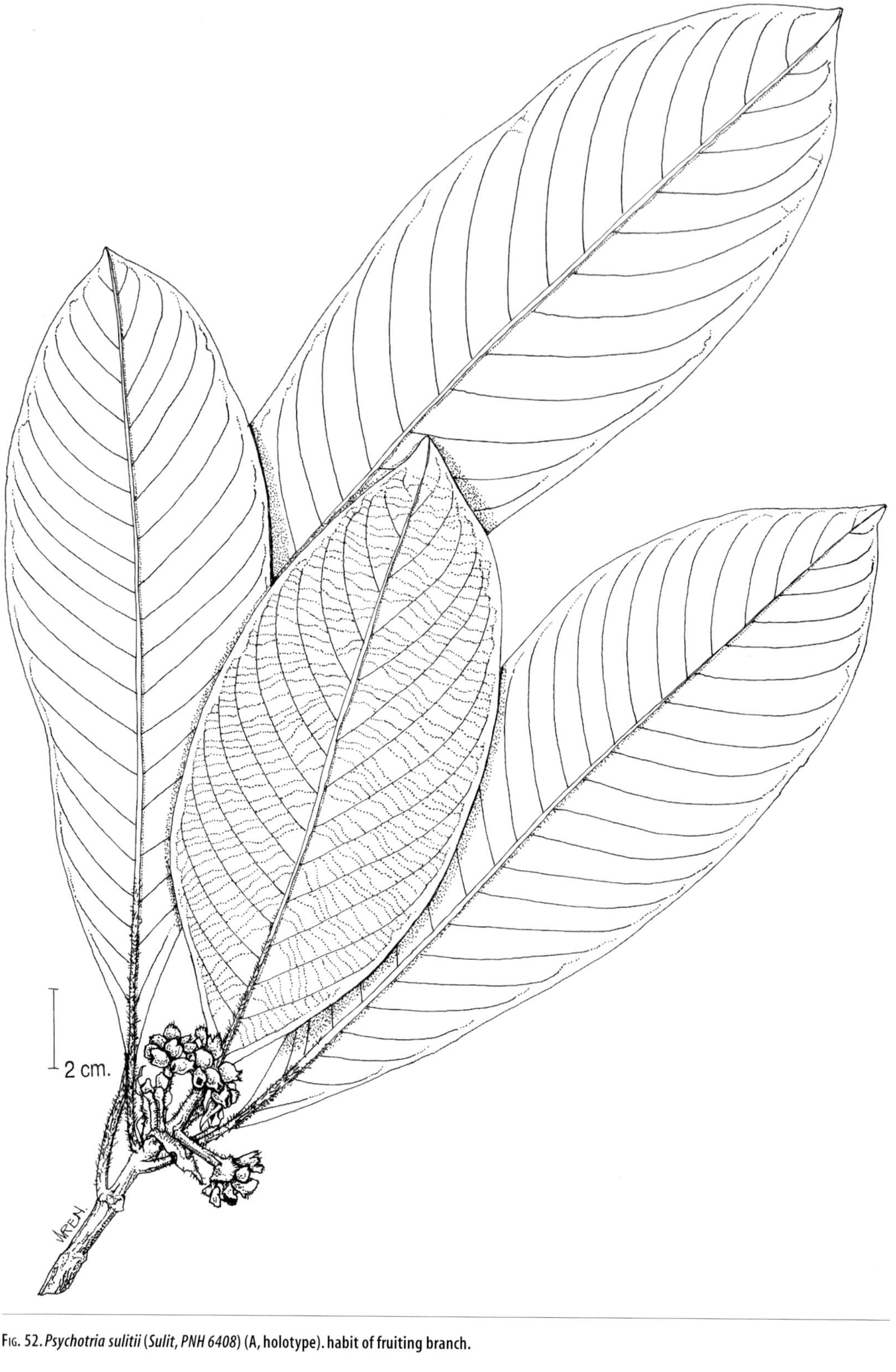

Fig. 52. *Psychotria sulitii* (*Sulit, PNH 6408*) (A, holotype). habit of fruiting branch.

Distribution.—Mt. Calbiga, Samar.

Ecology.—Forest; ca. 300 m.

Phenology records.—Flowering: May; fruiting: May.

Preliminary conservation assessment.—Critically Endangered (**CR B1ab(iii)**). Only collected once in 1948.

Distinguishing characteristics.—Leaf blades more than 17 cm long; inflorescence axes more than 3 mm wide; calyx pubescent.

Remarks.—This taxon is known only from the type collection. Merrill and Quisumbing recognized this as a potentially new species based.

46. Psychotria pulgarensis Sohmer & A.P. Davis, nom. nov. (**Fig. 53**). *Psychotria repens* Elmer, Leafl. Philipp. Bot. 4:1349. 1912. Non *Psychotria repens* (L.) L., Amoen Acad. 5:377. 1760. TYPE: PHILIPPINES. PALAWAN: Palawan Prov.: Mt. Pulgar, in moist, dense forest at about 80 m, Apr 1911, *Elmer 12991* (LECTOTYPE: L, designated here; ISOLECTOTYPES: A, BISH, BM, GH, K, MO, P, US, W).

Suffrutescent plant apparently creeping along the ground and rooting at the nodes, height unknown but probably < 20 cm high, young stems pubescent. Stipules valvate, lanceolate, 5–11 mm long, distinctly pubescent; apex cleft with each of the two tips drawn out into a long arista. Leaves: petioles (0.3–)0.4–0.7 cm long, pubescent; leaf blades elliptic to obovate, 4–10 × 2–4 cm, chartaceous to subcoriaceous, glabrous or pubescent on main and latral veins below; secondary veins 7–9 pairs; tertiary venation manifest to obscure; base cuneate to acute; apex obtuse to abruptly acute. Inflorescence/infructescence poorly known (only immature inflorescences seen), trichotomous, ca. 0.7 cm long, pubescent; principal axis, ca. 0.4 cm long, generally unbranched, each axis terminating in a cluster of sessile flowers/fruit; bracts linear, arista-like, ca. 4 mm long; bracteoles similar to the bracts, ca. 2 mm long. Flowers ± unknown (immature). Fruit unknown.

Other specimens examined: **Palawan. Palawan Prov.:** Mt. Pulgar, *Elmer 12868* (L); Balabac Island, Mt. Caunhayan, *Ramos & Edaño BS 49702* (UC); Balabac Island, without further data, *Ramos & Edaño BS 49793* (NY, UC).

Distribution.—Known only from two localities, Mt. Pulgar and Balabac Island, on Palawan.

Ecology.—Humid forest; ca. 80 m.

Phenology records.—Flowering: April; fruiting: unknown.

Preliminary conservation assessment.—Endangered (**EN B1ab(iii)**). Last collected in 1927.

Distinguishing characteristics.—Plant creeping, rooting at nodes; leaves nearly sessile; stipules densely pubescent; inflorescences trichotomous; flowers sessile.

47. Psychotria lancilimba Merr., Philipp. J. Sci. 17:451. 1921; Merr., Enum. Philipp. Fl. Pl. 3:557. 1923. (**Fig. 54**). TYPE: PHILIPPINES. LUZON. Rizal Prov.: Mt. Lumutan, Jul 1917, *Ramos & Edaño BS 29713* (LECTOTYPE: US, desiginated here).

Shrub or small tree, height unknown, young stems glabrous; Stipules valvate, ovate-deltate to narrowly triangular, 3–7(–11.5) mm long, glabrous; apex acute. Leaves: petioles 0.5–1.8 cm long, ± glabrous or sometimes sparsely pubescent; leaf blades narrowly elliptic to lanceolate, 7.5–16.8 × 1.7–3 cm, subcoriaceous, glabrous above, glabrous to sparsely pubescent on midrib and veins below; secondary veins 10–14(–17) pairs; tertiary venation rather prominent; base narrowly cuneate to acute; apex narrowly acuminate. Inflorescence/infructescence trichotomous, 2.8–3.3 cm long, pubescent; principal axis 0.8–3.8 cm long, unbranched, each axis terminating in a cluster os sessile flowers/fruits; bracts leaf-like, ca. 10 mm long, sometimes absent; bracteoles ?absent. Flowers very poorly known (immature), pedicels absent; calyx (incl. hypanthium) ca. 2.5 mm long, base pubescent, upper parts glabrous; calyx lobes shallowly triangular, ca. 0.5 mm long; corolla and other flower parts unknown. Fruit obovoid, 5–8 mm long, smooth, probably red, glabrous; calyx limb and lobes usually persistent; pedicels absent; pyrenes smooth on outer surface; endosperm ruminate.

Other specimens examined: **Luzon. Benguet Prov.:** Baguio, *Curran FB 5072* (NY). **Nueva Vizcaya Prov.:** Mt. Caraballo, *Loher s.n.* (UC). **Zambales Prov.:** without further data: *Ramos BS 5018* (US). **Rizal Prov.:** Mt. Irid, *Ramos & Edaño BS 48490* (A, BM, K, NY, UC), *BS 48526* (UC); Monte Batay, *Loher 6447* (K); Mt. Lumutan, *Ramos & Edaño BS 29629* (A, K, NY, P); Mt. Mapatag, San Andales, *Edaño BS 48737* (NY, UC).

Distribution.—Luzon.

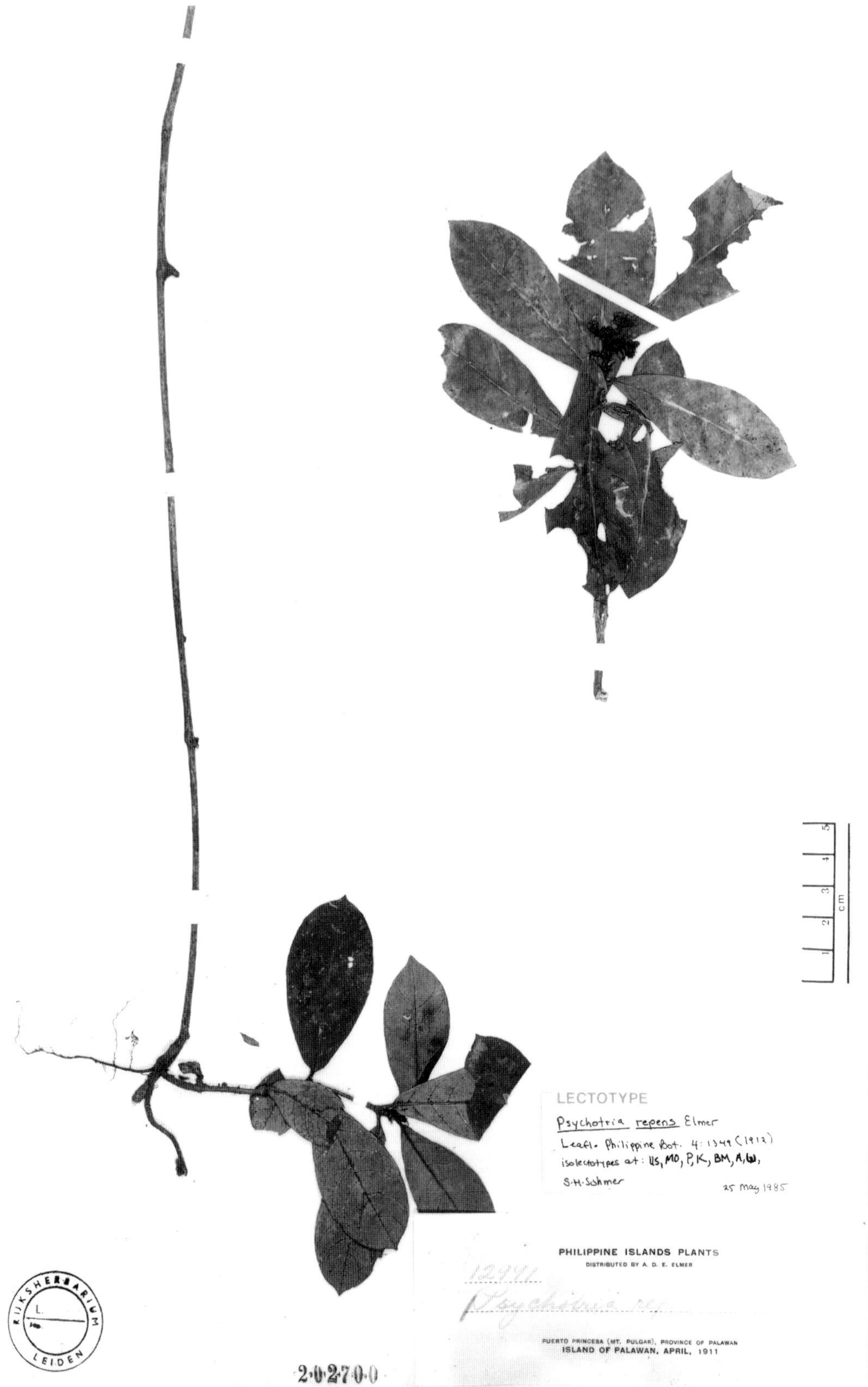

Fig. 53. *Psychotria pulgarensis* (*Elmer 12991*) (L, lectotype).

FIG. 54. *Psychotria lancilimba* (*Ramos & Edaño BS 29713*) (US, lectotype).

Ecology.—Lower montane forests; 600–1000 m.

Phenology records.—Flowering: July, November; fruiting: July.

Preliminary conservation assessment.—Endangered (**EN B1ab(iii)**). Last collected in 1926.

Distinguishing characteristics.—Leaf blades narrowly elliptic to lanceolate, glabrous, tertiary veins prominent; inflorescences trichotomous, each main axis terminated with a cluster of sessile flowers.

Remarks.—This is a taxon whose status still requires further clarification, as there are no known collections bearing flowers. However, based upon inflorescence and leaf characteristics it is distinct from other Philippine *Psychotria* species. Of the type material cited by Merrill (1921) only one specimen could be found in herbaria, namely *Ramos & Edaño BS 29713*, which is herein designated as the lectotype.

48. Psychotria pilosella Elmer, Leafl. Philipp. Bot. 1:353. 1908; Merr., Enum. Philipp. Fl. Pl. 3:561. 1923. (**Fig. 55**). *Grumilea pilosella* (Elmer) Merr., Philipp. J. Sci. 10:131. 1915. TYPE: PHILIPPINES. LUZON. Quezon Prov.: Mt. Banahão, in forested ravine at 850 m, May 1907, *Elmer 9213* (LECTOTYPE: US, designated here; ISOLECTOTYPES: A, K, L, MO, NY, W).

Psychotria heteromera Merr., Philipp. J. Sci. 17:449. 1921. TYPE: PHILIPPINES. LUZON. Quezon Prov.: Mt. Binuang in forest at about 1000 m, May 1917, *Ramos & Edaño BS 28773* (LECTOTYPE: K, designated here; ISOLECTOTYPES: BM, P, US).

Shrub or small tree, 1–7 m high, young stems ± glabrous to densely pubescent (hairs white *in vivo*, drying reddish brown). Stipules valvate, ovate to narrowly ovate, 4–20 mm long, densely pubescent or ciliate along margin; apex cleft. Leaves: petioles 1.4–2.9 cm long, pubescent; leaf blades elliptic, obovate to narrowly oblanceolate, 6–18.8 × 1.8–5 cm, chartaceous to subcoriaceous, usually with hairs above as well as below, at least on midrib (hairs usually crisped or bent, often thickened at the base); lateral veins 6–18 pairs; tertiary venation ± invisible to manifest; base usually narrowly acute, sometimes attenuate; apex acuminate to acute. Inflorescence/infructescence trichotomous, or sometimes monochotomous, 1.4–5(–6.2) cm long, pubescent to villous, rarely puberuolous to glabrous; principal axis of trichotomous inflorescences 0.7–2.7 cm long, principal axis (peduncle) of monochotomous inflorescences 0.4–2 cm long, generally unbranched or with 1 primary node, each axis or branch terminating in a cluster of sessile, or very rarely pedicellate, flowers/fruits; bracts long ovate, 4–6 mm long, margins serrate to laciniate, or absent (?caducous); bracteoles similar to bracts, ca. 2–5 mm long, or absent (?caducous). Flowers 4- or 5-merous; pedicels absent, or rarely present, then 0.5–3 mm long; calyx (incl. hypanthium) 1–7 mm long, external surface pubescent to densely villous, or sometimes glabrous; calyx lobes triangular, 0.5–3 mm long; corolla: white; corolla tube 2–7 mm long, glabrous, extremely villous at the throat within (and particularly the base of each anther); corolla lobes 2–5 mm long (usually less than 1 mm wide); stamens: filaments 1–2 mm long; anthers 1–2 mm long; style 1–4 mm long, lobes ca. 0.7 mm long. Fruit turbinate-globose, 6–12 mm long, smooth or ribbed, yellow, orange or red, sparsely to densely pubescent or glabrous; calyx limb and lobes usually persistent; pedicels absent; pyrenes ribbed to ribbed-ridged on outer surface; endosperm ruminate.

Distribution.—Widely distributed from Luzon through to Mindanao.

Ecology.—Humid forest; 50–1000 m.

Phenology records.—Flowering: January to June, December; fruiting: February, May to December.

Preliminary conservation assessment.—Near Threatened (**NT**). Last collected in 1996.

Distinguishing characteristics.—Inflorescences trichotomous; flowers sessile, arranged in tight clusters; an indumentum, which in its most developed form, consists of a blanket of relatively long bent or crisped hairs that cover twigs, both surfaces of the leaf blades, inflorescence axes, and the calyx (incl. hypanthium), ranging to a pubescence that sparsely covers these same organs. The hairs are usually a whitish color *in vivo* but dry a reddish brown.

Remarks.—This is a complex and variable taxon which has been reported in various guises throughout the Philippines, except Palawan. It consists of two somewhat variable subspecies. The subspecies usually can be distinguished with the following key.

KEY TO THE SUBSPECIES OF *PSYCHOTRIA* SUBSP. *PILOSELLA*

1. Hairs on leaf blades usually thickened at base, distributed evenly; leaf blades 9–18.5 cm long ____ **48a.** subsp. **pilosella**
1. Hairs on leaf blades usually not thickened at base, matted or sparse; leaf blades generally 6–18.8 cm long ____ **48b.** subsp. **erythrotricha**

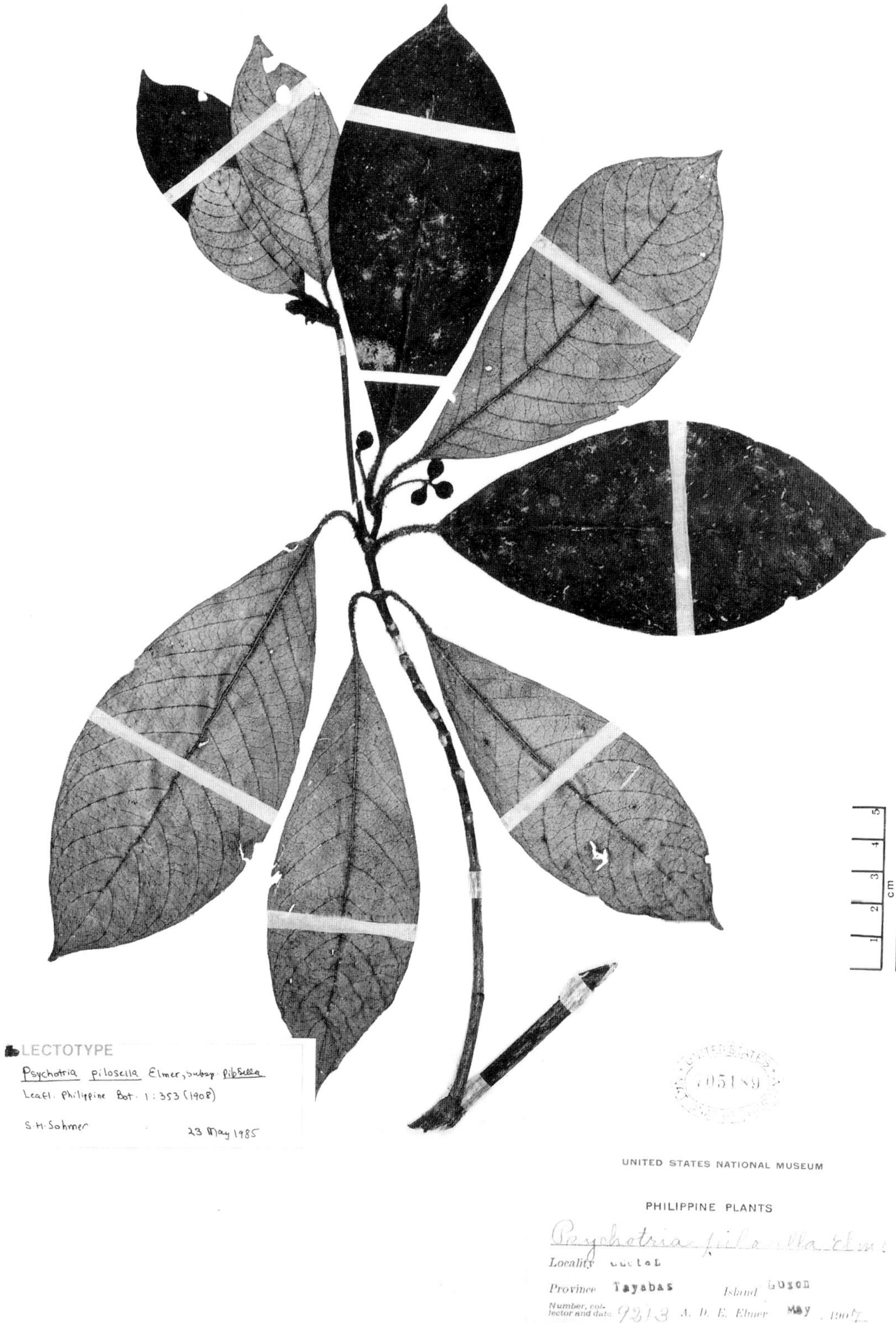

FIG. 55. *Psychotria pilosella* subsp. *pilosella* var. *pilosella* (*Elmer 9213*) (US, lectotype).

48a. Psychotria pilosella subsp. pilosella

Shrub, or small tree, 1–7 m tall, young stems usually pubescent, sometimes nearly glabrous. Stipules ovate, 1–2 cm long, pubescent. Leaves: petioles 0.5–3 cm long, densely pubescent; leaf blades broadly elliptic to oblanceolate, 9–18.5 × 3–7 cm, usually chartaceous but sometimes subcoriaceous, usually pubescent above and below, hairs evenly spaced along surface, usually noticeably thickened at base; lateral veins 11–18 per side; base obtuse to acute; apex abruptly acute or acuminate. Inflorescence 2.5–6.2 cm long. Flowers: calyx (incl. hypanthium) ca. 1 mm long (long styled), calyx lobes obscure, glabrous; corolla glabrous, villous within, corolla tube 2–7 mm long (long styled), corolla lobes 2–4 mm long (long styled). Fruit slightly to densely hairy.

Two varieties are recognized; the key can be used to separate them.

KEY TO THE VARIETIES OF *PSYCHOTRIA PILOSELLA* SUBSP. *PILOSELLA*

1. Leaf blades with hairs usually on both sides; corolla tube ca. 2 mm long (long styled flower) _______ **48aa.** var. **pilosella**
1. Leaf blades with hairs generally restricted to lower surface and midrib; corolla tube 5–7 mm long (long styled flower) _______ **48ab.** var. **samarensis**

48aa. Psychotria pilosella subsp. pilosella var. pilosella

Shrub, 1–2 m tall. Leaf blades generally with hairs evenly distributed on both surfaces. Corolla tube ca. 2 mm long (long styled).

Other specimens examined: **Luzon. Cagayan Prov.:** Claveria, Kilkiling, Mt. Taggat, *Reynoso, Fuentes & Majaducon PPI 11916* (BRIT). **Quezon Prov.:** Mt. Alzapan, *Ramos & Edaño BS 45606* (A, BM, K, NY, P, SING UC, US), *BS 45612* (NY, P, UC, US); Mt. Binuang, *Ramos & Edaño BS 28587* (A, P, K), *BS 28617* (MO, US), *BS 28602* (K); Infanta, *Robinson BS 9363* (K, NY); Infanta-Siniloan Trail, *Ramos & Edaño BS 29198* (A, K); Philipp. National Botanic Garden, forest near Garden & along Lalawinan River, *Sohmer 12016* (BISH, K), *12017* (BISH, K); Mt. Pular, *Ramos BS 19429* (BM, K, L). **Rizal Prov.:** Manila (Laguna de Bay), *Andersson s.n.* (S). **Laguna Prov.:** Majayjay, *Rivera & Duyag BS 49971* (UC). **Camarines Norte Prov.:** Paracale, *Ramos & Edaño BS 33615* (A, NY), *BS 33662* (A); Minasag River, *Edaño PNH 40217* (BM, CANB, K, L). **Camarines Sur Prov.:** Mt. Potianay, *Edaño BS 71919* (NY), *BS 75960* (NY); Kamugong River, *Edaño BS 75903* (NY); Mt. Bagacay, *Ramos & Edaño BS 33842* (A). **Catanduanes. Catanduanes Prov.:** Catanduanes Island, Boundry of Barangay Summitt, Viga and Barangay San Pedro, Gigmoto, *Reynoso & Majaducon PPI 24947* (BRIT, K); Catanduanes Island, So. Boradan, Barangay Summitt, *Reynoso & Majaducon PPI 24831* (BRIT). **Samar: Eastern Samar Prov.:** Borongon, Pinamgasan, *Castro 5826* (K); Sitio Cantinoc, Barangay 1, Balangiga, *Gaerlan & Chavez PPI 26347* (K).

Distribution.—Central to north central Luzon, and Samar.

Ecology.—Moist forest, including remnant dipterocarp forest; 150–1000 m.

Phenology records.—Flowering: January, February, May, June, December; fruiting: May to December.

Preliminary conservation assessment.—Near Threatened (**NT**). Last collected in 1996.

Distinguishing characteristics.—Leaf blades usually pubescent on both sides; corolla tube ca. 2 mm long.

Remarks.—*Psychotria heteromera* (Merrill 1921) definitely belongs here and matches the type of *P. pilosella* Elmer, except for the smaller leaves. Merrill states in his treatment of this taxon, 'It somewhat resembles *P. pilosella* Elmer, but is not closely related to that species.'

48ab. Psychotria pilosella subsp. pilosella var. samarensis Sohmer & A.P. Davis, var. nov. (Fig. 56).

Type: PHILIPPINES. Samar. Northern Samar Prov.: Mt. Cansayao, Catarman in primary dipterocarp forest at 200 m, 3 Apr 1951, *Sulit PNH 14346* (holotype: A; isotypes: K, L).

Arbores parvae 5–7 m altae sunt, laminis supra cum pilis in midnervo vel glabris, tubo corollae 5–7 cm longo (floribus longistylis).

Small tree, 5–7 m tall. Leaf blades with the hairs on upper surface restricted to midrib or nearly glabrous above. Corolla tube 5–7 mm long (long styled).

Paratypes: **Samar. Western Samar Prov.:** Mt. Mahagna, Turabucan, Oquendo, *Sulit PNH 14469* (A, L); Rawis, Sohoton, *Univ. San Carlos 442* (L).

Distribution.—Samar.

Ecology.—Moist forest; 50–250 m.

Phenology records.—Flowering: April; fruiting: unknown.

Preliminary conservation assessment.—Endangered (**EN B1ab(iii)**). Last collected in 1971.

Distinguishing characteristics.—Upper surface of leaf blades with hairs restricted to midrib; corolla tubes 5–7 mm long.

Fig. 56. *Psychotria pilosella* subsp. *pilosella* var. *samarensis* (*Sulit PNH 14346*) (A, holotype). Habit of fruiting branch.

48b. Psychotria pilosella subsp. **erythrotricha** (Elmer) Sohmer & A.P. Davis, comb. et stat. nov. (**Fig. 57**). *Psychotria erythrotricha* Elmer, Leafl. Philipp. Bot. 5:1891. 1913. *Grumilea erythrotricha* (Elmer) Merr., Philipp. J. Sci. 10:133. 1915. Type: PHILIPPINES. Mindanao: Agusan Prov.: Mt. Urdaneta at about 1200 m, Aug 1912, *Elmer 13486* (lectotype: K, designated here; isolectotypes: A, BISH, BM, GH, L, MO, P, UC, US, W).

Small tree or shrub, 1.5–5 m tall, young stems often pubescent. Stipules narrowly ovate, up to 1.5 cm long, pubescent on outer surface or at least the margins. Leaves: petioles 0.4–3.2 cm long, pubescent; leaf blades elliptic to broadly obovate, 6–18.8 × 2.5–5 cm, usually chartaceous, usually pubescent above and below, or only hirsute above on midrib, hairs not usually spread evenly along surface, and either matted or sparse, normally not thickened at base; lateral veins 8–18 per side; base obtuse to

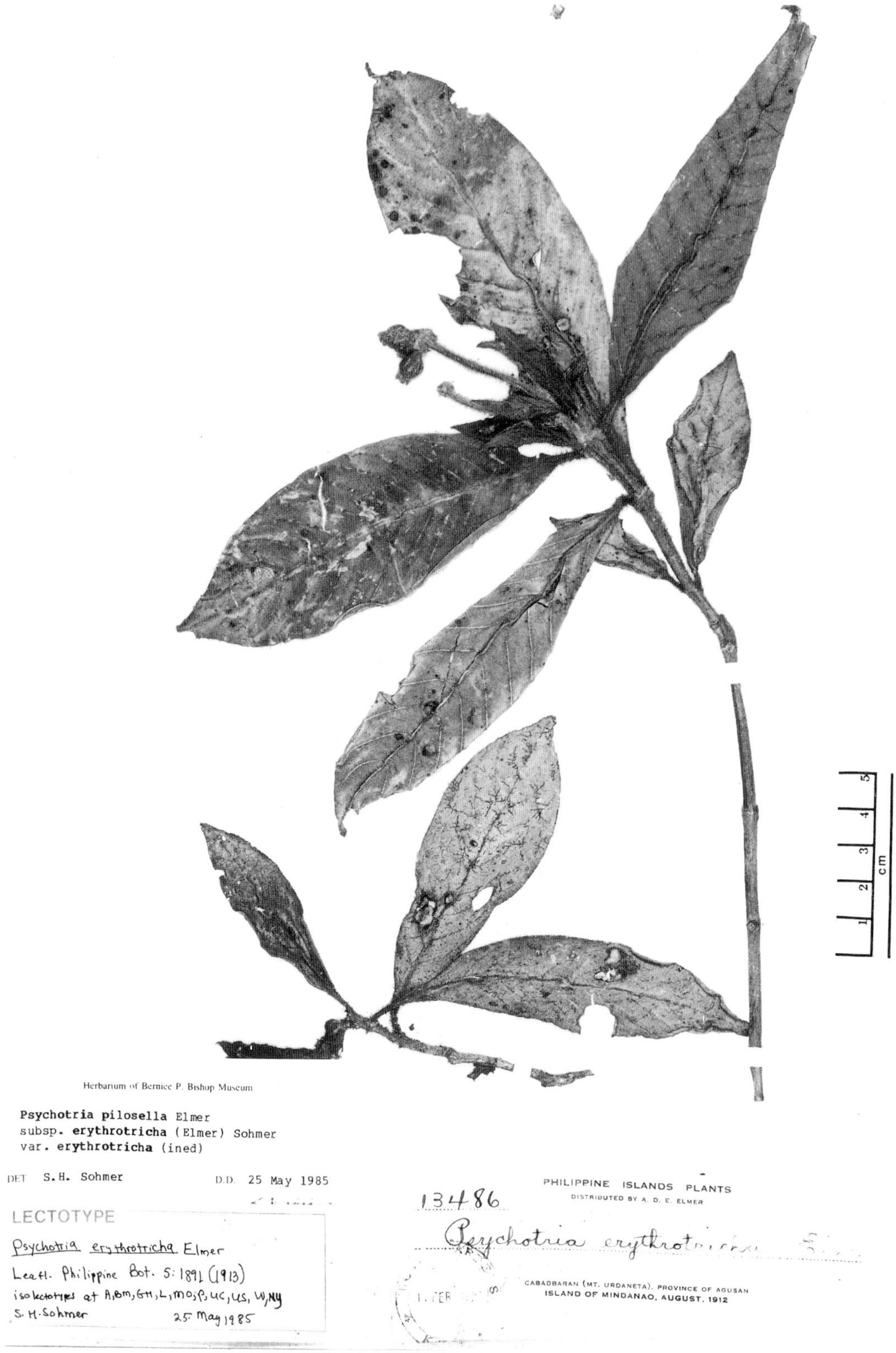

FIG. 57. *Psychotria pilosella* subsp. *erythrotricha* var. *erythrotricha* (*Elmer 13486*) (K, lectotype).

acute; apex broadly acute to acuminate. Inflorescence 1–6 cm long. Flowers: calyx (incl. hypanthium) 3–7 mm long, calyx lobes 0.5–3 mm long, usually pubescent; corolla glabrous or somewhat puberulent, villous within, corolla tube 4–7 mm long (short styled), corolla lobes 4–5 mm long (short styled). Fruit glabrous or lightly pubescent.

Remarks.—We have recognized this taxon principally on the type of hairs comprising the indumentum. As opposed to *P. pilosella* subsp. *pilosella*, the hairs do not seem to be prominently thickened at the base. If this proves to be an untenable character we would not hesitate to combine this taxon with *P. pilosella* subsp. *pilosella* as Merrill did (Merrill 1923:561).

There are two varieties for this subspecies.

KEY TO THE VARIETIES OF *PSYCHOTRIA PILOSELLA* SUBSP. *ERYTHROTRICHA*

1. Leaf blades densely hairy above as well as below, or at least hairy along the veins above _____ **48ba.** var. **erythrotricha**
1. Leaf blades usually glabrous above and often sparsely hairy below _____ **48bb.** var. **camarinensis**

48ba. Psychotria pilosella subsp. **erythrotricha** var. **erythrotricha**

Shrub, 2–5 m tall. Leaf blades densely hairy above, at least along veins.

Other specimens examined: **Luzon. Isabela Prov.:** Mt. Moises, *Clemens 17025* (NY, UC); Sierra Madre Mountains, Palanan, *Gutierrez PNH 78200* (A, K, L, SING), *PNH 78226* (A, K, L), *PNH 78244* (L). **Quezon Prov.:** without further data: *Oro FB 30972* (NY, SING). **Camarines Norte Prov.:** Bicol National Park, *Canicosa PNH 9763* (A), *Reynoso, Romero & Fuentes PPI 3689* (BRIT, K). **Samar. Western Samar Prov.:** Loquilocon, *Sulit PNH 6209* (L); Mt. Sohoton, Babasian Na Daku River, *Gutierrez et al. PNH 117680* (K); **Samar. Prov. unknown**: Mt. Concored, *Gutierrez et al. PNH 117192* (L); Mt. Comoral, *Gutierrez et al. PNH 117218* (L). **Panay. Antique Prov.:** Mt. Poras, Imparayan, Sibalom, *Reynoso, Sagcal & Fernando PPI 21047* (BRIT). **Leyte. Without further data:** *Wenzel 530* (A, BM, MO), *775* (US). **Mindanao. Surigao del Norte Prov.:** Dinagat Island, without further data, *Ramos & Convocar BS 83791* (A, SING), *BS 83888* (A); Bucas Grande Island, without further data, *Ramos & Pascasio BS 35049* (BM, K). **Surigao Prov.:** without further data: *Ramos & Pascasio BS 34702* (BM). **Agusan del Norte Prov.:** Mt. Hilonghilong, Cabadbaran, *Mendoza & Convocar PNH 10764* (A, L).

Distribution.—Widely distributed, from central Luzon through Samar, Panay and Leyte to Mindanao.

Ecology.—Moist forest, including disturbed forest; 300–1200 m.

Phenology records.—Flowering: April to May, August; fruiting: June to August, November.

Preliminary conservation assessment.—Near Threatened (**NT**). Last collected in 1995.

Distinguishing characteristics.—Leaf blades densely hairy above and below, or at least pubescent along veins above.

48bb. Psychotria pilosella subsp. **erythrotricha** var. **camarinensis** (Merr.) Sohmer & A.P. Davis, comb. et stat. nov. (**Fig. 58**). *Grumilea camarinensis* Merr., Philipp. J. Sci. 10:135. 1915. *Psychotria camarinensis* (Merr.) Merr., Enum. Philipp. Fl. Pl. 3:553. 1923. TYPE: PHILIPPINES. LUZON: Camarines Sur Prov.: Mt. Iriga, 29 Nov 1913, *Ramos BS 1552* (LECTOTYPE: BM, designated here; ISOLECTOTYPES: A, L, MO, P).

Shrub, 1.5–3 m tall. Leaf blades usually glabrous above or with hairs restricted to midrib.

Other specimens examined: **Luzon. Quezon Prov.:** Mt. Binuang, *Ramos & Edaño BS 28789* (A, K, US). **Laguna Prov.:** Mt. Banahao, *Duyag BS 76684* (NY). **Camarines Sur Prov.:** Mt. Isarog, *Edaño BS 76190* (A, K, NY); Mt. Potianay, *Edaño BS 75991* (NY). **Samar. Western Samar Prov.:** Mt. Calbiga, *Sulit & Conese PNH 6446* (A, L, PNH); Mt. Apoy, *Gutierrez et al. PNH 117463* (L). **Leyte. Leyte Prov.:** Biliran Island, Mt. Suiro, *Sulit PNH 21638* (L); Mt. Mamban, *Edaño PNH 12153* (A, L); Dagami, *Ramos BS 15357* (K); central divide, without further locality data, *Rosenbluth FB 16906* (K).

Distribution.—Samar, Leyte and Biliran Islands, and Quezon and Camarines Sur provinces of Luzon.

Phenology records.—Flowering: April to May; fruiting: February, August.

Ecology.—Forest, on slopes; 900–1500 m.

Preliminary conservation assessment.—Near Threatened (**NT**). Last collected in 1969.

Distinguishing characteristics.—Leaf blades usually glabrous above.

Remarks.—The status of this taxon may not be well understood due to its scattered distribution, and the variable nature of the characteristics that delimit it from the nominal variety: less hair or none at all on the upper surface of the leaf blades.

49. Psychotria crispipila Merr., Philipp. J. Sci. 1 (Suppl. 3): 240. 1906; Merr., Enum. Philipp. Fl. Pl. 3:554. 1923. (**Fig. 59**). *Grumilea crispipila* (Merr.) Merr., Philipp. J. Sci. 10:133. 1915. TYPE: PHILIPPINES. LUZON: Mountain Prov.: Mt. Data in mossy forest at 2200 m, Nov 1905, *Merrill 4490* (LECTOTYPE: K, designated here; ISOLECTOTYPE: US).

Psychotria fusca (Merr.) Merr., Enum. Philipp. Fl. Pl. 3:556. 1923. *Grumilea fusca* Merr., Philipp. J. Sci. 12:171. 1917. TYPE: PHILIPPINES. LUZON: Nueva Ecija Prov.: Mount Umingen, Aug 14 1916, *Ramos & Edaño BS 26232* (not seen).

Fig. 58. *Psychotria pilosella* subsp. *erythrotricha* var. *camarinensis* (*Ramos BS 1552*) (BM, lectotype).

Fig. 59. *Psychotria crispipila* (*Merrill 4490*) (K, lectotype).

Shrub or small tree, 1–3 m high, young stems pubescent. Stipules valvate, triangular to ligulate, 9–11.4 mm long, pubescent; apex acuminate to acute. Leaves: petioles 0.5–2.5 cm long, pubescent; leaf blades ovate-elliptic to obovate, 5.1–20 × 2.2–6.5 cm, subcoriaceous, pubescent below, glabrous or sometimes lightly scabrid above; secondary veins 7–16 pairs; tertiary venation manifest to rather prominent; base acute, infrequently narrowly acute; apex usually acute to nearly round. Inflorescence/infructescence trichotomous, 3–7.9 cm long, pubescent; principal axis 1.6–6.5 cm long, each axis with (0–)1–3 primary nodes, unbranched or with second order branches, each axis/branch terminating in a cluster of sessile or shortly pedicellate flowers/fruits, or flowers/fruits situated along the upper length of the middle (main) axis; bracts triangular to narrowly triangular, 1.4–2 mm long, or absent; bracteoles absent. Flowers 4- or 5-merous, pedicels absent, or sometimes present, then 0.3–2 mm long; calyx (incl. hypanthium) 2.5–3 mm long, external surface sparsely to densely pubescent on limb; calyx lobes ± obsolete; corolla: white; corolla tube ca. 5 mm long (short styled) or 3 mm long (long styled), glabrous, villous within; corolla lobes ca. 1 mm long (short styled) or 4 mm long (long styled); stamens: filaments ca. 2 mm long (short styled), ca. 1 mm long (long styled); anthers c 1 mm long (short and long styled); style ca. 2 mm long (short and long styled), lobes ca. 0.5 mm long (short and long styled). Fruit globose to turbinate, 7–10(–12) mm long, usually ribbed, red or black, glabrous to sparsely pubescent; calyx limb and lobes persistent, often pubescent; pedicels 0.4–2.2 mm long or absent; pyrenes smooth or with 1–3 sharp ridges on outer surface; endosperm ruminate.

Other specimens examined: **Luzon. Abra Prov.:** Mt. Posuey [Mt. Poosoy (?)], *Ramos BS 27038* (A, P, US). **Mountain Prov.:** Bontoc, slopes of Mt. Data at Bauko, *Lomibao 2241* (PNH); Mt. Data, *Clemens 16417* (UC), *17843* (UC), *18799* (UC), *Loher 1513* (K, US), *Ramos & Edaño BS 40275* (A, K, P, UC), *BS 40286* (L, P, SING, W); Mt. Pukis, *Ramos & Edaño BS 37785* (A, K, P, US). **Benguet Prov.:** Libolibo [Mt. Libolibo (?)], *Loher 1514* (K); Mt. Libung, *Steiner 2009* (L); Baguio, *Elmer 8347* (A, K, L, NY, US), *8794* (A, K, L, NY, US, W); Mt. Pulog, *Jacobs 7114* (A, CANB, K, L); Mt. Pulog, *Ramos & Edaño BS 40368* (A, P, SING); Bontoc Trail 88 to 59 K, *Clemens 17247* (UC); Pauai, *Clemens 9226* (BISH), *Santos BS 31775* (A, UC, US); "Haights in the Oaks," Pauai, *Mearns BS 4427* (L); 88 K. Resthouse, *Clemens 17248* (BM, UC); Mt. Tenglon, *Curran FB 4983* (US); Mt. Tonglon, *Ramos BS 5463* (P), *BS 5464* (L); Mt. Santo Tomas, *Elmer 6549* (K, NY), *Williams 1301* (K, NY), *1539* (NY), *1222* (K, NY). **Benguet Prov.:** without further data: *Ramos BS 5581* (US). **Ifugao Prov.:** Mt. Tabayoc, *Jacobs 7500* (A, CANB, K, L); Mt. Polis, *McGregor BS 19827* (US). **Quezon Prov.:** Sierra Madre Mountains, west-southwest of Baler, *Jacobs 7937* (A, CANB, K, L). **Rizal Prov.:** Bosoboso, *Ramos BS 2121* (NY). **Camarines Sur Prov.:** Mt. Potianay, *Edaño BS 75975* (NY). **Panay. Antique Prov.:** Culusai, Hangod Tubig, Mt. Madia-as, *Gaerlan & Fuentes PPI 5000* (BRIT).

Distribution—Luzon, most frequently in the northern and central provinces, and on Panay.
Phenology records.—Flowering: March, July, November; fruiting: January to March, June, September, October.

Ecology.—Drier types of montane forest; 1695–2500 m.

Preliminary conservation assessment.—Near Threatened (**NT**). Last collected in 1992.

Distinguishing characteristics.—Inflorescences trichotomous with clusters of sessile flowers; leaf blades ovate-elliptic to obovate, subcoriaceous and often strongly so, pubescent below, often with crisped hairs.

Remarks.—This species is closely related to *P. pilosella* Merr. as herein recognized, differing in the shape and indumentum of the leaves, usually being glabrous above, and the acute or obtuse leaf apex.

There are two distinct pyrene types for *P. crispipila*: (1) pyrenes obovate in outline, rather small (5.9 × 4.8 mm), ± smooth on the outer (dorsal) surface, and possessing short preformed germinations slits (PGSs), and (2) pyrenes broadly elliptic to almost circular in outline, larger (8.4 × 7.1 mm), with 1–3 sharp ridges on the dorsal surface, and lacking PGSs. Both pyrene types have a ruminate endosperm and a seed coat pigment that is soluble in 70% ethanol. Further detailed morpholgical study, with the aid of further collections, is required to ascertain whether *P. crispipila* can be separated into two species on the basis of the two pyrene types outlined above. The fruit of *P. crispipila* was unknown to Merrill (1906b: 240) when he described this species.

V—Membranifolia Species Group

Shrubs or small trees. Stipules valvate. ***Young shoots (branchlets) and leaves usually drying yellow-green. Leaf blades*** usually more than 3 cm wide; never more than 30 cm long; ***membranaceous or sometimes chartaceous***. Inflorescences with axes and branches, monochotomous (pedunculate). Flowers pedicellate. Fruits red or color unknown; ***pyrene inner (adaxial) surface concave; seed not producing a red color when placed in 70% ethanol (SCP negative); endosperm always entire; endosperm entire (non ruminate), pale.*** Comprising two species.

Note.—Text in ***bold italics*** denotes features useful for the placement of taxa within the respective species group.

KEY TO THE SPECIES OF THE MEMBRANIFOLIA SPECIES GROUP

1. Leaf blades obovate-elliptic; leaf apex acute, abruptly acute or obtuse; pyrenes smooth on upper surface ________ **50. P. balabacensis**
1. Leaf blades usually oblong-elliptic to oblanceolate; leaf apex apex acute, acuminate, or round with an abruptly acuminate tip; pyrenes weakly ribbed to smooth on outer surface ________ **51. P. membranifolia**

50. Psychotria balabacensis Merr., Philipp. J. Sci. 10:127. 1915; Enum. Philipp. Fl. Pl. 3:553. 1923. (**Fig. 60**). TYPE: PHILIPPINES. PALAWAN: Balabac, Jan 1913, *Escritor BS 21605* (LECTOTYPE: UC, designated here; ISOLECTOTYPES: BM, P, US).

Shrub, ca. 4 m high, young stems glabrous, drying yellow-green. Stipules valvate, ± ovate to triangular, 2–3 mm long, glabrous; apex acute to obtuse. Leaves: petioles 1–2 cm long, glabrous; leaf blades obovate-elliptic, 10–18.5 × 5.2–8 cm, membranaceous or sometimes chartaceous, drying yellow-green, glabrous above and below; secondary veins 7–10 pairs; tertiary venation manifest; base acute to cuneate; apex acute, abruptly acute or obtuse Inflorescence/infructescence monochotomous 2.5–7(–12) cm long, glabrous; principal axis (peduncle), 0.8–2(–5.2) cm long, with 1–2 primary nodes and opposite or verticillate branching at each node, the branches terminating in lax cymules; bracts absent (caducous); bracteoles ovate or deltate, ca. 0.3 mm long, often absent (?caducous). Flowers 4- or 5-merous, pedicels 1–3.8 mm long; calyx (incl. hypanthium) 1.1–1.8 mm long, external surface glabrous; calyx lobes ± absent, calyx limb truncate; corolla: yellowish white or yellowish green; corolla tube 2.5–3.2 mm long, glabrous, somewhat pubescent at throat within; corolla lobes ca. 0.9 mm long; stamens: filaments ca. 0.8 mm long; anthers ca. 1 mm long (long styled); style 2.5–3 mm long, lobes ca. 1 mm long (long styled). Fruit obovoid to almost globose, ca. 9 mm long, smooth, color unknown, glabrous; calyx limb and lobes persistent; pedicels 3–8 mm long; pyrenes smooth on outer surface, strongly concave on lower surface; endosperm entire.

Other specimens examined: **Camiguin. Cagayan Prov.:** Balatubat, *Edaño BS 79271* (A). **Mindoro. Oriental Prov.:** Pinmalayan, *Ramos BS 41046* (A,W). **Palawan. Palawan Prov.:** Bonabona, *Escritor BS 21541* (MO); Balabac Island, Mt. Caunayan, *Ramos & Edaño BS 49691* (NY, UC), *BS 49891* (UC); Ursula Island, without further data, *Kondo & Edaño PNH 36539* (BISH); Lumbucan Island, without further data, *Merrill BS 7192* (BM, L, K).

Distribution.—Recognized at present from several widely scattered island locations but principally from the southern end of the archipelago.

Ecology.—Moist forest; low elevation.

Phenology records.—Flowering: May, September; fruiting: September.

Preliminary conservation assessment.—Vulnerable (**VU B1ab(iii)**). Last collected in 1957.

Distinguishing characteristics.—All plant parts glabrous; young shoots usually drying yellow-green; leaf blades usually obovate, membranaceous or sometimes chartaceous, drying yellow-green, apex acute, abruptly acute or obtuse; fruit red, relatively large, outer (dorsal) surface of pyrenes smooth, inner (ventral) surface deeply concave; endosperm entire.

Remarks.—*Psychotria balabacensis* is very close to *P. leptothyrsa* but the pyrene features of these two species are clearly different. In *P. balabacensis* the outer (dorsal) surface of the pyrene is smooth, and the inner (ventral) surface is deeply concave; in *P. leptothyrsa* the outer (dorsal) surface is sharply ridged, and the inner flat. *Psychotria balabacensis* is endemic to the Philippines, whereas *P. leptothrysa* is a wide-spread taxon, which although described from the Celebes, is found throughout Melesia and the western pacific. An in-depth systematic study of *P. leptothrysa* is required.

51. Psychotria membranifolia Bartl. ex DC., Prodr. 4:522. 1830; Merr., Philipp. J. Sci 10:124. 1915; Merr., Enum. Philipp. Fl. Pl. 3:559. 1923. (**Fig. 61**). *Chassalia membranifolia* (Bartl. ex DC.) Elmer, Leafl. Philipp. Bot. 3:1032. 1911. TYPE: PHILIPPINES. LUZON: without further data, *Haenke s.n.* (HOLOTYPE: PR).

Psychotria membranifolia var. *elmeri* Merr., Philipp. J. Sci 10:124. 1915. TYPE: PHILIPPINES. LUZON: Tayabas (Quezon) Prov.: Antimonan, *Merrill 4083* (LECTOTYPE: K, designated here; ISOLECTOTYPES: NY).

Psychotria chasalioides Merr., Philipp. J. Sci. 10:125. 1915. TYPE: PHILIPPINES. SIBUYAN: Romblon Prov.: Mt. Giting-Giting, Apr 1910, *Elmer 12314* (LECTOTYPE: MO, designated here; ISOLECTOTYPES: A, BISH, L, NY, US).

Shrub or tree, 1–6 m high, young stems glabrous, drying yellow-green. Stipules valvate, delate-ovate to ± triangular, 2–5 mm long, glabrous; apex entire, slightly cleft or serrate. Leaves: petioles (0.8–)1–6.2 cm long, glabrous; leaf blades oblong-elliptic to oblanceolate, (6.7–)10.3–26.5 ×(3–)3.4–10.7 cm, membranaceous, drying green to yellow-green, glabrous above and below, rarely pubescent on veins below; secondary veins (6–)9–11(–13) pairs; tertiary venation manifest to quite prominent; base obtuse to acute; apex acute, acuminate, or round with an abruptly acuminate tip. Inflorescence/infructescence monochotomous, 1.2–6.5 cm long, glabrous or puberulent; principal axis (peduncle) 0.5–3.7(–4.5) cm long, with 1–2 primary nodes and opposite or verticellate branching at each node,

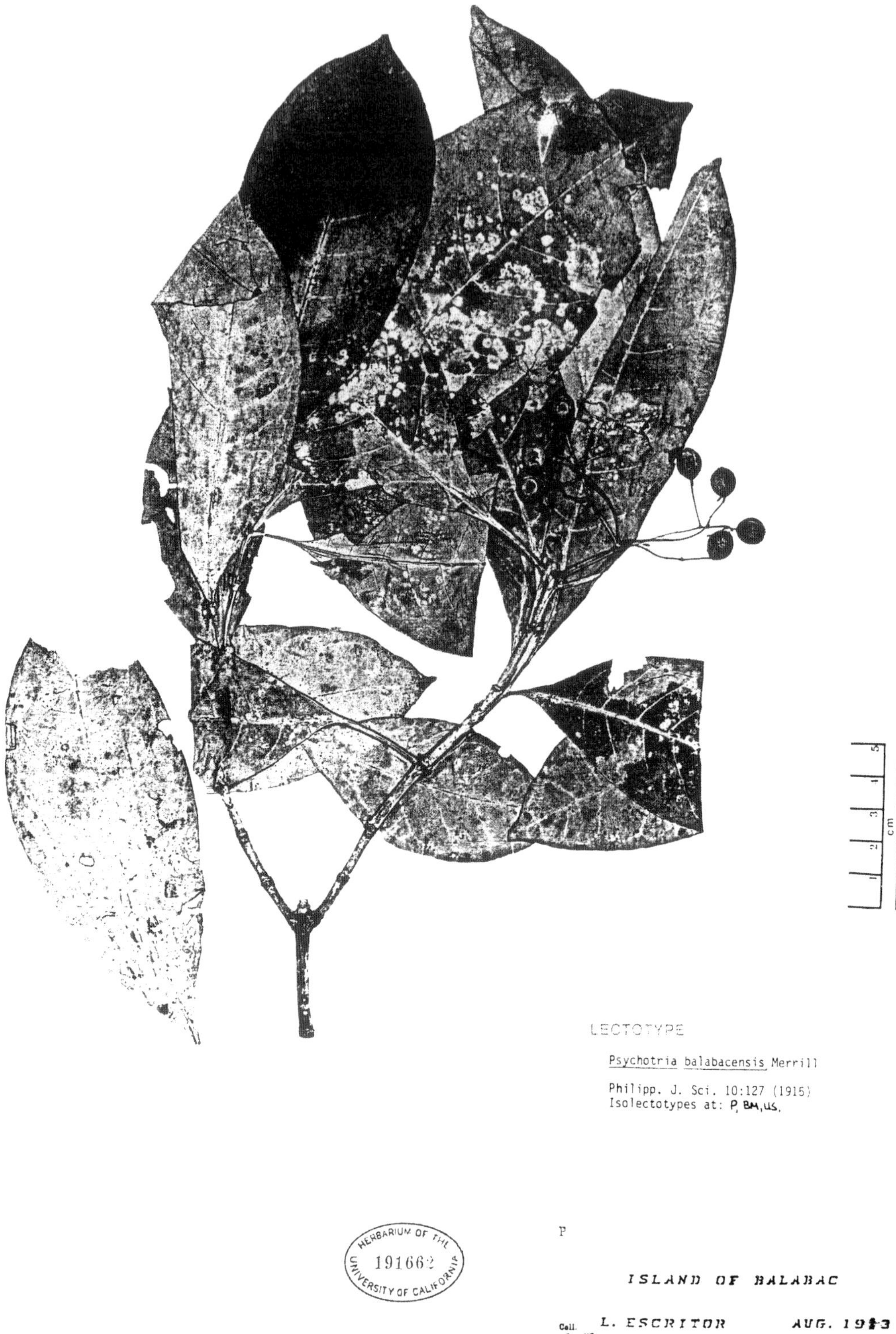

FIG. 60. *Psychotria balabacensis* (*Escritor BS 21605*) (UC, lectotype).

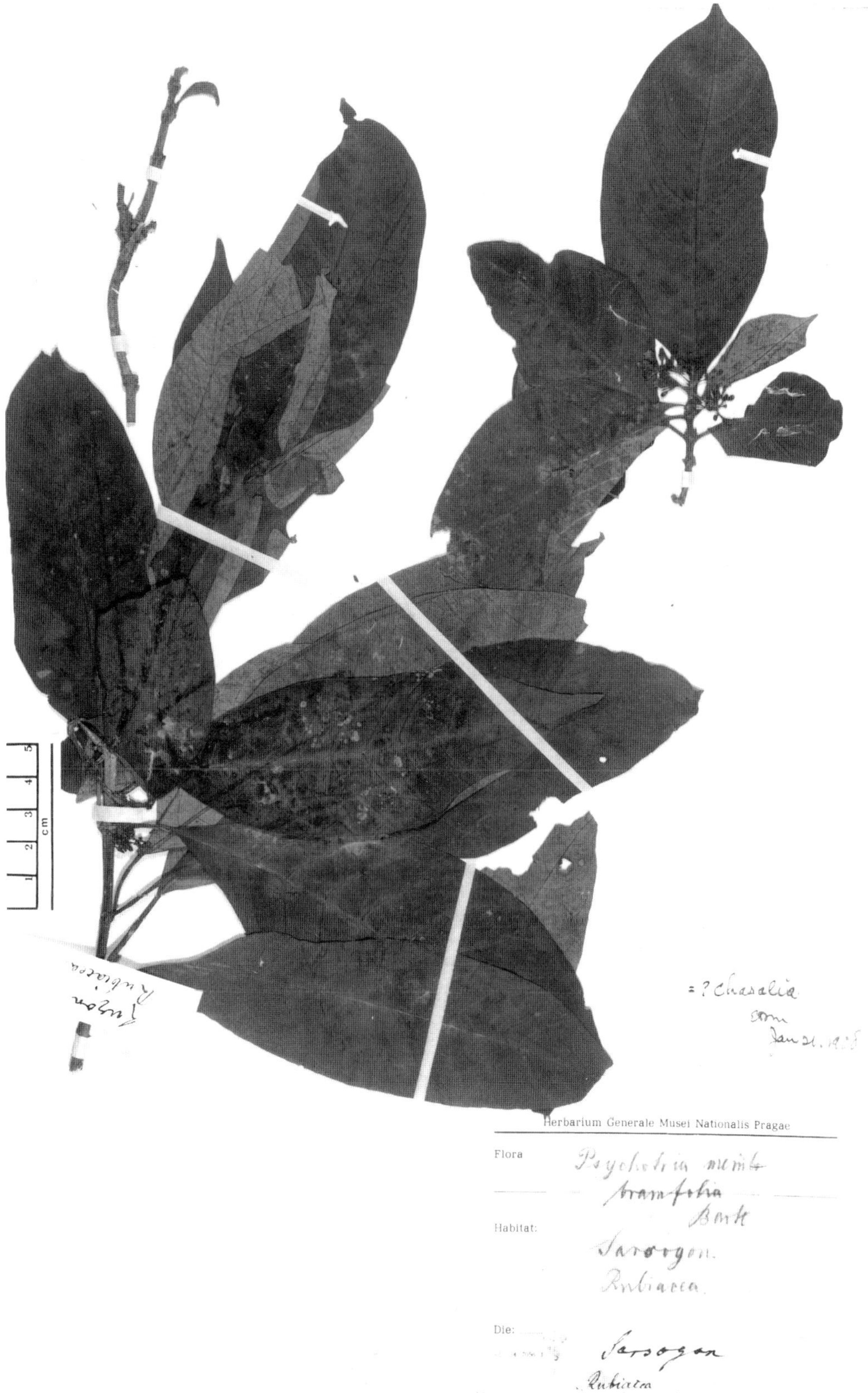

Fig. 61. *Psychotria membranifolia* (*Haenke s.n.*) (PR, holotype).

the branches terminating in lax cymules; bracts ovate to deltate, 0.3–0.7 cm long, or sometimes with leaf-like bracts, ca. 2 cm long; bracteoles ± ovate, 0.3–0.5 mm long. Flowers 4-merous, pedicels 1–2 mm long; calyx (incl. hypanthium) 1–1.5 mm long, external surface glabrous; calyx lobes obscure to ± absent, frequently truncate; corolla: white or yellowish white; corolla tube (1.5–)2.2–3 mm long, glabrous, villous at throat within; corolla lobes ca. 1 mm long; stamens: filaments obscure (long styled); anthers ca. 1 mm long; style ca. 3 mm long (long styled), lobes ca. 0.2 mm long. Fruit globose to obovoid, 5–9 mm long, smooth or ribbed, red, glabrous; calyx limb and lobes not persistent, or persistent but obscure; pedicels 1–2.4 mm long; pyrenes weakly ribbed to smooth on outer surface, weakly to strongly concave on lower surface; endosperm entire.

Other specimens examined: **Luzon. Ilocos Norte Prov.:** Bangui to Claveria, *Ramos BS 33090* (A, K, P); Mt. Palimlim, *Ramos BS 33371* (L). **Ilocos Norte Prov.:** without further data: *Ramos BS 7640* (L). **Kalinga-Apayao Prov.:** Mt. Magnas, *Edaño PNH 19807* (K, L). **Cagayan Prov.:** Gonzaga, *Edaño BS 78222* (A), *BS 78271* (A), *BS 78285* (A); Pinagteponan River, *Edaño BS 78349* (A); Mt. Babatngin, *Edaño BS 79397* (A); Palaui Island, Barangay San Vicente, Santa Ana, *Barbon, Garcia & Romero PPI 18837* (BRIT). **Isabela Prov.:** Malunu, *Warburg 11897* (NY). **Nueva Ecija Prov.:** Mt. Umingan, *Ramos & Edaño BS 26424* (A, US); Palanan, *Ridsdale, Dejan & Baquiran ISU 271* (K). **Quezon Prov.:** Casiguran, *Ramos & Edaño BS 45200* (UC), *BS 45550* (NY, UC, W); Umiray, *Ramos & Edaño BS 29005* (US); Lucban, *Elmer 7954* (A, K, L), *9185* (A, K, L, MO), *McGregor BS 47367* (UC); Mt. Binuang, *Ramos & Edaño BS 28491* (K), *Ramos & Edaño BS 28853* (K); Santa Lucia, Mt. Cristobal, Dolores, *Reynoso & Majaducon PPI 21856* (BRIT); Quezon National Park, *Reynoso et al. PPI 3872* (BRIT); Atimonan, *Merrill 4083* (K, NY); Kinabuhayan, *Sulit FB 30050* (UC); Mt. Tulaog, *Ramos & Edaño BS 29105* (US); Mt. Banahaw, Barangay Kinabuhayan, Dolores, *Reynoso & Majaducon PPI 21662* (K). **Pampanga Prov.:** Mt. Pinatubo, Camp Stotsenburg, *Elmer 22302* (L, MO, NY). **Rizal Prov.:** Mt. Irid, *Ramos & Edaño BS 48559* (UC); Montalban, *Loher 13394* (UC), *s.n.* (UC); Mt. Mapatag, San Andales, *Edaño BS 48664* (UC, US), *BS 48794* (UC)); Balut Island, *Merrill 5416* (NY). **Laguna Prov.:** Los Banos, *Elmer 8195* (K, NY), *17585* (A, K, L, MO, NY); Majayjay, *Duyag BS 76694* (NY), *Rivera & Duyag BS 75018* (UC); Mt. Maquiling [?Mt. Makiling], *Bartlett 15679* (A, BISH, NY, RSA, UC); Mt. Makiling, UPLB College, *Gaerlan et al. PPI 10408* (BRIT); Mt. Banahão, *Duyag BS 76656* (NY), *Ocampo BS 27986* (US). **Laguna Prov.:** without further data: *Canicosa 729* (US), *Mabesa FB 25380* (A, P, US), *Tamesis FB 11967* (K, P). **Camarines Norte Prov.:** Parcale, *Ramos & Edaño BS 33658* (NY), *BS 33740* (P). **Camarines Sur Prov.:** Buhi, *Vidal 404* (L); Agosais, *Edaño BS 76121* (NY); Kamugong River, *Edaño BS 75871* (NY); Sarapan, *Edaño BS 76145* (NY). **Camarines Sur Prov.:** without further data: *Vidal 3050* (A, K). **Albay Prov.:** Mayon Volcano, *Mendoza PNH 18360* (K, L), *Ramos & Edaño BS 75719* (NY, UC). **Sorsogon Prov.:** Bulusan Lake, Barangay San Roque, Bulusan, *Garcia & Fernando PPI 22916* (BRIT); without further data: *Ramos BS 23539* (BM, L, US). **Catanduanes. Catanduanes Prov.:** So. Boradan, Barangay Summit, *Reynoso & Majaducon PPI 24820* (BRIT); without further data: *Ramos & Edaño BS 75474* (UC). **Marinduque. Marinduque Prov.:** So Matalim, Barangay Dampulan, *Romero & Chavez PPI 29193* (K). **Mindoro. Mindoro Oriental Prov.:** Mt. Halcon, *Edaño PNH 3304* (A), *Merrill 5576* (NY); San Teodoro, Sibuang, *Ridsdale 843* (L); Mindoro, Prov unknown: Mt. Ilong, *Edaño PNH 12266* (PNH); Mindoro, without further data, *Merritt FB 11382* (L). **Sibuyan. Romblon Prov.:** Magallanes, *Elmer 12314* (A, BISH, L, MO, NY, US); Madiwang, Tampayan, *Reynoso, Fuentes & Rivers PPI 14010* (K); Camp 1 above Magdiwang on ridge leading to Mayos Peak, *Argent & Reynoso PPI 8919* (K); Camp 3, above Magdiwang on ridge leading to Mayos Peak, *Argent & Reynoso PPI 89130* (K); Punong River, *Sohmer 12441* (K); *Sohmer 12443* (K); Punong Mountain, *Sohmer 12458* (K), *Sohmer 12470* (K), *Sohmer 12484* (K); trail from Sitio Agsalay, Barrio to Punong River, *Sohmer 12378* (K), *Sohmer 12380* (K), *Sohmer 12379* (K); left bank of Punong River, near Jaosan trail, *Sohmer 12421* (K); *Sohmer 12428* (K), *Sohmer 12429* (K); Mt. Guiting-guiting, *Reynoso & Gaerlan et al. PPI 21481* (BRIT, K); Mt Guiting-guiting, *Reynoso, Fuentes & Rivers PPI 14030* (K). **Masbate. Masbate Prov.:** Masbate, Barangay Tugbo, Barbon, *Garcia & Fernando PPI 12536* (BRIT). **Samar. Northern Samar Prov.:** Catubig River, *Edaño BS 24859* (A, US), *Ramos BS 24574* (A, BISH, L, US); Oquendo, *Sulit PNH 14503* (A). **Eastern Samar Prov.:** Oras, *Castro & Anonuevo PNH 5693* (A, L, PNH), *PNH 5730* (A), *PNH 5736* (A); Tagaslian Borongan, *Castro & Anonuevo PNH 5795* (A, PNH). **Western Samar Prov.:** Mt. Malingon, Barangay Lokilokon, Paranas, *Reynoso, Sagcal & Garcia PPI 7330* (BRIT); Sitio Kaguyanan, Barangay Lokilokon, Paranas, *Reynoso, Sagcal & Garcia PPI 7384* (BRIT); Loquilocon, *McGregor BS 43772* (UC); *Wright, Sulit & Conese PNH 6179* (A, L, SING); Mt. Concord, *Gutierrez et al. PNH 117023* (L); Rawis, Sohoton, *Univ. San Carlos 209* (L), *245* (L). Prov. unknown: Mt. Apey, near summit, *Gutierrez et al. PNH 117464* (L); Balanyiga, *Madulid et al. PNH 118346* (L). Samar, without further data, *Ramos BS 17549* (BM, K, L). **Biliran. Leyte Prov.:** without further data: *McGregor BS 18809* (BISH, US). **Leyte. Leyte Prov.:** Tongonan, Ormoc, *Barbon, Alvarez & Fuentes PPI 8706* (BRIT); Mt. Janagdan, Ormoc, *Edaño PNH 12067* (A, L); Mt. Mamban, Ormoc, *Edaño PNH 12096* (A, L). **Leyte. Without further data:** *Wenzel 280* (L, MO) *702* (A, BM, MO), *1200* (BM, MO). **Camotes/Cebu. Cebu Prov.:** Poblacion Alcoy, Argao, *Madulid et al. PPI 20189* (BRIT); without further data: *Ramos BS 41610* (BM, K, UC), *Ramos BS 41611* (L). **Bohol. Bohol Prov.:** Bilar, Logarita, *Reynoso, Alvarez & Fuentes PPI 7833* (BRIT), *Reynoso, Alvarez & Fuentes PPI 7866* (BRIT); Bilar, *Ramos BS 43337* (A, UC); Valencia, *Ramos BS 43127* (A, UC). Bohol, without further data, *Cuming 1863* (K). **Panay. Capiz Prov.:** Libacão, *Martelino & Edaño BS 35437*; Jamindan, *Ramos & Edaño BS 30848* (K, UC), *BS 30958* (A, BM, K, P); Mt. Kinablan, *Edaño BS 46048* (UC). **Mindanao. Surigao del Norte Prov.:** Dinagat Island, Cambinliw, Santiago, Loreto, *Gaerlan, Sagcal & Fernando PPI 4588* (BRIT); Dinagat Island, without further data, *Ramos & Convocar BS 83886* (A), *Ramos & Pascasio BS 35263* (L); Siargao Island, Poneas Islet, Sitio Dabo, Sunkoi Hill, *Stone, Barbon & Garcia PPI 9442* (BRIT); Siargao Island, without further data, *Ramos & Pascasio BS 34864* (UC); Placer, *Wenzel 2630* (A, K, MO, UC), *2749* (A, K, MO, UC), *3196* (UC), *3226* (UC);Mt. Kabatuan, *Mendoza & Convocar PNH 10472* (A, L). **Agusan del Norte Prov.:** Mt. Urdaneta, Cabadbaran, *Elmer 13389* (A, BISH, K, L, MO, NY, P, UC, US); Mt. Hilonghilong, *Mendoza & Convocar PNH 10616* (L); Butuan, *Weber 1008* (A, BM, GH, K, MO, P, UC), *Weber s.n.* (L, NY). **Zamboanga del Norte Prov.:** Siuya, *Frake PNH 38033* (L). **Zamboanga del Sur Prov.:** Malangas, *Ramos & Edaño BS 36786* (A, US), *BS 36790* (BISH); San Ramon, *Hallier 4721* (L). **Palawan. Palawan Prov.:** Municipality of Quezon, Sumindap river, Barangay Bundog, *Pipoly & Romero et al. PPI 38135*

(BRIT); Bancud, *Anonuevo PNH 13503* (A, L, PNH); Malasgao River, Aborlan, *Edaño PNH 14119* (A); Panalingajan River, *Edaño BS 77379* (A); Mt. Gatung, Mantalingajan Mt. Range, District of Tindogan, *Pipoly & Romero et al. PPI 38041* (BRIT).

Distribution.—Found on nearly all the islands and island groups of the Philippines, although extirpated in many lowland areas. Occurring in New Guinea (Sohmer, 1988) and probably many other places in Malesia.

Ecology.—Moist forest, mostly in primary forest (incl. mixed dipterocarp forest, and swamp forest) but also found in secondary forest (e.g. emerging from logged-over forest); quite often found near rivers; coral limestone, limestone, clay, and ultramafic rocks; 50–1300(–1500) m.

Phenology records.—Flowering: February to May, August; fruiting: March to August, October to December.

Preliminary conservation assessment.—Least Concern (**LC**). Last collected in 1996.

Distinguishing characteristics.—Leaf blades membranaceous, oblong-elliptic to oblanceolate, drying green or yellow-green, apex often abruptly acuminate; pyrenes smooth or ridged on outer (dorsal) surface, often strongly concave on inner surface, endosperm entire.

Remarks.—This is the most widely distributed taxon of Philippine *Psychotria* with the possible exception of *P. luzoniensis.* It is probably not nearly as widely distributed as it once was, however, because its primary habitat, lower elevation moist forest, is the most severely degraded of all forest types in the Philippines. Today very little of such forest remains. There is considerable variation within this species. The variation of the inflorescence ranges from those with a rather thick peduncle, numerous secondary branches, and compressed internodes, to those with relatively thin peduncles and elongated internodes. The inflorescence is either puberulent or completely glabrous. The basic characteristic that ties most of these collections together in the herbarium is the very thin, membranaceous leaves. Herbarium specimens of *Chassalia curviflora* Thwaites can look very similar to specimens of *P. membranifolia* and may sometimes be confused, as they both have membranaceous leaves. Unlike *Psychotria*, however, *Chassalia* species usually have pale, papery (chartaceous) stipueles, relatively long curved corolla tubes, and pyrenes with a deep excavation on the inner surface of each pyrene.

Psychotria membranifolia may be an artificial concept (i.e. comprise more than one species/taxon), and is found in other parts of Malesia, including New Guinea and the Indonesian archipelago. The variability within this species is more than expected from a single taxon, and there appears to be an intergradation into *P. leptothyrsa* Miq. The principal characteristics distinguishing these two taxa are the texture and shape of the leaves. In addition, *P. leptothyrsa* never has thickened peduncles and main axes with compression between the branches. *Psychotria leptothyrsa* is itself widely distributed and highly variable. These taxa are obviously related, particularly in the nature of the pyrenes, which possess preformed germination slits (as in most Philippine *Psychotria*), homogeneous (non ruminate) endosperm and a seed coat pigment which is insoluble in 70% ethanol (SCP negative). After studying them throughout Malesia, we are certain that only careful field work will allow an understanding of this *Psychotria* group. A taxon described from New Britain, *P. damasiana* Sohmer, is morphologically very similar to *P. membranifolia* in leaf size, shape, texture, and venation. They can be separated on the basis of their stipule morphology: *P. damasiana* has calyptrate stiuples whereas *P. membranifolia* are valvate.

Two specimens from Samar (*Reynoso & Majaducon PPI 21983* (K), and *Gaerlan & Chavez PPI 26351* (K)) have pyrenes with a smooth outer (dorsal) surface (like *P. balabacensis*) but otherwise conform to *P. membranifolia*.

VI.—Magnifolia Species Group

Shrubs or small trees. ***Stipules imbricate*** or unknown, ***large*** (e.g. 3–6 cm long, but sometimes up to 10 cm long). ***Leaf blades*** more than 3 cm wide, and usually more than 9 cm wide; ***usually more than 24 cm long and often more than 30 cm long;*** chartaceous to subcoriaceous, sometimes coriaceous. Inflorescences with axes and branches, monochotomous (pedunculate) or trichotomous. Flowers pedicellate. Fruits red or color unknown; pyrene inner (adaxial) surface flat; seed producing a red color when placed in in 70% ethanol, due to a soluble seed coat pigment (SCP positive), rarely SCP negative; endosperm ruminate, or unknown, usually dark. Comprising four species.

Note.—Text in ***bold italics*** denotes features useful for the placement of taxa within the respective species group.

KEY TO THE SPECIES OF THE MAGNIFOLIA SPECIES GROUP

1. Inflorescence axes fused into a thick axis from the top of which a congested mass of numerous cymules are borne; petioles < 1 cm long ______ **52. P. conglomeratifolia**
1. Inflorescence axes not fused, inflorescences either trichotomous or monochotomous, with cymules borne on secondary branches or clustered on widely separated nodes, not congested; petioles 1.5–6 cm long.
  2. Leaf blades with 15–20 pairs of lateral veins ______ **97. P. casiguraensis***
  2. Leaf blades with 20–30(–36) pairs of lateral veins.
    3. Flowers or fruits borne directly on the inflorescence axes or in cymules distinctly clustered at widely separated nodes on these axes; leaf blades 36.5–49(–55) cm long ______ **53. P. magnifolia**
    3. Flowers and fruits borne on secondary or tertiary branches and not distinctly clustered at nodes directly on the primary axes; leaf blades 24.5–38.5 cm long.
      4. Leaf blades oblanceolate to narrowly obovate; petioles < 1 cm long; leaf base ± attenuate; inflorescences monochotomous ______ **54. P. alzapaniensis**
      4. Leaf blades broadly obovate; petioles 3–6 cm long; leaf base acute, narrowly acute or cuneate; inflorescences trichotomous or sometimes monchotomous ______ **55. P. amplissima**

*Taxon included here because leaf blade size grades into this group, but is treated in the Luzoniensis Species Group.

52. Psychotria conglomeratiflora Sohmer & A.P. Davis, sp. nov. (**Fig. 62**). TYPE: PHILIPINES: SAMAR: Western Samar Prov.: Bagacay, bank of Kansulaban Creek, in dipterocarp forest at ca. 200 m, 26 Apr 1948, *Sulit PNH 6306* (HOLOTYPE: PNH; ISOTYPES: A, L, PNH).

Frutices 1–2 m alti glabri sunt, stipulis 2.3–5 cm longis apice obtuso ad acuto, laminis 9.2 × 25.8–13 × 37 cm coriaceis oblanceolati-oblongis apice late rotundato basi truncata ad obtusa quando sicis subrubre subviride badiis, petiolis minus que 1 cm longis, inflorescentia 8.5–16.5 cm longa axibus tribus in pedunculo connato grandi 4–10 cm longo, floribus multis congestis apicalibus, fructibus 8–9 mm longis obovoideis valde liratir, pyrenis in dorso valde liratis, endospermo ruminato.

Shrub, 1–2 m high, young stems glabrous. Stipules imbricate, elliptic to obovate, 2.3–5 cm long, glabrous; apex obtuse to acute. Leaves: petioles 0.6–1 cm long, glabrous; leaf blades oblanceolate-oblong, 25.8–37 3 9.2–13 cm, coriaceous, drying reddish greenish brown, glabrous above and below; secondary veins 20–27 pairs; tertiary venation pattern manifest to quite prominent (particulalry on the upper surface); base truncate to obtuse, or subauriculate; apex obtuse to broadly rounded. Inflorescence/infructescence appearing monochotomous, 8.5–16.5 cm long, glabrous; principal axis (peduncle) apparently formed by fusion of 3 from a trichotomous inflorescence, greatly thickened, 4–10 cm long, with numerous flowers congested and arising directly on the distal end of the inflorescence, parted but not obviously branched; bracts and bracteoles absent. Flowers unknown. Fruit obovoid, 8–9 mm long, strongly ridged, color unknown, glabrous; calyx limb and lobes not persistent; pedicels 3–8 mm long; pyrenes strongly ridged on outer surface; endosperm ruminate.

PARATYPE: **Samar. Borongon Prov.:** Tagaslian, *Castro PNH 5834* (A).

Distribution.—Samar.

Ecology.—Primary dipterocarp forest; 150–200 m.

Phenology records.—Flowering: unknown; fruiting: April to June.

Preliminary conservation assessment.—Critically Endangered (**CR B1ab(iii)**) or possibly extinct (**EX**). Last collected in 1948.

Distinguishing characteristics.—Leaf blades oblanceolate-oblong, large, 25.8–37 cm long, apex broadly rounded, base obtuse to truncate, lateral veins 20–27 per side; inflorescence with one stout peduncle formed by the fusion of the three axes of a trichotomous inflorescence.

Remarks.—This is one of the most distinctive species of *Psychotria* in the Philippines, distinguishable from the other species in the Magnifolia Species Group by the nature of the inflorescence and the shape of the leaves. It is known from only two collections made in 1948. Hopefully it still exists, but as both collections were made in primary lowland dipterocarp forest, it is highly likely that it has disappeared.

53. Psychotria magnifolia Merr., Philipp. J. Sci. 17:453. 1920; Merr., Enum. Philipp. Fl. Pl. 3:558. 1923. (**Fig. 63**). TYPE: PHILIPPINES. LUZON: Quezon Prov.: Mt. Binuang, along small streams in damp forests, 11 May 1917, *Ramos & Edaño BS 28828* (LECTOTYPE: US, designated here; ISOLECTOTYPES: A, K, P).

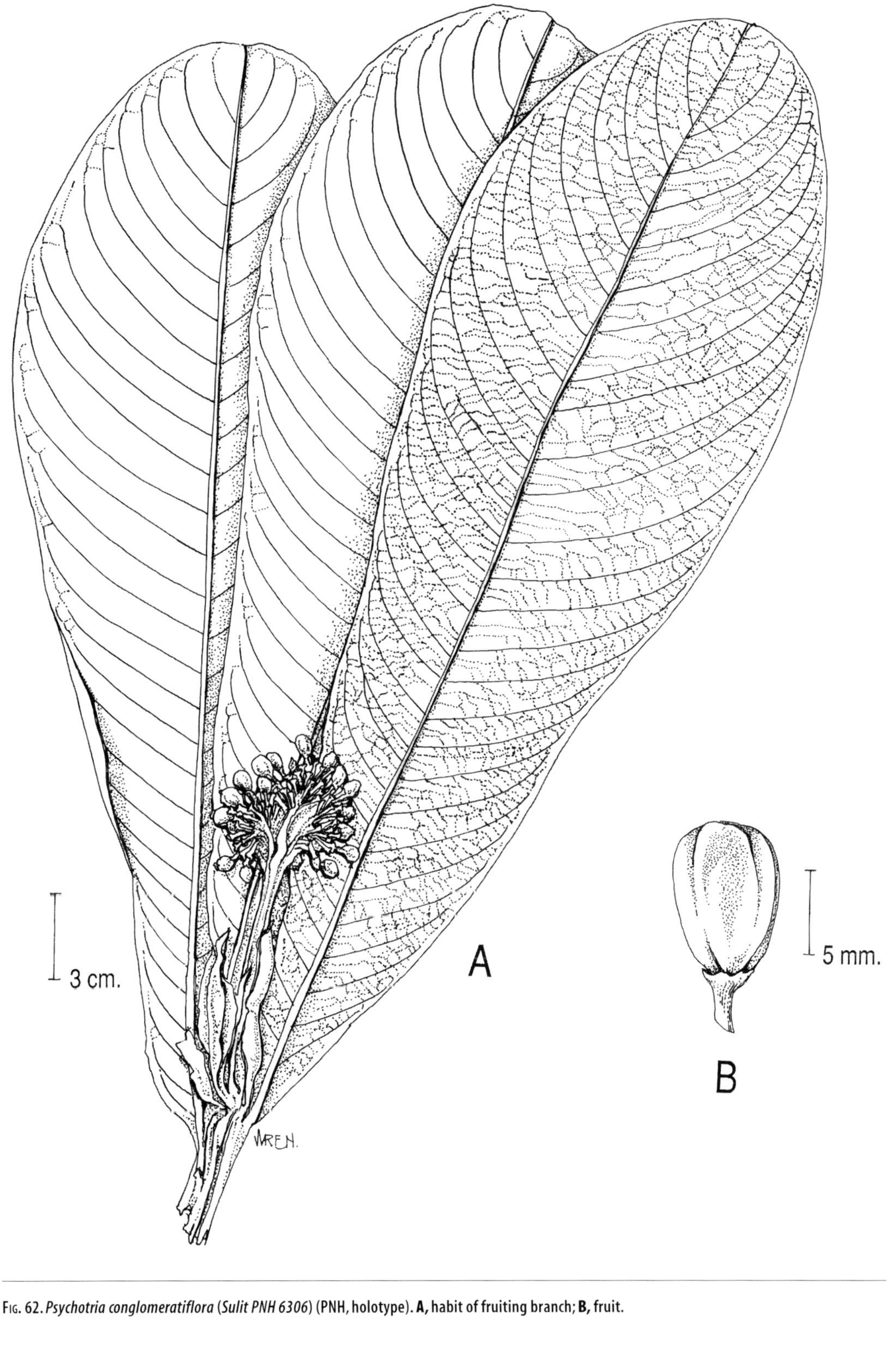

Fig. 62. *Psychotria conglomeratiflora* (*Sulit PNH 6306*) (PNH, holotype). **A,** habit of fruiting branch; **B,** fruit.

FIG. 63. *Psychotria magnifolia* (*Ramos & Edaño BS 28828*) (US, lectotype).

Shrub, 3–4 m high, young stems glabrous. Stipules probably imbricate, leaf-like, broadly ovate, 3–6 cm long, glabrous; apex narrowly acute to acuminate. Leaves: petioles 1.5–2 cm long, stout (4.5–5 cm wide), glabrous; leaf blades broadly oblanceolate to obovate, 36.5–49(–55) × 14.5–18.5(–22) cm, sub-coriaceous, drying dark brown, glabrous above and below; secondary veins 24–30 pairs; tertiary venation manifest; base acute to obtuse; apex obtuse to to broadly acute. Inflorescence/infructescence rather poorly known, trichotomous, (11.5–)13–20 cm long, glabrous; principle axis (9.4–)11.5–16.5 cm long, each axis with 1–3 primary nodes and opposite branching at each node, the nodes bearing clusters of several to many pedicellate flower/fruits; bracts and bracteoles unknown. Flowers poorly known, ?5-merous, pedicels ca. 6 mm long; calyx (incl. hypanthium) ca. 3 mm long, external surface glabrous; calyx lobes tooth-like; corolla: color unknown; corolla tube ca. 3 mm long, glabrous, densely pubescent at throat within; corolla lobes ca. 5 mm long; stamens unknown; style unknown. Fruit poorly known, subglobose to ellipsoidal, 7–9 mm long, probably ribbed, color unknown, glabrous, calyx limb and lobes persistent but inconspicuous; pedicels 5.9–6.5 mm long; pyrenes immature, unknown.

Other specimens examined: **Luzon. Rizal Prov.:** locality unknown: *Loher 15011* (UC).

Distribution.—Central and north central Luzon.

Ecology.—Moist forest; reported along streams; elevation unknown.

Phenology records.—Flowering: unknown; fruiting: May.

Preliminary conservation assessment.—Extinct (**EX**). Last collected in 1917.

Distinguishing characteristics.—Leaf blades oblanceolate-obovate, very large, reported to reach 55 cm long, lateral veins more than 20 per side and with a conspicuous marginal vein; petioles relatively short, 1.5–2 cm long; stipules large, 3–6 cm long; inflorescence/infructescence with several to many flowers and fruits clustered directly at the primary nodes.

Remarks.—*Psychotria magnifolia* is certainly one of the most conspicuous species of Philippine *Psychotria*, due to the large leaves alone. The inflorescence/infructescence with flowers/fruits clustered directly at the nodes is another conspicuous characteristic. *Psychotria magnifolia* is morphological close to *P. alzapaniensis* or *P. amplissima* Merr., and they share the characteristic of large leaves. The inflorescence structure of *P. magnifolia* is, however, unlike the other two species.

54. Psychotria alzapaniensis Sohmer & A.P. Davis, sp. nov. (**Fig. 64**). TYPE: PHILIPPINES. LUZON. Quezon Prov.: Mt. Alzapan, on mossy forest slopes at ca. 1600–1700 m ("5000 ft"), 20 May 1925, *Ramos & Edaño BS 45688* (HOLOTYPE: BM; ISOTYPES: K, NY, UC).

Frutex 1 m altus est, stipulis 5–10 cm longis glabris foliaceis imbricatis apice acuminato, laminis 9.5 × 35–11.5 × 38.5 cm latis et longis chartaceis vel subcoriaceis quando siccis subviridi-badiis oblanceolatis ad anguste obovatis apice obtuso vel rotundato basi attenuata nervis lateralibus 24–28(–36) in dimidio quoque nervulis tertiis subprominentibus, petiolis minus que 1 cm longis, inflorercentia monochotoma 10–12 cm longa, pedunculo 8–10 cm longo nodis primariis 3–4 cum ramis verticillatis brevibus puberulis cum cymulis terminalibus plerumque sine ramulis, calycibus cum hypanthio at tubo omnibus 1.5 × 2 mm, lobis obscuris, fructibus ignotis.

Shrub, ca. 1 m high, young stems glabrous. Stipules imbricate, almost leaf-like, ± elliptic, 5–10 cm long, glabrous; apex narrowly acute to acuminate. Leaves: petioles 0.4–1 cm long, rather stout (ca. 4 mm wide), glabrous; leaf blades oblanceolate to narrowly obovate, 35–38.5 × 9.5–11.5 cm, chartaceous to subcoriaceous, drying greenish brown, glabrous, secondary veins 24–28(–36) pairs; tertiary venation manifest; base ± attenuate; apex obtuse to rounded. Inflorescence/infructescence monochotomous, 10–12 cm long, glabrous to puberulous; principal axis (peduncle) 8–10 cm long, with 3–4 primary nodes and verticillate branching at each node, branches generally rather short, terminating in cymules; bracts absent; bracteoles deltate to ovate, ca. 0.4 mm long. Flowers poorly known, pedicels ca. 2 mm long; calyx (incl. hypanthium) ca. 2mm long, external surface glabrous; calyx lobes obscure or absent, calyx limb truncate; corolla: color unknown; corolla tube ca. 3.5 mm long, otherwise unknown; stamens unknown style; unknown. Fruit unknown.

PARATYPES: None—known only from the type collection.

Distribution.—Known only from Mt. Alzapan, Quezon Province, Luzon.

Ecology.—Forest, including 'mossy forest'; ca. 1600–1700 m.

Phenology records.—Flowering: April to June.

Preliminary conservation assessment.—Critically Endangered (**CR B1ab(iii)**), or possibly Extinct (**EX**). Only collected once in 1925.

Distinguishing characteristics.—Leaves oblanceolate to narrowly obovate, over 35 cm long, lateral veins over 24 per side, petioles short, less than 1 cm long; stipules large and leaf-like, 5–10 cm long; inflorescences monochotomous, cymules borne on short secondary branches.

Fig. 64. *Psychotria alzapaniensis* (*Ramos & Edaño BS 45688*) (BM, lectotype).

Remarks.—It is rarely good practice to describe a species from a single collection, and the one desribed here also lacks mature flowers and fruits. However, the large leaves distinguish it from most other species of Philippine *Psychotria*, and the leaf shape, short petioles, and gross morphology of the inflorescence separate it from the other large-leaved members of the Magnifolia Species Group. A specimen collected at sea level in the Izabela Province (*Ridsdale, Dejan & Baquiran ISU 34*) resembles *P. alzapaniensis* but differs due to its fewer veined leaves, slightly longer petioles, and smaller, fewer flowered infloresecences. We have been unable to authoritatively determine this specimen, and for the time being place it with above species, as *P.* cf. *alzapaniensis*.

55. Psychotria amplissima Merr., Philipp. J. Sci. 17:444 1921; Merr., Enum. Philipp. Fl. Pl. 3:552. 1923. (**Fig. 65**). TYPE: PHILIPPINES. LEYTE: Leyte Prov.: Dagami, along streams in forests, 9 Aug 1912, *Ramos BS 15198* (LECTOTYPE: US, designated here).

Shrub or small tree, 1–2m high, young stems glabrous. Stipules imbricate, almost leaf-like, ± elliptic-obovate, 2.6–5.2 cm long; apex broadly acute to acute, entire or becoming cleft. Leaves: petioles 3–6 cm long, glabrous or puberulous (on lower surface); leaf blades broadly obovate, 24.5–38.5 × 10–15 cm, coriaceous to subcoriaceous, drying brown, reddish brown or greenish, glabrous above, softly puberulous below; secondary veins 20–28 pairs; tertiary venation prominent; base acute, cuneate or narrowly acute; apex broadly rounded to obtuse. Inflorescence/infructescence trichotomous or monochotomous, (5–)8–16.5 cm long, glabrous; principal axis (sometimes the peduncle) (5–)12–13 cm long, each axis with 2–4 primary nodes and verticillate branching at each node, branches bearing several to many cymules; bracts triangular, ca. 2 mm long, but often absent; bracteoles absent. Flowers 5-merous, pedicels 2–5 mm long; calyx (incl. hypanthium) 1–2 mm long, external surface glabrous; calyx lobes deltate to triangular, 0.5–1 mm long; corolla: color unknown corolla tube 2.5–3 mm long, glabrous, villous within at throat; corolla lobes 2–2.5 mm long; stamens: filaments ca. 2 mm long (long styled); anthers ca. 1 mm long; style ca. 2 mm long, lobes ca. 0.2 mm long (long styled). Fruit subglobose to obovoid, 6–8 mm long, ± smooth, red, glabrous; calyx limb and lobes persistent, rather obvious; pedicels 2–5.2 mm long; pyrenes smooth on outer surface; endosperm ruminate.

Other specimens examined: **Samar. Northern Samar Prov.:** Catubig River, *Ramos BS 24445* (A, K, P). **Western Samar Prov.:** Mt. Mahagna, Tarabucan, *Sulit PNH 14496* (A, BM, L, PNH); Mt. Sohoton, Babasian Na Daku River, *Gutierrez et al. PNH 117759* (K, L). **Leyte. Leyte Prov.:** locality unknown: *Wenzel 1628* (A, BM, GH, MO), *1751* (A, BM, MO, NY). **Mindanao. Agusan Prov.:** Diuata Mtns., *Ramos & Convocar BS 83745* (A, SING). **Surigao Prov.:** Sukailung, *Mendoza & Convocar PNH 10231* (A, L, PNH).

Distribution.—Leyte, Samar and Mindanao.

Ecology.—Moist Forest; sometimes near streams; low elevation (e.g. 150 m).

Phenology records.—Flowering: August; fruiting: February to April.

Preliminary conservation assessment.—Endangered (**EN B1ab(iii)**). Last collected in 1970.

Distinguishing characteristics.—Leaf blades 24.5–38.5 cm long, lateral veins more than 20 per side, petioles 3–6 cm long; inflorescences or infructescences trichotomous or monochotomous, with secondary branches up to 5 cm long in fruit.

Remarks.—This is, or was, one of the most commonly encountered members of this species group in the Philippines. It is distinguished from *P. alzapaniensis* and *P. magnifolia* by the branched, spreading nature of the inflorescence. Since the habitat appears to be humid lowland forest, it is unlikely that many populations of this species remain.

VII.—Gitingensis Species Group

Shrubs or small trees. Stipules valvate or imbricate. Leaf blades usually more than 3 cm wide, but sometimes less; usually more than 5 cm long, only very rarely more than 30 cm long, chartaceous to subcoriaceous, or coriaceous. ***Inflorescences with axes and branches, monochotomous (pedunculate), rarely trichotomous.*** Flowers pedicellate, rarely sessile. Fruits usually red, sometimes yellow, purple or black; pyrene inner (adaxial) surface flat; seed producing a red color when placed in 70% ethanol, due to a soluble seed coat pigment (SCP positive), rarely SCP negative; endosperm ruminate, often strongly so, usually dark. Comprising 17 species.

Note.—Text in ***bold italics*** denotes features useful for the placement of taxa within the respective species group.

FIG. 65. *Psychotria amplissima* (*Ramos BS 15198*) (US, lectotype).

KEY TO THE SPECIES OF THE GITINGENSIS SPECIES GROUP

1. Plants pubescent, the leaves usually prominently pubescent below (sometimes glabrous in *P. rubiginosa*); inflorescence axes often densely pubescent.
  2. Lateral veins 11–17 per side ____ **56. P. rubiginosa**
  2. Lateral veins 18–20 per side ____ **57. P. phanerophlebia**
1. Plants usually glabrous, at most somewhat pubescent on veins below; inflorescence axes usually glabrous.
  3. Stipules imbricate, narrowly triangular, several times longer than broad in bud, often twisted in bud, usually 0.9–4.5 cm long.
    4. Fruits (6–)9–11 mm long, obovoid; petioles 5.1–6 cm long ____ **58. P. boholensis**
    4. Fruits 5–8 mm long, globose-turbinate or obovoid and sometimes diplococcus; petioles 0.8–5 cm long.
      5. Stipules 3–4.5 cm long; leaf blades (15–)21.5–26 cm long, lateral veins 14–20 per side ____ **59. P. iridensis**
      5. Stipules usually 0.9–1.2 cm long or smaller; leaf blades 11–18.7 cm long, lateral veins 11–15 per side.
        6. Stipules cleft at apex; petioles 0.8–1.5 cm long; inflorescence peduncle 1.5–3.5 cm long ____ **60. P. palawanensis**
        6. Stipules not cleft at apex; petioles 2–4.4 cm long; inflorescence peduncle (3.3–)4–6.8 cm long ____ **61. P. quisumbingiana**
  3. Stipules usually valvate, ovate, one-half as broad as long or broader than long in bud, usually not twisted, 0.2–1(–1.5) cm long.
    7. Fruits oblong-ellipsoid, 8–15 mm long, prominently ribbed when dry ____ **62. P. manillensis**
    7. Fruits usually obovoid, turbinate-obovoid, obovoid-globose, or globose, rarely sub-ellipsoid, 5–12 mm long, with or without prominent ribbing when dry.
      8. Calyx (incl. hypanthium and limb) 2.8–4.7 mm long in flower, and calyx limb 2.5–3 mm long in fruit (also *P. crassifolia*); fruits 8.5–12 mm long ____ **63. P. alabatensis**
      8. Calyx (incl. hypanthium and limb) < 3 mm long in flower, and calyx limb < 2 mm long in fruit; fruits 5–10 mm long.
        9. Infructescence peduncle usually < 3 cm long.
          10. Secondary branches of inflorescence/infructescence 2–5 cm long, spreading, lax ____ **64. P. merrittioides**
          10. Secondary branches of inflorescence/infructescence 0.7–2 cm long, often < 0.5 cm long, ± contracted, not spreading.
            11. Leaf blades (5.1–)7–8.8 cm wide, lateral veins 10–13 pairs, not prominent ____ **66. P. panayensis**
            11. Leaf blades 4.9–7 cm wide, lateral veins (13–)16–17 pairs, prominent ____ **67. P. frakei**
        9. Infructescence peduncle infructescence usually > 3 cm long.
          12. Fruits globose, 8–9 mm long ____ **65. P. foxworthyi**
          12. Fruits other shapes, very rarely globose, 8 mm long .
            13. Stipules persistent and conspicuous on lower nodes ____ **56b. P. rubiginosa** var. **jacobsii**
            13. Stipules early falling or conspicuous on lower nodes.
              14. Leaf blades usually < 3 cm wide ____ **68. P. mindanaensis**
              14. Leaf blades usually . 3 cm wide.
                15. Primary branches of infructescence often reflexed, bracts usually fused and sheathing the inflorescence axes ____ **69. P. pyramidata**
                15. Primary branches of infructescence never reflexed; bracts, where known, not sheathing the inflorescence axes.
                  16. Leaf blades 6.2 11 cm wide ____ **70. P. sorsogonensis**
                  16. Leaf blades 4–6 cm wide.
                    17. Leaf blades subcoriaceous to ± coriaceous, drying a shiny reddish brown; corolla tubes 8–9 mm long; fruits (6–)8–10 mm long ____ **71. P. crassifolia**
                    17. Leaf blades chartaceous to subcoriaceous, usually drying greenish brown to brown, rarely shiny; corolla tubes 3.5–9 mm long; fruits 5–8.5 mm long ____ **72. P. gitingensis**

56. Psychotria rubiginosa Elmer, Leafl. Philipp. Bot. 1:25. 1906; Merr., Philipp. J. Sci. 1(Suppl. 1): 135. 1906; Merr., Enum. Philipp. Fl. Pl. 3:562. 1923. (**Fig. 66**). *Grumilea rubiginosa* (Elmer) Merr., Philipp. J. Sci. 10:131. 1915. Type: PHILIPPINES. Luzon: Bataan Prov.: Mt. Mariveles, Nov 1904, *Elmer 6718* (lectotype: K, designated here).

Shrub or small tree, 1–6 m high, young stems pubescent to puberulous. Stipules valvate, usually with distinct 'shoulders' at the base, ovate to ± lanceolate, 0.7–1.2 cm long, usually pubescent; apex partially to distinctly cleft, the lobes often almost arisitate. Leaves: petioles (1.1–)1.5–4 cm long, pubescent to puberulous; leaf blades oblong-elliptic to oblanceolate, (12–)19–33 × (3.4–)4.1–11.6 cm, coriaceous, pubescent to puberulous below (hairs with bulbous bases); secondary veins 11–17 pairs; tertiary venation manifest; base acute to narrolwy acute, or attenuate; apex acuminate, abruptly acuminate, or

Fig. 66. *Psychotria rubiginosa* var. *rubiginosa* (*Elmer 6718*) (K, lectotype).

long-acuminate. Inflorescence/infructescence monochotomous, (9–)13–28 cm long, densely pubescent to pubescent (hairs drying reddish brown); principal axis (peduncle) (4.7–)5–13.5(–15.7) cm long, with 3–6 primary nodes and verticillate branches at each node, secondary branches terminating in cymules; bracts ovate to triangular, 2–2.5 mm long, sometimes absent (caducous); bracteoles similar to the bracts, ca. 1.5 mm long. Flowers 4- or 5-merous, pedicels 1–2 mm long or absent; calyx (incl.

hypanthium) 1.9–2.3 mm long, external surface pubescent to puberulous, or glabrous; calyx lobes broadly ovate, or obscure and tooth-like; corolla: white; corolla tube 2–3 mm long, glabrous, ?villous at throat within; corolla lobes ca. 1.5 mm long (short styled); stamens: filaments ca. 2 mm long; anthers ca. 1.5 mm long; style ca. 2 mm long, lobes ± unknown. Fruit globose-turbinate or obovoid, 7–8 mm long, ridged or ribbed, reported yellow but probably turning orange or red at maturity, glabrous; calyx limb and lobes usually persistent but insignificant; pedicels 1–2 mm long, or absent; pyrenes ribbed or ridged on outer surface; endosperm ruminate.

Distribution.—Luzon; one specimen record from Cebu.

Ecology.—Humid forest; ridges and ravines; 500–1100 m.

Phenology records.—Flowering: January to May; fruiting: March, May, August.

Preliminary conservation assessment.—Near Threatened (**NT**). Last collected in 1993.

Distinguishing characteristics.—Leaf blades oblanceolate, coriaceous; inflorescences monochotomous, peduncle 5–13.5 cm long.

Remarks.—This species was published twice in the same year, by Elmer (1906) and Merrill (1906a). In this revision we assume that Elmer's description of *P. rubignosa* was first and that it has priority. If this is not the case this species should be cited as *P. rubiginosa* Elmer ex Merr.

There are to be two varieties of this taxon. The following key serves to distinguish them.

KEY TO THE VARIETIES OF *PSYCHOTRIA RUBIGNOSA*

1. Inflorescence axes, and often undersurface of leaf blades, pubescent to densely pubescent; stipules usually not persistent ____ **56a.** var. **rubiginosa**
1. Inflorescence axes and leaf blades glabrous; stipules ± persistent ____ **56b.** var. **jacobsii**

56a. Psychotria rubiginosa var. rubiginosa

Shrub or small tree, 1–6 m tall, generally pubescent. Stipules caducous to semi-persistent, 1–1.2 cm long, pubescent. Leaves: petioles 1.5–4 cm long, pubescent; leaf blades oblong-elliptic to oblanceolate, 19–33 × 4.1–11.6 cm, pubescent below; secondary veins 11–17 per side; base acute to narrowly acute; apex acuminate to abruptly acuminate. Inflorescence densely pubescence (hairs drying reddish brown); main axis (peduncle) 5–13.5 cm long, with 4–6 nodes. Flowers 4–5-merous, pedicels 1–2 mm long.

Other specimens examined: **Luzon. Ilocos Norte Prov.:** Mt. Palimlim, *Ramos BS 33261* (K). **Zambales Prov.:** Acoje Mine concession area, Santa Cruz, *Ridsdale 1400* (K). **Rizal Prov.:** Mt. Pinagbayanan, *Edaño BS 48777* (K, P, NY, UC); Mt. Angilo, *Ramos BS 40779* (A, P); Mt. Irid, *Ramos & Edaño BS 48481* (UC); Montalban, *Loher 13340* (UC), *Loher 1518* (K), *s.n.* (UC); Mt. Canumay, *Ramos BS 13776* (K, L, P); Mt. Tokduanbanoy, *Ramos & Edaño BS 48621* (NY, UC); Limutan, *Loher 1516* (K, US), *Loher 1517* (K); **Rizal Prov.:** without further data: *Loher 14301* (A), *Loher 6389* (K), *s.n.* (UC). **Bataan Prov.:** Mt. Marivales, *Reynoso, Alvarez & Fernando PPI 3932* (BRIT, K); Mt. Mariveles, Lamao River, *Meyer FB 2778* (K, NY, US), *Whitford 243* (K, P, US). **Luzon. District unknown:** 'district of Morong', *Vidal 3070* (K). **Cebu. Cebu Prov.:** Tebunan, Talamban, *Reynoso, Alvarez & Fuentes PPI 7905* (BRIT, K). **PHILIPPINES. Without further data:** *Loher 6394* (K), *6433* (K).

Distribution.—Luzon and one specimen record from Cebu.

Ecology.—Forest; ridges and ravines; 600–900 m.

Phenology records.—Flowering: January, March to May; fruiting: March, May, August.

Preliminary conservation assessment.—Near Threatened (**NT**). Last collected in 1993.

Distinguishing characteristics.—Most plant parts pubescent; leaf blades oblanceolate, 19–33 cm long, coriaceous; inflorescences monochotomous, peduncle usually long, usually 5–13.5 cm long, axes conspicuously pubescent.

56b. Psychotria rubiginosa Elmer var. **jacobsii** Sohmer & A.P. Davis, var. nov. (**Fig. 67**). TYPE: PHILIPPINES. LUZON: Quezon Prov.: Sierra Madre Mountains north-northeast of Dingalan, low mixed mossy primary forest at 1000–1100 m, 18 Mar 1968, *Jacobs 7847* (HOLOTYPE: L; ISOTYPES: A, CANB, K).

Frutices vel arbores parvae 3–4 m alti sunt, stipulis 0.7–0.8 cm longis et in latitudine fere aequalilus glabris, persistentibus et in nodis inferis conspicuis, laminis 3.4 × 12–5 × 19 cm latis et longis coriaceis infra glabris elliptici-lanceolatis longe acuminatis basi acuminata vel attenuato nervis lateralilus 12–16 in dimideo quoque, petiolis 1.1–2.1 cm longis, inflorescentia monochotoma 9.5–13.7 cm longa, pedenculo 4.7–9 cm longo, nodis primariis 3–6 et cum ramis verticillatis 2–5 cm longis ascendentilus eis 1–2-ramosis, cymulis terminalibus cum bracteis, floribus sessilibus vel in pedicellis minutis.

Shrub or small tree, 3–4 m tall, glabrous. Stipules usually semi-persistent and conspicuous, 0.7–0.8

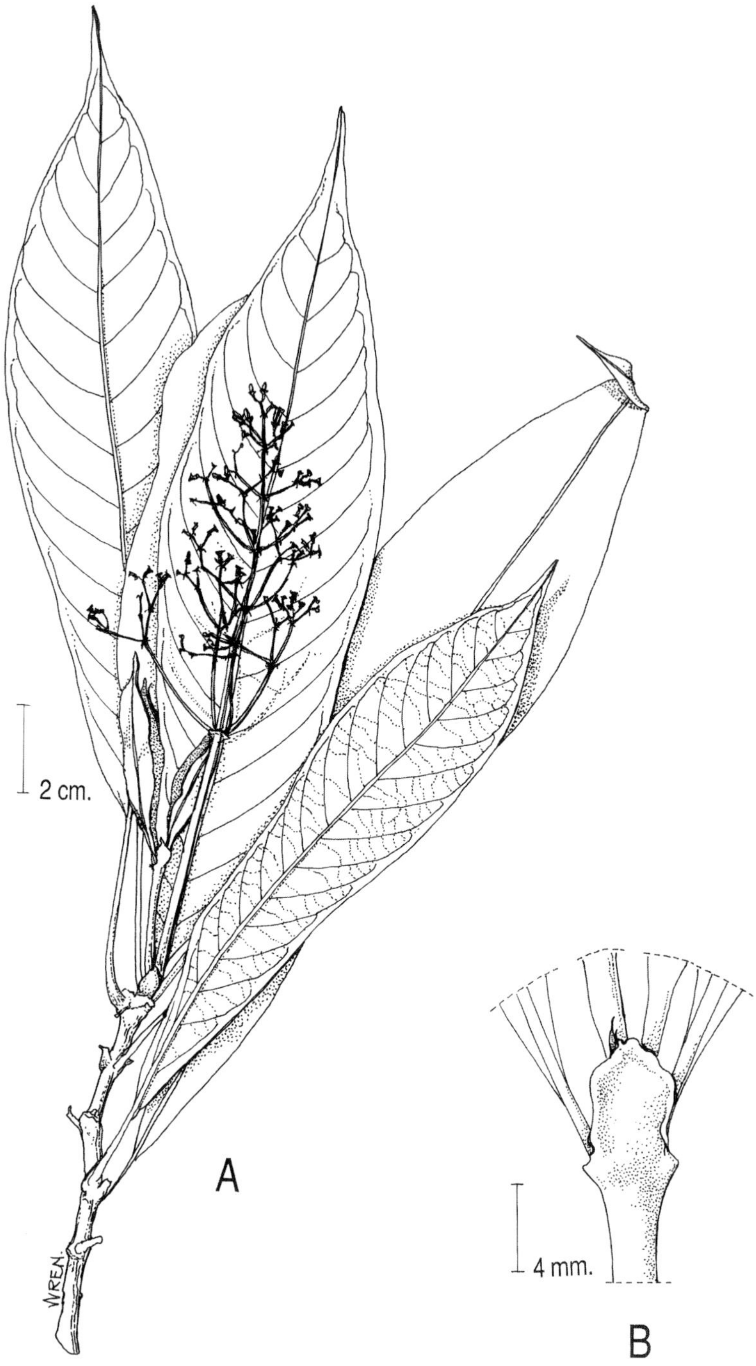

Fig. 67. ***Psychotria rubiginosa*** **var.** ***jacobsii*** **(*Jacobs 7847*) (L, holotype). A, habit of flowering branch; B, stipule.**

cm long, glabrous. Leaves: petioles 1.1–2.1 cm long, glabrous; leaf blades elliptic-lanceolate, 12–19 × 3.4–5 cm, glabrous below; secondary veins 12–16 per side; base narrowly acute to attenuate; apex long-acuminate. Inflorescence glabrous; main axis (peduncle) 4.7–9 cm long, with 3–6 nodes. Flowers poorly known (immature), sessile or pedicels 0.1–0.4 mm long.

PARATYPES: **Luzon. Ilocos Norte Prov.:** Mt. Tagbi, Nueva Era, *Vendivil & Reynoso PNH 121680* (K), *PNH 121681* (K). **Negros. Negros Occidental Prov.:** Mt. Dumalabdab, Talissay mun., *Stone et al. PPI 6366* (BRIT, K).

Distribution.—On eastern side of north central Luzon.

Ecology.—Humid, mixed forest; 1000–1100 m.

Phenology records.—Flowering: February, March; fruiting: unknown.

Preliminary conservation assessment.—Vulnerable (**VU B1ab(iii)**). Last collected in 1992.

Distinguishing characteristics.—Leaf blades elliptic-lanceolate, 12–19 cm long, apex long-acuminate; stipules prominent and persistent on lower nodes; inflorescences monochotomous.

Remarks.—The persistent and conspicuous stipules on the lower nodes immediately sets this taxon apart from others; the generally glabrous condition at first belies the relationship with var. *rubiginosa*. However, upon closer study it becomes apparent that there is more than a casual or coincidental relationship between the two taxa, especially as intermediate specimens have been found. There are some specimens of var. *rubiginosa* that are less pubescent than most of the other specimens cited for that variety, e.g. *Vendivil & Reynoso PNH 121680* (K), and *Vendivil & Reynoso PNH 121681* (K), *Stone et al. PPI 6366* (K). This taxon is named after Dr. Marius Jacobs, who was an indefatigable worker with the Rijksherbarium in Leiden for world conservation until his untimely death.

57. Psychotria phanerophlebia Merr., Philipp. J. Sci. 5:243. 1910; Merr., Enum. Philipp. Fl. Pl. 3:561. 1923. (**Fig. 68**). *Grumilea phanerophlebia* (Merr.) Merr., Philipp. J. Sci. 10:131. 1915. TYPE: PHILIPPINES. LUZON: Cagayan Prov.: near Pamplona, Mar 1909, *Ramos BS 7404* (LECTOTYPE: US, designated here; ISOLECTOTYPE: K).

Shrub or small tree, height unknown, young stems densely pubescent. Stipules unknown. Leaves: petioles 3–3.5 cm long, pubescent; leaf blades oblanceolate to oblong-oblanceolate, 21.3–27.2 × 6.9–8.2 cm, coriaceous, densely pubescent below; secondary veins 18–20 pairs; tertiary venation manifest to rather prominent; base acute; apex broadly acute to obtuse. Inflorescence/infructescence monochotomous, 2.8–4.1 cm long, pubescent to densely pubescent; principal axis (peduncle) 1.6–3 cm long, with 2–4 primary nodes and opposite or verticillate branching at each node, secondary branches terminating in rather dense cymules; bracts ± aristate, 2.5–3 mm long; bracteoles similar to the bracts, ca. 0.7 mm long. Flowers 5-merous, pedicels 1–3 mm long; calyx (incl. hypanthium) 3–5.2 mm long, external surface puberulous (at base) to glabrous; calyx lobes triangular to narrowly triangular, 1.5–2 mm long; corolla and other flower parts uknown. Fruit unknown; pyrenes ± unknown; endosperm reported to be ruminate.

Other specimens examined: Known only from the type collection.

Distribution.—Pamplona, Luzon.

Ecology.—Unknown.

Phenology records.—Flowering: unknown, fruiting: March.

Preliminary conservation assessment.—Extinct (**EX**). Only collected once in 1909.

Distinguishing characteristics.—Leaves densely pubescent, lateral veins numerous and prominent; inflorescences apparently monochotomous.

Remarks.—This unique taxon is known only from the type collection.

58. Psychotria boholensis Merr. ex Sohmer & A.P. Davis, sp. nov. (**Fig. 69**). TYPE: PHILIPPINES. BOHOL: Bohol Prov.: Kalingohan (?), rocky forest, 29 Jan 1923, *Ramos BS 43201* (HOLOTYPE: PNH; ISOTYPES: BM, UC, W).

Frutices (1–)2–6 m alti glabri sunt, stipulis in gemma contortis imbricatis glabris 1.7–2.2 cm longis acuminatis multiplo longioribus quam latis, laminis coriaceis glabris oblongo-ellipticis vel late ellipticis (7.5 × 17.5–) 9 × 21.5–13 × 25.5 cm latis et longis nervis lateralilus 14–18 in dimidio nervis tertiis prominentibus apice rotundato cum puncto brevi obtuso vel acuto basi acuta obtusa cuneatave, petiolis 5.1–6 cm long, infructescentia monochotoma glabra, pedunculo (1.9–)4–4.8 cm longo nodis 2–3 cum ramis oppositis vel verticillatis, fructibus ut dicetur albis glabris obovoideis (6–)9–11 mm longis, pyrenis sublaevibus vel in dorso cum costis minimis, endospermio valde ruminato.

Shrub or small tree, (1–)2–6 m high, glabrous throughout. Stipules imbricate (with one somewhat overlapling the other in bud), twisted in bud, triangular to narrowly triangular, or ± lanceolate, 1.7–2.2 cm long, glabrous; apex acute. Leaves: petioles 5.1–6 cm long, glabrous; leaf blades oblong-elliptic to broadly elliptic, (17.5–)21.5–25.5 × (7.5–)9–13 cm, coriaceous, glabrous; secondary veins 14–18 pairs; tertiary venation manifest to rather prominent; base acute, obtuse, or cuneate; apex round to very broadly acute, with a short obtuse or acute tip. Inflorescence/infructescence monochotomous, 4.5–8 cm long, glabrous; principal axis (peduncle) (1.9–)4–4.8 cm long, with 2–3 primary nodes and opposite

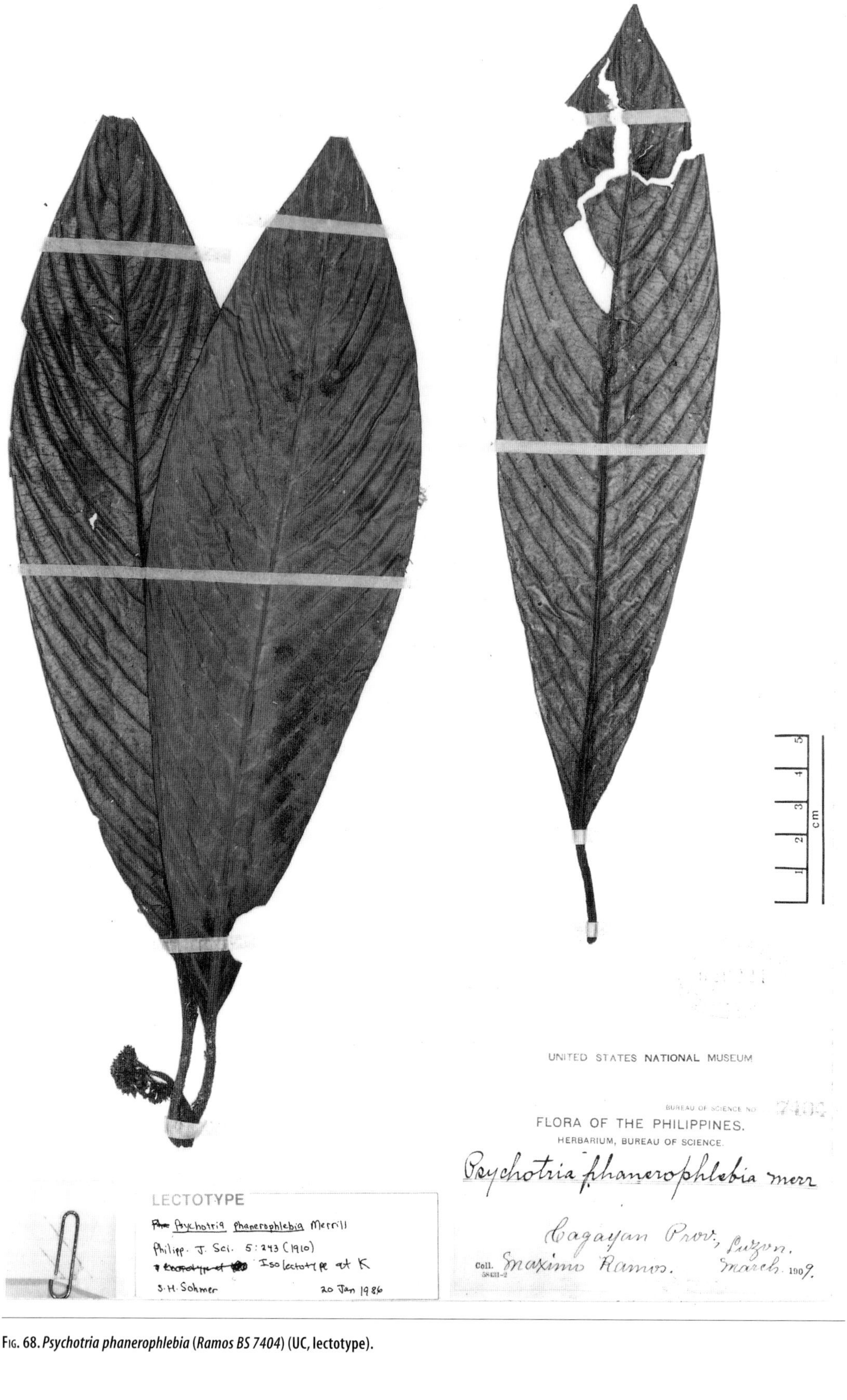

Fig. 68. *Psychotria phanerophlebia* (*Ramos BS 7404*) (UC, lectotype).

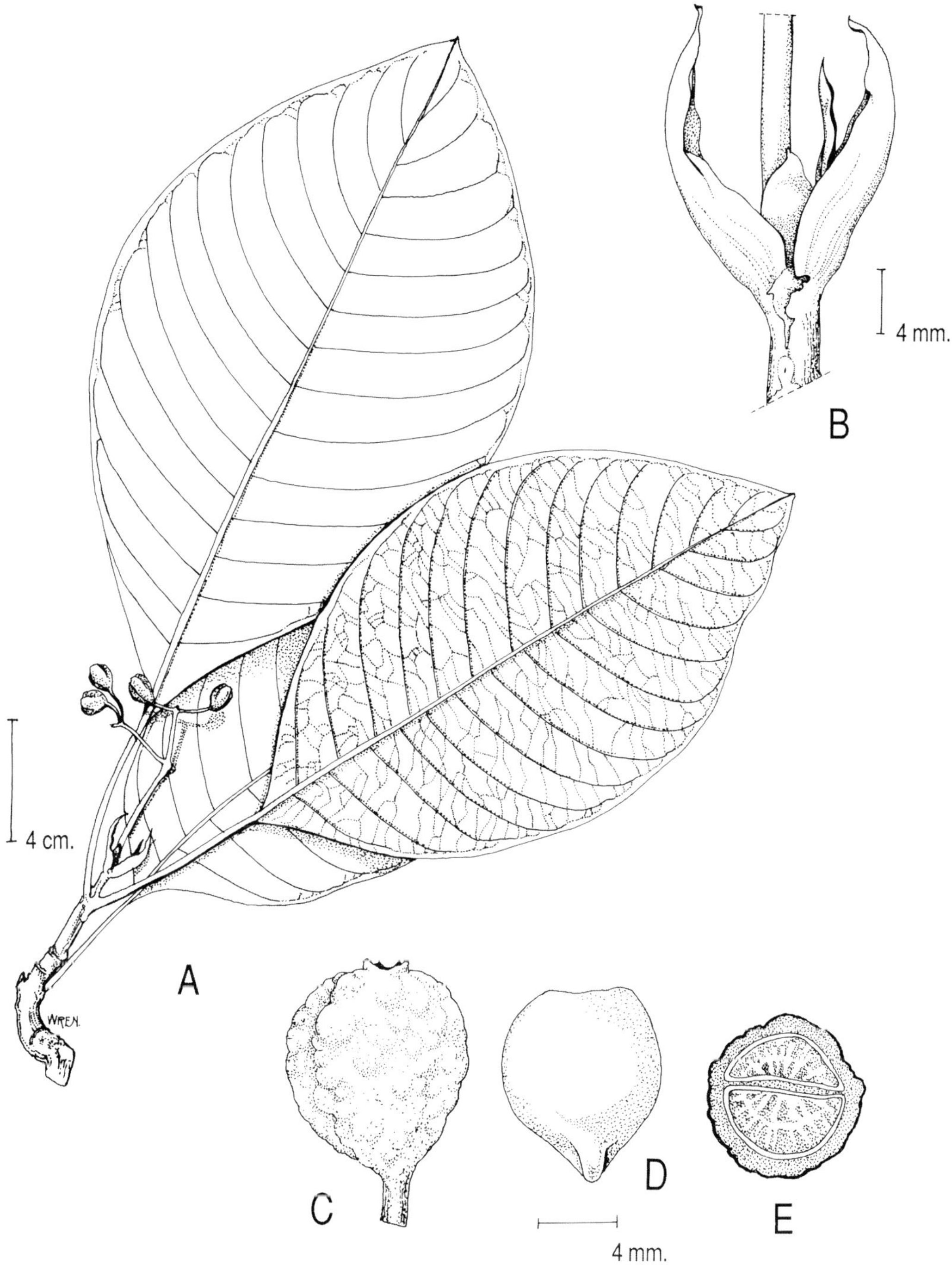

FIG. 69. *Psychotria boholensis* (*Ramos BS 43201*) (PNH). **A,** habit of fruiting branch; **B,** stipule; **C,** fruit; **D,** pyrene; **E,** fruit in cross section.

or verticillate branching at each node, sconday branches terminating in lax cymules; bracts ovate to triangular, ca. 1.5 mm long; bracteoles unknown (?caducous). Flowers unknown. Fruit obovoid, (6–) 9–11 mm long, ± smooth or ridged (due to edge of pyrene, not a feature of the dorsal surface of the pyrene), orange or red, glabrous; calyx limb and lobes persistent but inconspicuous; pedicels ca. 2 mm long; pyrenes ± smooth on outer surface; endosperm strongly ruminate.

PARATYPES: **Bohol. Bohol Prov.:** Bilar, *Ramos BS 43145* (K, PNH, UC); Magsaysay Camp, Logarita, Bilar, *Reynoso, Alvarez & Fuentes PPI 7825* (BRIT).

Distribution.—Bohol.

Ecology.—Relatively dry or well-drained forest, including secondary forest; recorded on limestone (*Reynoso, Alvarez & Fuentes PPI 7825*); 600–700 m.

Phenology records.—Flowering: unknown; fruiting: January, March, August, October.

Preliminary conservation assessment.—Endangered (**EN B1ab(iii)**). Last collected in 1993.

Distinguishing characteristics.—Leaf blades broadly elliptic, relatively large; fruits relatively large (usually 9–11 mm long), reported to be white but a recent collection (*Reynoso, Alvarez & Fuentes PPI 7825* records fruit color as yellow-green. At maturity the fruit would probably be orange or red.

Remarks.—This taxon was recognized in the herbarium by Merrill. The label on the PNH specimen appears to have been annotated by Merrill as a new species. We concur that this is an undescribed taxon and have designated the PNH specimen as holotype.

59. Psychotria iridensis Sohmer & A.P. Davis, sp. nov. (**Fig. 70**). TYPE: PHILIPPINES. LUZON: Rizal Prov.: Mt. Irid, on damp forest slopes, 13 Nov 1926, *Ramos & Edaño BS 48443* (HOLOTYPE: A; ISOTYPES: BM, K, NY, UC).

Frutices 2–4 m alti glabri sunt, stipulis imbricatis 3–4.5 cm longis multiplo longioribus quam latis, laminus subcoriaceis vel coriaceis late ellipticis vel obovato-ellipticis (8 × 15–)10.5 × 21.5–10.2 × 26 cm latis et longis nervis lateralibus 14–20 in dimidio quoque apice obtuso vel acuto interdum abrupto acuto basi obtusa, petiolis 3–5 cm longis, inflorescentia monochotoma vel trichotoma, pedunculo 4–9 cm longo, floribus 5-meris dimorphicis, pedicellis brevissimis, hypanthio et tubo calycis 1.4–2 mm longis, lobis 5 indistinctis, corollis albis intra in fauce villosis tubo 2 mm longo lobis 1.5 mm longis, fructibus ut dicetur luteis sed versimiliter aurantiacis vel rubris obovoideis 6–8 mm longis, pyrenis in dorso laevibus vel paulum costatis, endospermio ruminato.

Shrub or small tree, 2–4 m high, ± glabrous throughout. Stipules imbricate, very narrowly triangular, 3–4.5 cm long, glabrous; apex acute. Leaves: petioles 3–5 cm long, glabrous; leaf blades broadly elliptic or obovate-elliptic, (15–)21.5–26 × (8–)10.2–10.5 cm, subcoriaceous to coriaceous, glabrous; secondary veins 14–20 pairs; tertiary venation manifest to prominent; base cuneate to obtuse; apex obtuse to acute, sometimes abruptly acute. Inflorescence/infructescence monochotomous or sometimes trichotomous, 9.5–16.5 cm long, glabrous to minutely puberulous; principal axis (usually the peduncle) 4–9 cm long, with 2–4 primary nodes and verticillate branching at each node, secondary branches terminating in cymules; bracts and bracteoles unknown (?caducous). Flowers 5-merous, pedicels 0.5–1.2 mm long; calyx (incl. hypanthium) 1.4–2 mm long, external surface glabrous; calyx lobes shallowly triangular to tooth-like, ca. 0.3–0.4 mm long; corolla: white; corolla tube ca. 2 mm long, glabrous, villous at throat within; corolla lobes ca. 1.5 mm long; stamens: filaments ca. 2 mm long; anthers ca. 1 mm long; style ca. 2 mm long, lobes ca. 0.5 mm long (short styled). Fruit obovoid, 6–8 mm long, very weakly ribbed to ± smooth, reported to be yellow but probably orange or red at maturity, glabrous; calyx limb and lobes not persistent; pedicels 2–4 mm long; pyrenes very weakly ribbed to ± smooth on outer surface; endosperm ruminate.

PARATYPES: **Luzon. Ifugao Prov.:** Brgy Batad, Mt. Bobor, *Barbon & Pedersen PPI 12940* (BRIT). **Rizal Prov.:** Mt. Angilo, *Ramos BS 40745* (A, K); Mt. Irid, *Ramos BS 41879* (A, K, L, P, US, W). **PHILIPPINES. Without further data:** *Loher 6402* (K).

Distribution.—Luzon in the provinces of Ifugao, Rizal and Quezon.

Ecology.—Forest, including disturbed and possibly secondary forest; 1000–1700 m.

Phenology records.—Flowering April; fruiting: February, April, November.

Preliminary conservation assessment.—Vulnerable (**VU B1ab(iii)**). Last collected in 1994.

Distinguishing characterstics.—Leaf blades broadly elliptic, relatively large, usually over 20 cm long, lateral veins 14–20 per side; stipules large and prominent, more than 3 cm long.

Remarks.—This taxon is notable in this group by its large, glabrous leaf blades. It bears some resemblance to *P. rubiginosa*.

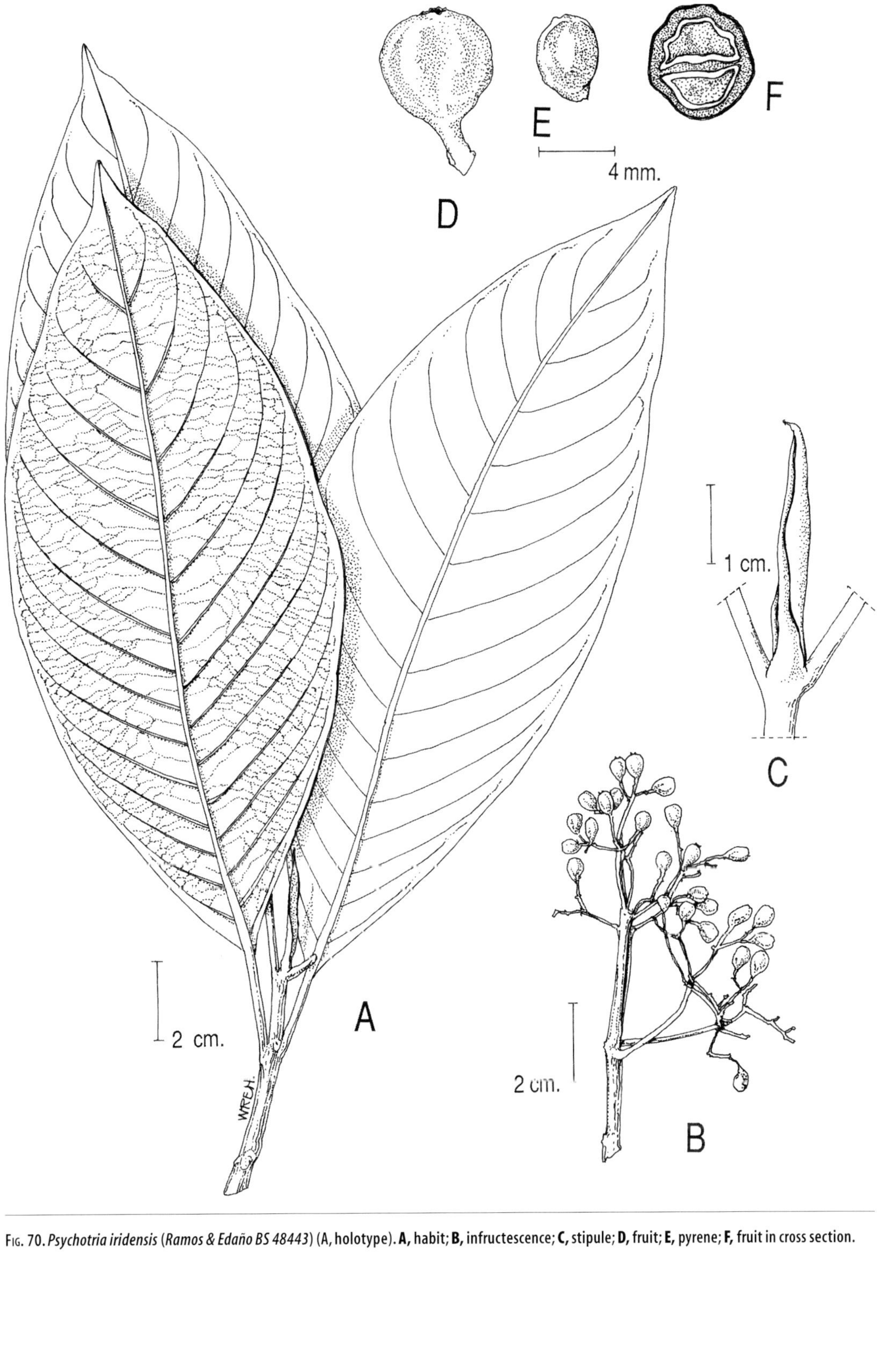

Fig. 70. *Psychotria iridensis* (*Ramos & Edaño BS 48443*) (A, holotype). **A,** habit; **B,** infructescence; **C,** stipule; **D,** fruit; **E,** pyrene; **F,** fruit in cross section.

60. Psychotria palawanensis Elmer, Leafl. Philipp. Bot. 4:1348. 1912; Merr., Enum. Philipp. Fl. Pl. 3:560. 1923. (**Fig. 71**). *Grumilea palawanensis* (Elmer) Merr., Philipp. J. Sci. 10:134. 1915. TYPE: PHILIPPINES. PALAWAN: Palawan Prov.: Mt. Pulgar, in a slightly wooded flat at 76 m, Mar 1911, *Elmer 12737* (LECTOTYPE: A, designated here; ISOLECTOTYPES: BISH, BM, GH, K, L, MO, US, W).

Shrub, probably 1–2 m high, ± glabrous throughout. Stipules ?imbricate, ± triangular to ± lanceolate, 0.8–1.5 cm long, glabrous but margin ciliate; apex cleft. Leaves: petioles 0.8–1.5 cm long, glabrous; leaf blades subobovate to subelliptic, 11–17 × 4.5–6.6 cm, subcoriaceous, usually glabrous, infrequently pubescent on veins below, secondary veins 11–14 pairs; tertiary venation manifest to obscure; base narrowly acute or obtuse; apex abruptly acuminate or shortly acute. Inflorescence/infructescence monochotomous, 2.5–9.5 cm long, glabrous; principal axis (peduncle) 1.5–3.5 cm long, with 1–3 primary nodes and usually with verticillate branching at each node, secondary branches terminating in cymules; bracts deltate to triangular, ca. 0.5 mm long; bracteoles ± broadly ovate, ca. 0.2 mm long. Flowers unknown. Fruit globose-turbinate or subobovoid, 5.5–7 mm long, very weakly ribbed to ± smooth, reported to be black at maturity, glabrous; calyx limb and lobes persistent but inconspicuous; pedicels 1–2 mm long, or pedicels absent; pyrenes slightly ribbed on outer surface; endosperm ruminate.

Other specimens examined: **Palawan. Palawan Prov.:** Tagimbung, *Olsen 220* (L).

Distribution.—Palawan.

Ecology.—Forest, incl. those on well drained soils; low elevation (e.g. 76 m).

Phenology records.—Flowering: March; fruiting: March.

Preliminary conservation assessment.—Critically Endangered (**CR B1ab(iii)**). Last collected in 1961.

Distinguishing characteristics.—Stipules cleft; infructescences/inflorescences monochotomous, with a short peduncle and verticillate branching, branches short, bracts conspicuous.

Remarks.—This taxon is really no better known today than when Elmer (1912) described it. It bears some resemblance, at least in the structure of its infloresence, to *P. pyramidata* and *P. wenzelii*.

61. Psychotria quisumbingiana Sohmer & A.P. Davis, sp. nov. (**Fig. 72**). TYPE: PHILIPPINES. BOHOL: Bohol Prov.: Serilla, along stream in the forest, 15 Sep 1923, *Ramos BS 43188* (HOLOTYPE: UC; ISOTYPES: BM, K).

Arbores vel frutices 3–8 m altae glabrae sunt, stipulis imbricatis in gemma contortis 9 mm longis multo longioribus quam latis acuminatis, laminis coriaceis ellipticis vel elliptico-obovatis 5.6 × 12.5–8 × 18.7 cm latis et longis nervis lateralis 11–15 in dimidio quoque nervis tertiis non prominentibus apice acuto vel obtuso vel abrupte acuminato basi acuta vel cuneata, petiolis 2–4.4 cm longis, infuctescentia monochotoma 11.5–18 cm longa, pedunculo (3.3–)4–6.8 cm longo, nodis 2–4, ramulis verticillatis, fructibus ut dicetur luteis sed versimiliter in maturite rubris obovoideis vel duplococcaceis 6 mm longis, pyrenis in dorso laevibus vel minute costatis, endospermio ruminato.

Small tree or shrub, 3–8 m high, glabrous throughout. Stipules imbricate, twisted in bud, ± narrowly triangular, ca. 0.9 cm long, glabrous; apex acute. Leaves: petioles 2–4.4 cm long, glabrous; leaf blades elliptic to elliptic-obovate, 12.5–18.7 × 5.6–8 cm, subcoriaceous, glabrous; secondary veins 11–15 pairs; tertiary venation obscure; base acute or cuneate; apex acute or obtuse or abruptly acuminate. Inflorescence/infructescence monochotomous, 11.5–18 cm long, glabrous; principal axis (peduncle) (3.3–)4–6.8 cm long, with 2–4 primary nodes and verticillate branching at each node, secondary branches terminating in cymules; bracts deltate to triangular, ca. 1 mm long; bracteoles unknown (?caducous). Flowers unknown. Fruit obovoid and often ± diplococcus, 5.8–7 mm long, weakly ribbed, reported to be yellow but probably orange or red at maturity, glabrous; calyx limb and lobes not persistent; pedicels 1–10 mm long; pyrenes smooth or minutely ribbed on outer surface; endosperm ruminate.

PARATYPE: **Bohol. Bohol Prov.:** Valencia, *Ramos BS 43140* (BM, K, UC).

Distribution.—Bohol Island.

Ecology.—Humid forest; recorded near stream; 300–700 m.

Phenology records.—Flowering: unknown; fruiting: August to October.

Preliminary conservation assessment.—Critically Endangered (**CR B1ab(iii)**) or possibly Extinct (**EX**). Last collected in 1923.

Distinguishing characteristics.—Glabrous throughout; stipules long acuminate; inflorescences monochotomous, peduncles 4–6.8 cm long; fruit yellow (probably ripening orange or red), ca. 9 mm long.

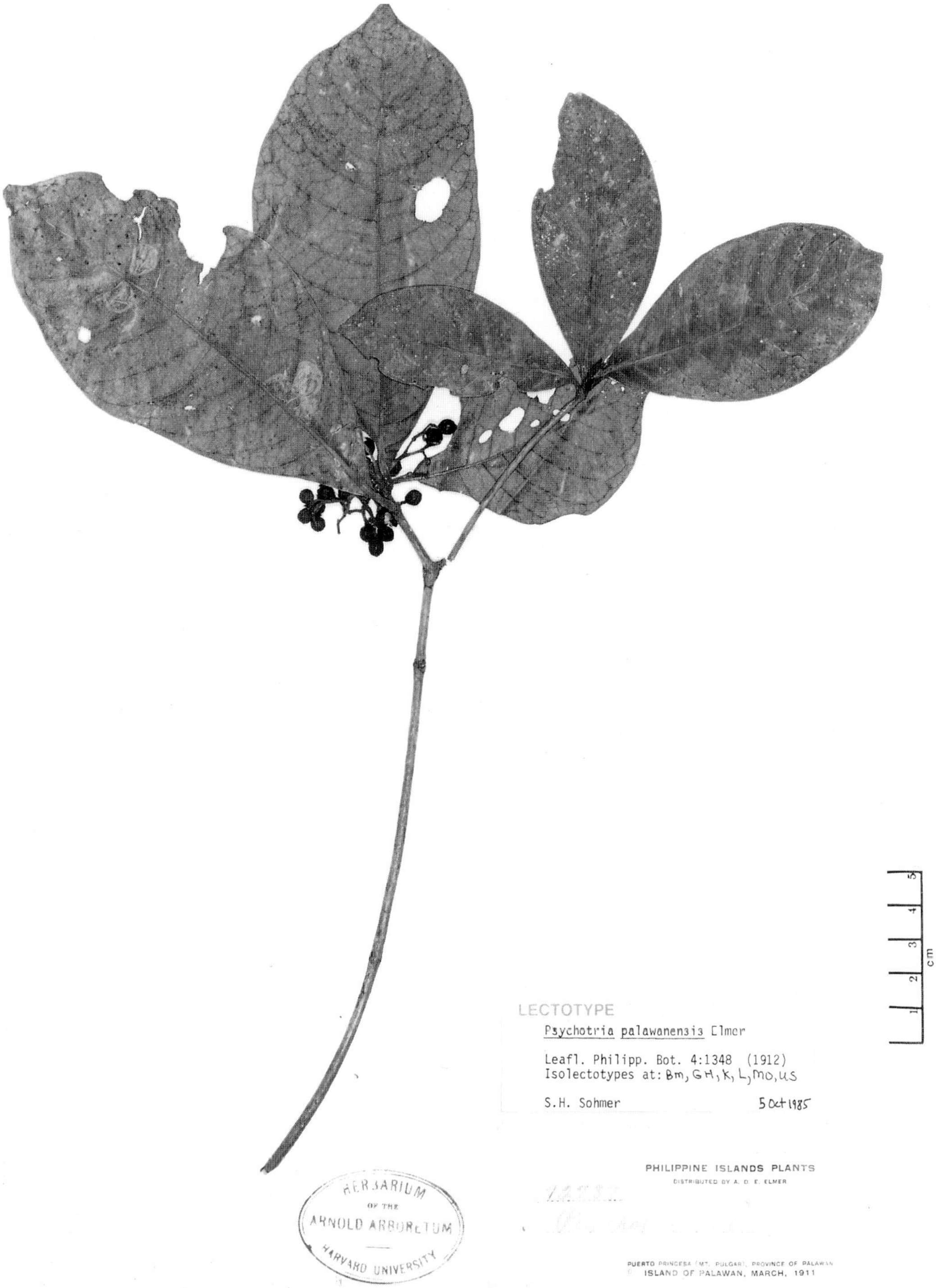

Fig. 71. *Psychotria palawanensis* (*Elmer 12737*) (A, lectotype).

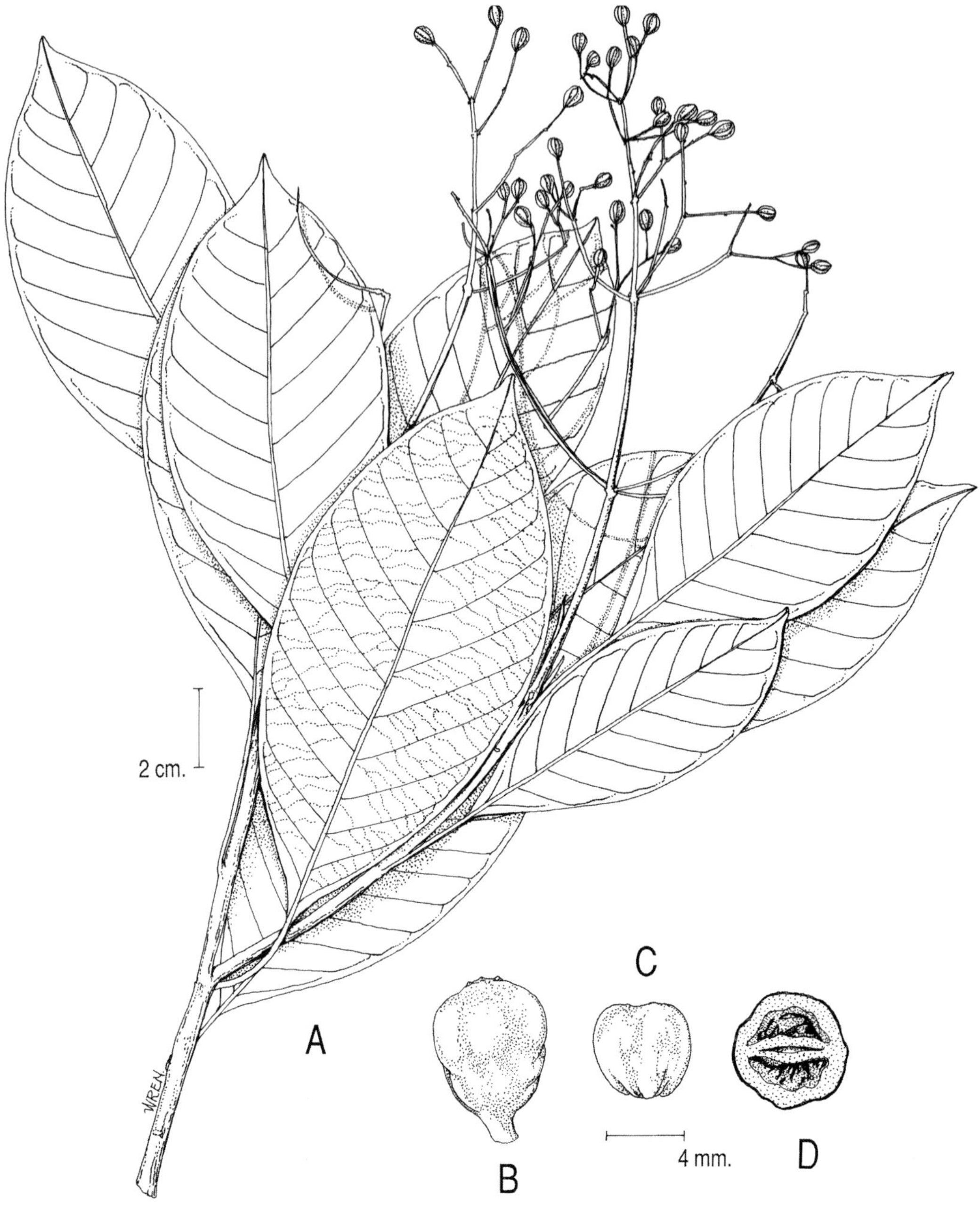

FIG. 72. *Psychotria quisumbingiana* (*Ramos BS 43188*) (UC, holotype). **A,** habit of fruiting branch; **B,** fruit; **C,** pyrene; **D,** fruit in cross section.

Remarks.—The long-acuminate stipules, glabrous condition, and peduncles 4–6.8 cm long distinguish this taxon from other taxa in the Gitingensis Species Group.

62. Psychotria manillensis Bartl. ex DC., Prodr. 4:522. 1830; Merr., Enum. Philipp. Fl. Pl. 3:559. 1923. (**Figs. 73a, 73b**). TYPE: PHILIPPINES. LUZON: Sorsogon, without precise locality, *Haenke s.n.* (HOLOTYPE: PR).

Psychotria propinqua (Merr.) Merr., Enum. Philipp. Fl. Pl. 3:561. 1923. *Grumilea propinqua* Merr., Philipp. J. Sci. 12:170. 1917. TYPE: PHILIPPINES. Luzon: Quezon Prov.: Mt. Dingalan, 27 Jan 1916, *Ramos & Edaño BS 26570* (LECTOTYPE: K, designated here; ISOLECTOTYPE: NY).

Psychotria liukiuensis Hatus., J. Jap. Bot. 13:674. 1937. TYPE: JAPAN. Liukiu: Syuri, Okinaurajima, 1 Jun 1924, *Tashiro s.n.* (not seen).

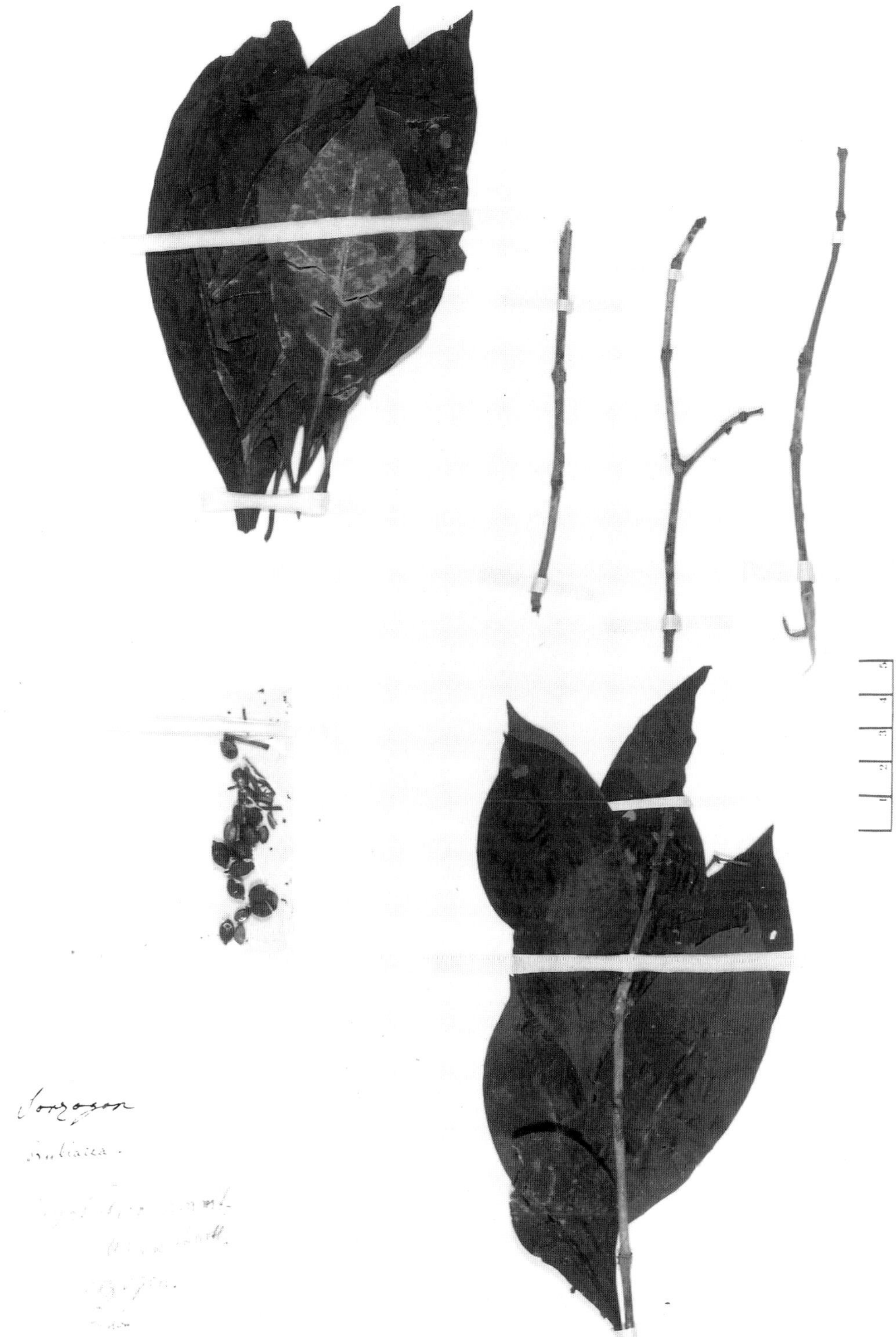

Fig. 73a. *Psychotria manillensis* (*Haenke s.n.*) (PR, holotype).

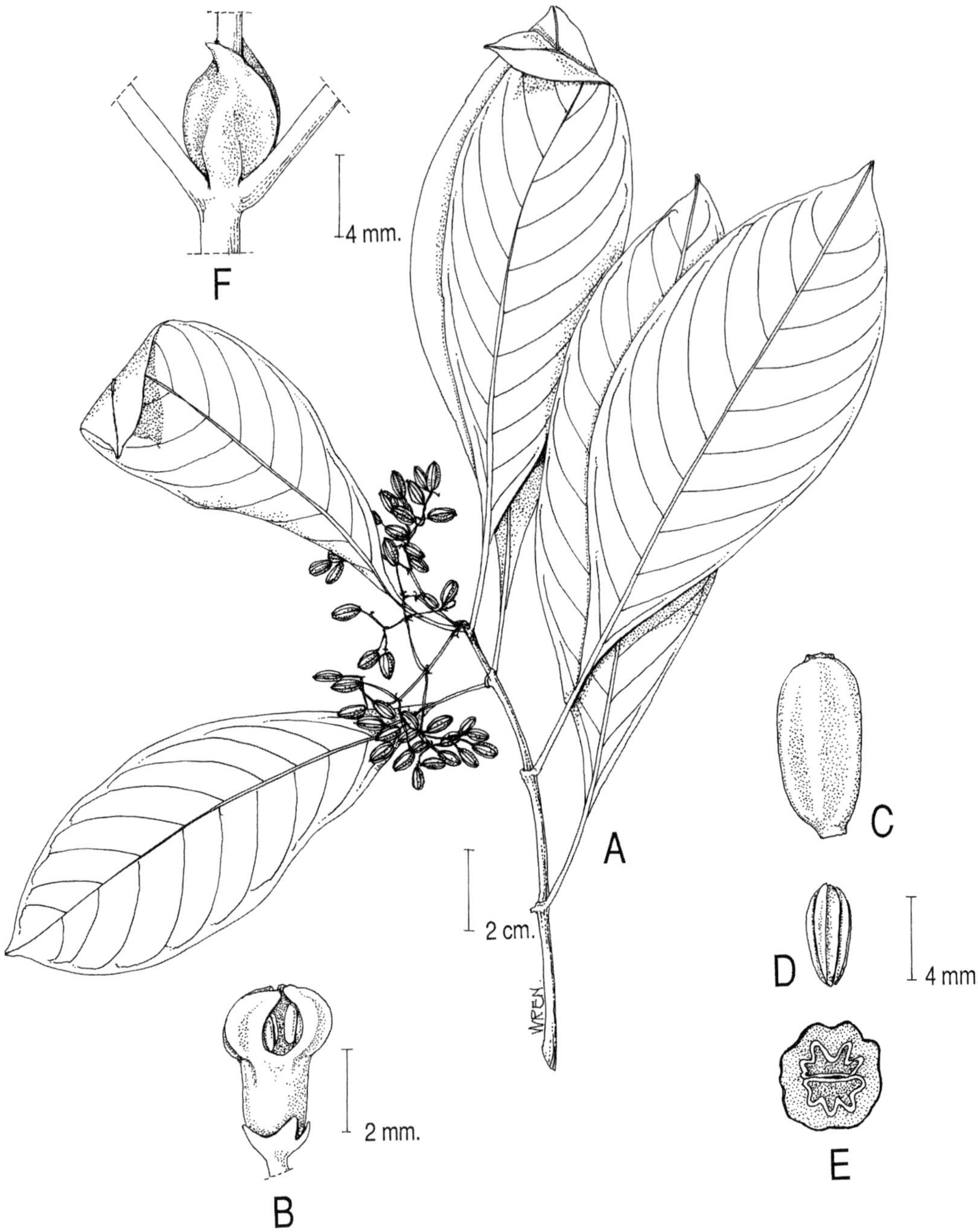

Fig. 73b. ***Psychotria manillensis*** **(*Ridsdale SMHI 1648*) (L). A, habit of fruiting branch; B, flower; C, fruit; D, pyrene; E, fruit in cross section; F, stipule.**

Shrub or small tree, 1–2(–6) m high, young stems glabrous. Stipules valvate, ovate to triangular, or sometimes ± lanceolate, 0.5–1 cm long, glabrous or infrequently pubescent, margin sometimes ciliate; apex acute or acuminate. Leaves: petioles (0.5–)1.1–2.5(–3.4) cm long, glabrous, or rarely pubescent; leaf blades elliptic to elliptic-obovate, 10–18(–20.5) × 3.7–7.3 cm, subcoriaceous, usually glabrous above and below, infrequently puberulent to somewhat scabrous below; secondary veins (6–)8–12 pairs; tertiary venation manifest to obscure; base narrowly acute to cuneate; apex abruptly acute to acuminate. Inflorescence/infructescence monochotomous but sometimes appearing trichotomous due to a very short principal axis, (2.5–)6–9.5 cm long, glabrous or infrequently puberulous;

principal axis (peduncle) (0.1–)1–5.1 cm long, with 1–2 primary nodes and opposite branching at each node, the main axis and secondary branches in small clusters of (1–5) flowers/fruits; bracts triangular to narrowly triangular, 1.7–3(–5.8) mm long, margins serrate-laciniate to ± laciniate, pubescent (?hairs or colleters); bracteoles triangular, 0.7–2.5 mm long, margins serrate-laciniate to ± laciniate, pubescent (?hairs or colleters), bracteoles also subtending the flower and usually the fruit, paired and often united at the base. Flowers 4–6-merous, pedicels 0.1–2 mm long or absent; calyx (incl. hypanthium) 2.9–3.7 mm long, external surface glabrous; calyx lobes deltate, 0.3–0.6 mm long, margins ciliate; corolla: white; corolla tube 4 mm long, glabrous, villous at throat within; corolla lobes ca. 2 mm long; stamens: filaments ca. 1 mm long; anthers ca. 1 mm long (short styled); style ca. 2.5 mm long, lobes ca. 0.5 mm long (short styled). Fruit oblong-ellipsoid, 8–15 mm long, ribbed to conspicuously ribbed, red, glabrous; calyx ± persistent or obscure; pedicels 0.1–3(–4) mm long, or pedicels absent; pyrenes distinctly ribbed on outer surface; endosperm entire or rarely very weakly ruminate.

Other specimens examined: **Itbayat. Batanes Prov.:** Mt. Iraya, *del Rosario PNH 79916* (PNH), *Ramos BS 79917* (A), *BS 80090* (A, SING), *Hatusima & Sato 28730* (PNH), *Ramos BS 80669* (A); Mt. Iraya, Deohango, *Madulid et al. PPI 1509* (BRIT, K); *Madulid et al. PPI 1528* (BRIT, K). **Batan. Batanes Prov.:** Mt Matarem, Barangay San Vicente, *Madulid, Gaerlan et al. PPI 23761* (BRIT, K); along road from Itbud to Mahatao (south to north), *Madulid & Gaerlan et al. PPI 23843* (BRIT, K); from Puereto to CENRO, *Madulid & Gaerlan et al. PPI 23878* (BRIT, K); Mt. Mariveles, *Leiberg 6008* (K). **Dalupiri. Without further data:** *Bartlett 15104* (A, NY, PNH, UC). **Camiguin. Batanes Prov.:** Mt. Malabsing, *Edaño BS 79217* (A); Camiguin Volcano, *Edaño BS 79151* (A); Mt. Mapolapola, *Edaño BS 79354* (A, SING). **Batanes Prov.:** without further data: *Fenix BS 3648* (NY); *Quisumbig et al. PNH 79408* (PNH). **Luzon. Cagayan Prov.:** Barangay San Juan, Santa Praxedes, *Garcia, Fuentes & Romero PPI 18431* (BRIT, K); So. Salpad, Barangay Balatubat, Calayan, *Fuentes & Fernando PPI 37183* (BRIT); Claveria, Kilkiling, Mt Taggat, *Reynoso, Fuentes & Majaducon PPI 11932* (K); Abulug Riv, *Weber 1561* (K, P). **Cagayan Prov.:** without further data: *Ramos BS 13957* (P). **Kalinga-Apayao Prov.:** Mt. Magnas, *Edaño PNH 19786* (L, PNH). **Ilocos Norte Prov.:** Bangui, *McGregor BS 43541* (A, UC, W); Mt. Pico de Loro, *Edaño PNH 18022* (K, L); Mt. Bubungkayo, from Santo Nino to Miner's Camp, *Iwatsuki et al. P-351* (L); Mt. Darna, *Edaño PNH 18035* (L, PNH), *PNH 18127* (L); Sitio Nalamus, Barangay Dampig, Pagudpod, *Garcia & Fernando PPI 25226* (BRIT); Mangoratao, *Gaerlan et al. PPI 10394* (BRIT, K); Dumalneg, Cabel, Simuba Mts., *Gaerlan et al. PPI 9899* (BRIT, K). **Ilocos Sur Prov.:** Burgos, *Ramos BS 32724* (L, NY). **Nueva Ecija Prov.:** Mt. Umingan, *Ramos & Edaño BS 26343* (K, US). **Bulucan Prov.:** Angat Watershed Area, Barangay San Lorenzo, Norzagaray, *Garcia et al. PPI 15167* (BRIT). **Quezon Prov.:** Karlagan, *Fox PNH 9101* (L, PNH); Alabat Island, *Ramos & Edaño BS 48313* (UC); So. Juan, Mt. Banashaw, Barangay Kinabuhayan, Dolores, *Reynoso & Majaducon PPI 21737* (BRIT, K); *Reynoso & Majaducon PPI 21740* (BRIT); Mt. Banashaw, Barangay Kinabuhayan, Dolores, *Reynoso & Majaducon PPI 21642* (BRIT, K); Mt. Cristobal, Retritan, Barangay Santa Lucia, *Reynoso & Majaducon PPI 27210* (BRIT). **Rizal Prov.:** Mt. Irid, *Ramos & Edaño BS 48404* (NY, UC), *BS 48505* (UC), *BS 48561* (UC); Montalban, *Loher s.n.* (UC); Mt. Mapatog, San Andales, *Edaño BS 48738* (UC); Mt. Tokduanbanoy, *Ramos & Edaño BS 48629* (NY, UC, W). **Bataan Prov.:** Mt. Mariveles, Lamao River, *Borden FB 1771* (BM, K, NY), *Collector unknown FB 1480* (NY, US), *Elmer 6758* (K, NY), *Williams 560* (NY); Mt. Mariveles, lower eastern slope, *Bartlett 14637* (A). **Cavite Prov.:** Mt. Palay Palay/Mataas na Gulod National Park, Ternate, *Medecilo, Romero & Agoo PPI 29314* (BRIT); Tala, Ternate, *Reynoso, Sagcal & Fuentes PPI 17116* (BRIT). **Laguna Prov.:** Mt. Maquiling [Mt. Makalin], *Forestry School FB 20137* (L, P, US), *Foxworthy 16* (US), *32* (A), *FB 25706* (P), *Holman 87* (A), *Robinson BS 9726* (L), *BS 17185* (BM, L, US), *Sulit PNH 8495* (A, L), *Villamil FB 21393* (US), *Gaerlan et al. PPI 10406* (BRIT, K); on ridge above Maralas Creek, *Sulit PNH 9718* (A, L, PNH, SING). **Laguna Prov.:** without further data: *Valderama 10* (US). **Camarines Sur Prov.:** Sitio Kalagikhik, Barangay Guijalo, *Reynoso, Sagcal & Fuentes PPI 14799* (BRIT, K); Barangay Guijalo, *Reynoso, Sagcal & Fuentes PPI 14680* (BRIT). **Catanduanes. Catanduanes Prov.:** Kasalugan Riv, *Ramos & Edaño BS 75286* (NY). **Catanduanes Prov.:** without further data: *Ramos BS 30347* (NY, P). **Mindoro. Mindoro Oriental Prov.:** Puerto Galera, *Ramos BS 46354* (L, NY, UC); Pinamalayan, *Ramos BS 43134* (A, L, P, US); Mt Naapug, Mansalay, *Conklin PNH 17507* (A, L, PNH); Mt. Yagaw (eastern slope), Mansalay, *Sulit & Conklin PNH 16765* (A, PNH), *PNH 16963* (A, L, PNH), *PNH 17608* (A, L), *PNH 17680* (A, K, L), *PNH 17720* (A, L, PNH); Mt. Yagaw, south-eastern slope, *Conklin PNH 18670* (A, L, PNH). **Masbate. Masbate Prov.:** 7-R Ranch, Sitio Kalunakan, Barangay Matipuron, Milagros, *Barbon, Garcia & Fernando PPI 12724* (BRIT). **Panay. Iloilo Prov.:** Miagoo, *Vidal 3053* (K). **Mindanao: S Cotabato Prov.:** Cotabato, *Edaño PNH 1479* (A,L) *vel aff.* **Palawan. Palawan Prov.:** Mt. Balaghag, *Edaño BS 77773* (A); Quezon, *Reynoso PNH 87848* (L, PNH); Mt. Cleopatra Ranges, So. Calabayog, Barangay Tanabag, Puerto-Princesa, *Reynoso & Majaducon PPI 24379* (BRIT); Balabac Island, without further data, *Ramos & Edaño BS 49760* (UC) ; St. Paul's Bay, Cabayugan Village, regrowth around meadows at 50 m elevation, 7 May 1984, *Ridsdale SMHI 1648* (K, L). **Philippines. Without further data:** *Loher 6400* (K).

Distribution.—Widely distributed throughout the Philippines and also found in Japan (Nansei-shoto [Ryuku Islands]) and Taiwan.

Ecology.—Forest: primary forest (including remnant dipterocarp forest), disturbed forest, regrowth and regenerating forest, at the margins of forest (e.g. near meadows and fields), and secondary forest; recorded on limestone; 50–900 m.

Phenology records.—Flowering: March to September; fruiting: January August, November.

Preliminary conservation assessment.—Least Concern (**LC**). Last collected in 1996.

Distinguishing characteristics.—Lateral veins of leaf usually forming closed loops at margin; inflorescences monochotomous; bract and bracteole margins serrate-laciniate to ± laciniate; flowers subsessile; fruit usually ellipsoid and strongly ribbed.

Remarks.—This was one of the first taxa described from the Philippines. The type material apparently used by De Candolle (l.c.) is at Prague (PR). Dr. Jiri Sojak, the Curator at PR in 1986, was of the opinion that specimen *Haenke s.n.* was the material seen by De Candolle and this is therefore cited here as the holotype. *Psychotria manillensis* is probably closely related to *P. angulata* Korthals, a species described from Borneo. It is possible that these two species could be conspecific, in which case *P. manillensis* would have priority. *Psychotria manillensis* was collected with some frequency during the Philippine Plant Inventory Project (PPI), which is explained by its ability to grow in forests that are disturbed, regenerating or secondary. Unlike many *Psychotria* species it does not have a strong requirement for primary forest.

63. Psychotria alabatensis Sohmer & A.P. Davis, sp. nov. (**Fig. 74**). TYPE: PHILIPPINES. ALABAT: Quezon Prov.: without further data: Sep–Oct 1926, *Ramos & Edaño BS 48006* (HOLOTYPE: K; ISOTYPES: A, BM, K, NY, UC).

Ut videtur frutices vel arbores parvae sunt, stipulis valvatis plerumque glabris et 0.4–1.1 cm longis plus minusve triplo longis quam latis acuminatis apici plus minusve fissi, laminis chartaceis vel subcoriaceis supra glabris infra glabris vel paulum secus costam et nervas secundarias pubescentibus ellipticis vel elliptio-obovatis (4.3 × 12–) 6.7 × 17.5–9.6 × 24 cm latis et longis nervis lateralibus 11–15 et infra prominentibus apice acuminato vel acuto, petiolis 2–4.5 cm longis, infructescentia monochotoma, (6.2–)9.6–23 cm longis, pedunculo (3.2–)5–14.2 cm longo, fructibus ut dicetur luteis ut videtur aurantiaci-rubris laevibus vel minute costatis glabris obovoidis vel subellipsoideis 8.5–12 mm longis, pyrenis in dorso laevibus, endospermio ruminato. [$modify E, Fig. 89]

Tree, ca. 10 m high; young stems glabrous. Stipules valvate, triangular to ± lanceolate, 0.4–1.1 cm long, glabrous; apex acuminate, with a definite or obscure cleft. Leaves: petioles 2–4.5 cm long, glabrous; leaf blades elliptic to elliptic-obovate, (12–)17.5–24 × (4.3–)6.7–9.6 cm, chartaceous to subcoriaceous, glabrous above, and glabrous or somewhat pubescent along midrib, lateral veins and sometimes surface, below; secondary veins 11–15 pairs; tertiary venation manifest to rather prominent; base narrowly acute to acute; apex obtuse or abruptly acuminate with the tip often curved. Inflorescence/infructescence monochotomous, (6.2–)9.6–23 cm long, glabrous, or sparsely pubescent to pubescent; principal axis (peduncle) (3.2–)5–14.2 cm long, with 2–4 primary nodes and opposite branching at each node, secondary branches terminating in lax cymules; bracts triangular, ca. 3 mm long; bracteoles deltate to triangular, or ovate, ca. 0.9 mm long. Flowers largely unknown (buds only), pedicels 0.1–0.8 mm long or absent; calyx (incl. hypanthium) 2.8–4.7 mm long, glabrous; calyx lobes broadly ovate to deltate, 0.6–1.2 mm long, margins appearing sparsely pubescent (probably due to presence of colleters); corolla: white, otherwise unknown; other flower parts unknown. Fruit obovoid to subellipsoid, 8.5–12 mm long, smooth or weakly striated, yellow to orange-red, glabrous; calyx limb and lobes persistent, obvious, 2.5–3 mm long, often pubescent (colleters); pedicels 0.7–1.2 mm long, or pedicels absent; pyrenes smooth on outer surface; endosperm ruminate.

PARATYPES: **Luzon. Quezon Prov.:** Umiray, *Ramos & Edaño BS 28962* (A, K, P); Alabat Island, without further data, *Ramos & Edaño BS 48361* (BM, NY, UC); Alabat Island, Sitio Canumay, Villa Norte to Barangay Bacung, *Gaerlan & Romero PPI 23618* (K). **Catanduanes. Catanduanes Prov.:** Mt. Nagpakdit, *Ramos & Edaño BS 75166* (NY, SING,UC).

Distribution.—Alabat and Catanduanes Islands, and Quezon Province, Luzon.

Ecology.—Moist forest; low elevation, including sea level.

Phenology records.—Flowering: March; fruiting: May, June, September to October.

Preliminary conservation assessment.—Endangered (**EN B1ab(iii)**). Last collected in 1996 (Alabat Island).

Distinguishing characteristics.—Leaves elliptic or elliptic-obovate, relatively large, 17.5–24 cm long, venation prominent below; inflorescences monochotomous, infructescences usually 10–14.2 cm long; fruits relatively large, 9–12 mm long.

Remarks.—This taxon is certainly different from other members of this group. The peduncles, usually over 9 cm long, and the relatively large fruits, distinguish it from other taxa.

64. Psychotria merrittioides Sohmer & A.P. Davis, sp. nov. (**Fig. 75**). TYPE: PHILIPPINES. LUZON: Kalinga-Apayao Prov.: Apayao subprov., Mt. Magnas, summit of mountain at 2900 m, 9 Jun 1953, *Edaño PNH 19774* (HOLOTYPE: PNH; ISOTYPES: K, L, PNH).

Frutices 3 m alti sunt, stipulis 0.5–1 mm longis imbricatis puberulis laminis 2 × 4.6–4 × 8.5 cm latis et longis chartaceis glabris obovati-ellipticis apice breve acutis vel breve acuminatis vel obtusis basi acuta vel cuneata in sicco subgriseo-subluteo-viridibus

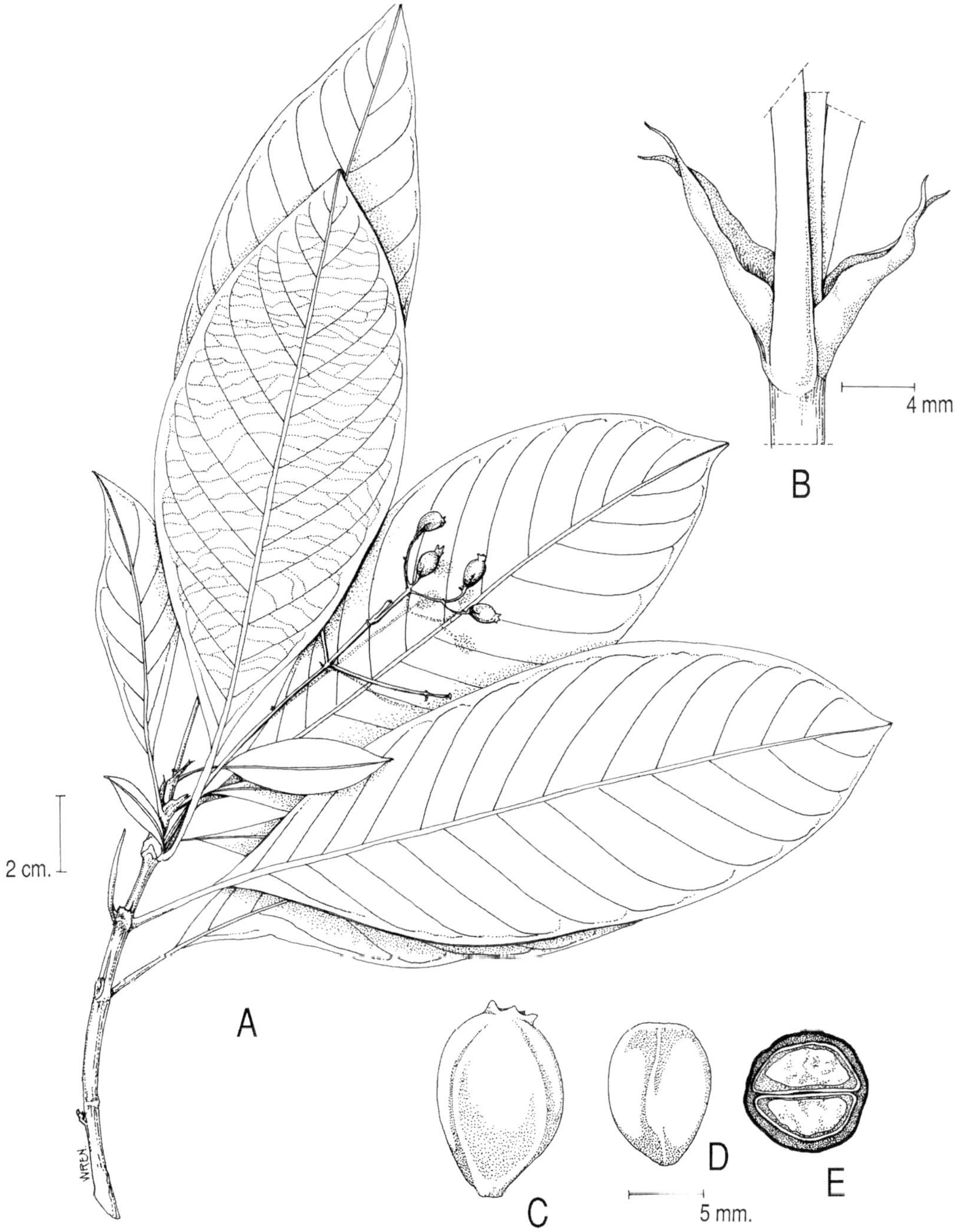

Fig. 74. *Psychotria alabatensis* (*Ramos & Edaño BS 48361*) (K). **A,** habit of fruiting branch; **B,** stipule (*Ramos & Edaño BS 48006*) (K, holotype); **C,** fruit (*Ramos & Edaño BS 28962*) (A); **D,** pyrene; **E,** fruit in cross section.

nervis lateralibus 7–9 in dimidio quoque, petiolis 1–2.2 cm longis, inflorescentia monochotoma 6 cm longa axibus cum 2–3 nodis primariis eis cum ramis verticillatis, pedunculo (0.1–)1–3.6 cm longo glabro, nodis cum bracteis minutis inflorescentia 20–30 flores ferentibus, floribis apicalibus 4–5-meris heteromorphicis, pedicellis 6–8 mm longis, calycibus et hypanthio 2 mm longo in apice dilatatis tubo 1 mm longo, lobis obscuris, corollis immaturis, infructescentia monochotoma interdum trichotoma 6–8 cm longa cum foliis sustentibus subaequalibus, fructilus paucis, axi principali 0–3.6 cm longo cum 1–3 nodeis inferis ramis verticillatis in illis superis ramis oppositis, ramulis cum 1–2 cymules et bracteolis minimis, fructibus 7–10 mm longis turbinati-obovoideis et in sicco valde costatis, pedicellis longis gracilibus 5–15 mm longis, pyrenis in dorso liratis, endospermo ruminato.

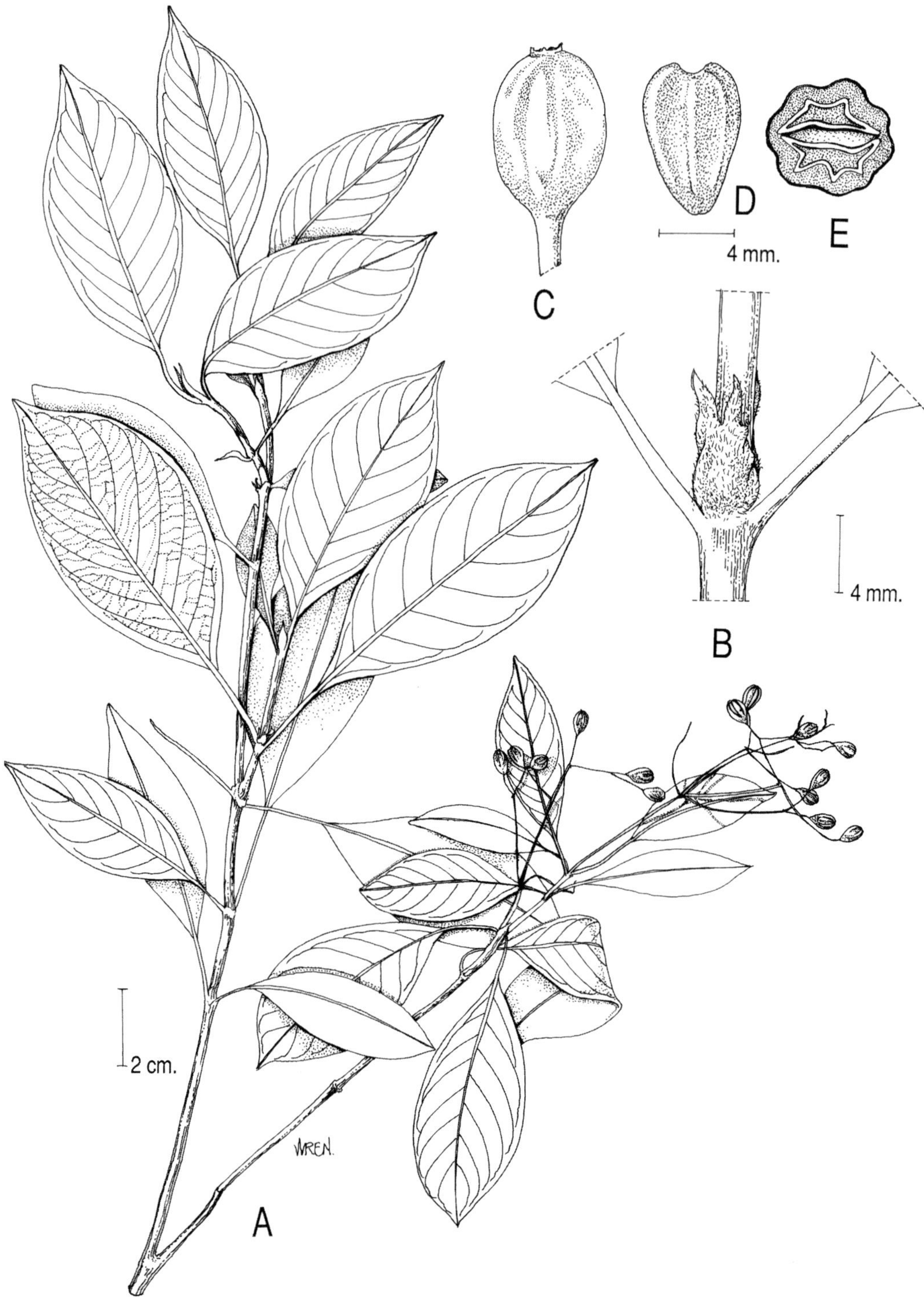

FIG. 75. *Psychotria merrittioides*. (*Edaño PNH 19774*) (PNH, holotype). **A,** habit of vegetative and fruiting branches; **B,** stipule; **C,** fruit; **D,** pyrene; **E,** fruit in cross section.

Shrub, ca. 3 m high, young stems glabrous. Stipules valvate, ± lanceolate-triangular, 0.5–1 cm long, puberulent to sparsely pubescent; apex acute. Leaves: petioles 1–2.2 cm long, glabrous; leaf blades elliptic to obovate-elliptic, 4.6–8 × 2–4 cm, chartaceous, drying grayish yellowish green, glabrous above and below; secondary veins 7–9 pairs; tertiary venation rather prominent; base acute to cuneate; apex abruptly short acute or short-acuminate, or obtuse. Inflorescence/infructescence monochotomous but sometimes appearing trichotomous due to a very short principal axis, 6–8 cm long, glabrous; principal axis (peduncle) (0.1–)1–3.6 cm long, with 1–3 primary nodes and verticillate (lower) or opposite (upper) branching at each node, terminating in lax cymules (with 20–30 flowers/fruits); bracts often caducous, ± ovate, ca. 1 mm long, apex distinctly apicaulate; bracteoles ± ovate, 0.5–0.7 mm long, apex distinctly acuminate. Flowers 4- or 5-merous, pedicels 6–8 mm long; calyx (incl. hypanthium) ca. 2 mm long, glabrous; calyx lobes ± deltate, ca. 0.6 mm long, or obscure; corolla and other flower parts unknown. Fruit turbinate-obovoid, 7–10 mm long, strongly ribbed or ridged (when dry), color unknown, glabrous; calyx limb and lobes persistent or falling, sometimes rather obvious; pedicels 5–15 mm long; pyrenes distinctly ridged on outer surface; endosperm ruminate.

PARATYPES: None—known only from the type collection.

Distribution.—Known only from Mt. Magnas, Luzon.

Ecology.—Forest; high altitude (e.g. ca. 2900 m).

Phenology records.—Flowering: unknown; fruiting: June.

Preliminary conservation assessment.—Critically Endangered (**CR B1ab(iii)**). Only collected once in 1953.

Distinguishing characteristics.—Leaves small, 4.6–8.5 cm long; inflorescences monochotomous, sometimes trichotomous, lax, as long or longer than the subtending leaves; fruit turbinate-obovoid.

Remarks.—The lax nature of the inflorescence distinguishes this taxon from all other groups. It closely resembles *P. merrittii* and it is for that reason that the specific epithet *merrittioides* was chosen. In *P. merrittii*, however, the inflorescence is much shorter than the subtending leaves, and the inflorescences are nearly always trichotomous.

65. Psychotria foxworthyi Sohmer & A.P. Davis, sp. nov. (**Fig. 76**). TYPE: PHILIPPINES. PALAWAN: without further data, Mar–Apr 1906, *Foxworthy BS 657* (HOLOTYPE: NY; ISOTYPES: GH, K).

Frutices vel arbores parvi sunt (?), stipulis licet imbricatis 0.8–1.1 cm longis 5 mm latis ovatis vel triangularis glabris, laminis 4.6 × 15–6 × 17.5 cm latis et longis subcoriaceis oblanceolatis apice auto basi cuneata vel obtusa nervis lateralilus 12–14 in dimidio quoque nervis tertiis in reticulo sub-prominenti, petiolis 1–2.3 cm longis glabris, inflorescentia monochotoma 4–5 cm longa, pedunculo 0.8–2.9 cm longo, nodis primaries 3 cm ramis verticillatis, ramis 0.7–2 cm longis in apice cum 1–2 cymulis floribus subsessilibus, infructescentia 6 cm longa monochotoma, pedunculo 1.5 cm longo, fructibus 8–9 mm diaimetro globosis laevibus, pyrenis in dorso rugosis vel laevibus, sine costis vel liris, endospermo valde ruminato.

Shrub or small tree, height unknown, young stems glabrous. Stipules valvate, ovate to ± triangular, 0.8–1.1 cm long, glabrous; apex obtuse to ± acute. Leaves: petioles 1–2.3 cm long, glabrous; leaf blades elliptic-oblanceolate to oblanceolate, 15–17.5 × 4.6–6 cm, subcoriaceous, glabrous above and below; secondary veins 12–14 pairs; tertiary venation manifest to prominent; base cuneate to obtuse; apex acute to rarely obtuse. Inflorescence/infructescence monochotomous, 4–7.4 cm long, glabrous; principal axis (peduncle) 0.8–2.9 cm long, with 3 primary nodes and verticillate branching at each node, terminating in cymules; bracts unknown (?caducous); bracteoles ovate to triangular, ca. 0.6 mm long. Flowers ± unknown (very immature). Fruit globose, 8–9 mm long, very smooth (mesocarp very thick, fleshy), color unknown, glabrous; calyx limb and lobes not persistent; pedicels absent; pyrenes with a distinct median keel on outer surface; endosperm strongly ruminate.

PARATYPES: None—known only from the type collection.

Distribution.—Known only from Palawan.

Ecology.—Unknown.

Phenology records.—Flowering: March, April; fruiting: March, April.

Preliminary conservation assessment.—Critically Endangered (**CR B1ab(iii)**), or possibly Extinct (**EX**). Only collected once in 1906.

Distinguishing characteristics.—Leaf blades 15–17.5 cm long; inflorescences/infructescences

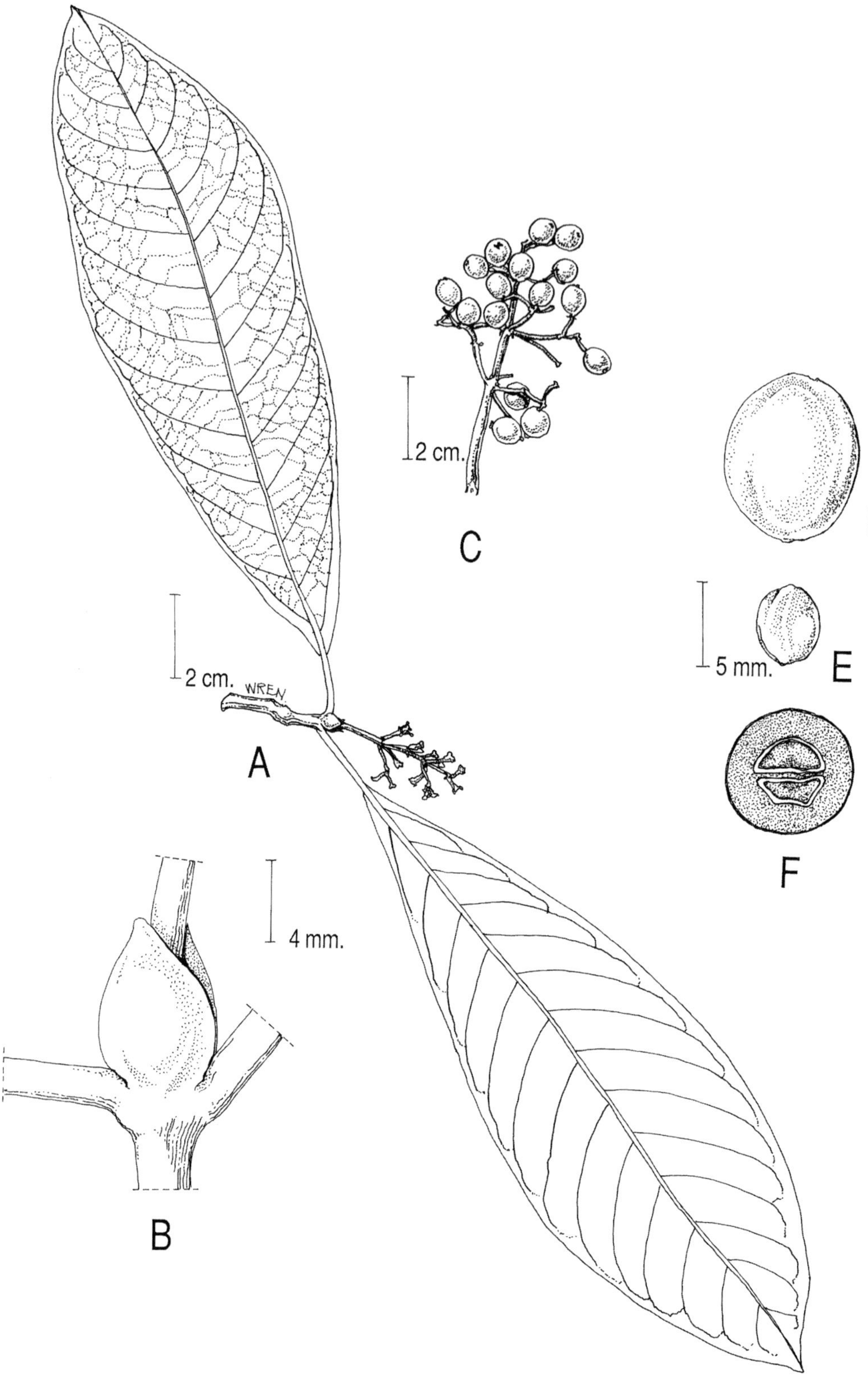

FIG. 76. *Psychotria foxworthyi* (*Foxworthy BS 657*) (NY, holotype). **A,** habit of branch; **B,** stipule; **C,** infructescence; **D,** fruit; **E,** pyrene; **F,** fruit in cross section.

monochotomous, short, 6 cm long or less; stipules ca. 1 cm or less long, about half as broad as long; fruits perfectly globose, 8 mm in diameter, smooth; pyrenes keeled on outer (dorsal) surface, otherwise smooth, and endosperm strongly ruminate.

Remarks.—This taxon is well marked in its group by the large, perfectly globose fruits, which, with the relatively short and well defined inflorescences/infructescences, distinguish it from other taxa.

66. Psychotria panayensis Merr., Philipp. J. Sci. 17:456. 1921; Merr., Enum. Philipp. Fl. Pl. 3:560. 1923. (**Fig. 77**). TYPE: PHILIPPINES. PANAY: Capiz Prov.: Jamindan, moist forests along small streams, Apr–May 1918, *Ramos & Edaño BS 31198* (LECTOTYPE: K, designated here; ISOLECTOTYPE: US).

Small tree or shrub, 1–4 m high, young stems glabrous to puberulous. Stipules valvate, triangular to narrowly triangular, 0.6–0.8 cm long, glabrous; apex acute. Leaves: petioles 1.5–3.5 cm, glabrous; leaf blades oblong-elliptic to elliptic-obovate, (12–)15–21.5 × (5.1–)7–8.8 cm, chartaceous, khaki-brown when dry; glabrous above and below but midrib below glabrous to puberulous; secondary veins 10–13 pairs; tertiary venation manifest; base acute to attenuate; apex abruptly acute to very short acuminate, or obtuse. Inflorescence/infructescence monochotomous or sometimes trichotomous, 3–4.4 cm long, glabrous to puberulous; principal axis (usually the peduncle) (0.2–)0.5–2 cm long, with 1–3 primary nodes and verticillate branching at least at lower nodes, secondary branches branching once more and terminating in cymules; bracts absent (?caducous); bracteoles few or absent, ± ovate, ca. 0.4 mm long. Flowers 4- or 5-merous, pedicels 0.9–2 mm long; calyx (incl. hypanthium) 2–3 mm long, puberulous; calyx lobes shallowly triangular, 0.1–0.3 mm long; corolla: white; corolla tube ca. 2 mm long, external surface puberulent, villous at throat within; corolla lobes ca. 2 mm long; stamens: filaments ca. 1 mm long; anthers ca. 1 mm long (long styled); style ca. 3 mm long, lobes ca. 1 mm long (long styled). Fruit obovoid, 5–7 mm long, with obscure striations, reported to be yellow but probably red or black at maturity, glabrous; calyx limb and lobes persistent; pedicels 1.2–4 mm long; pyrenes smooth on outer surface; endosperm reported to be entire.

Other specimens examined: **Luzon. Camarines Sur Prov.:** Botol River, *Edaño BS 76287* (NY), *vel aff.* **Panay. Capiz Prov.:** Jamindan, *Ramos & Edaño BS 31335* (A, K, US). **Mindoro. Prov unknown:** Mt. Ilong, *Edaño PNH 3576* (A, L, PNH) *vel aff.*

Distribution.—Limited to Panay but with possible records from Camarines Sur province on Luzon and Mt. Ilong on Mindoro.

Ecology.—Moist forest; recorded along small streams; elevation unknown.

Phenology records.—Flowering: February to May; fruiting: unknown.

Preliminary conservation assessment.—Endangered (**EN B1ab(iii)**). Last collected in 1948.

Distinguishing characteristics.—Leaves relatively large; inflorescences monochotomous or sometimes trichotomous; outer (dorsal) surface of pyrenes smooth, endosperm recorded as non ruminate.

Remarks.—The nature of the inflorescence is not well understood, and the endosperm needs to be investigated to verify that it is non ruminate (Merrill 1921). The dubious collections cited here provide inconclusive evidence for this trait. This is another example of how the lack of field knowledge and availability of adequate collections provide a less than adequate concept for a taxon. *Psychotria panayensis* shares certain features with *P. sorsogonensis*, and we believe that future collections of these taxa may show that the condition of the endosperm in the former taxon is ruminate.

67. Psychotria frakei Sohmer & A.P. Davis, sp. nov. (**Fig. 78**). TYPE: PHILIPPINES. MINDANAO: Zamboanga del Norte Prov.: without precise locality, 2 Jan 1958, *Frake PNH 38038* (HOLOTYPE: A; ISOTYPE: L).

Arbores parvae 3–5 m altae glabrae sunt, stipulis valvatis glabris 0.4–0.5 cm longis apice acuto vel breve acuminato integro, laminis subcoriaceis supra glabris infra glabris vel in costa sparse pubescentibus ellipticis vel subobovatis saepe asymmetrices 4.9 × 12.4–7 × 17.5 cm latis et longis nervis lateralibus (13–)16–17 in dimidio quoque infra nervulis nonnihil prominentibus apice abrupte acuto vel acuminato basi obtusa vel cuneata, petiolis 1.2–2.5 cm longis, inflorescentia monochotoma glabra (3–)3.6–5.4 cm longa, pedunculo (1.5–)2–3.3 cm longo, floribus 5-meris, calycibus cum hypanthio et tubo 2 mm longis lobis obscuris glabris, corolla licet alba extra glabra intra in fauce villosa tubo 2 mm longo lobis 3 mm longis, infructescentia 4.5–5 cm longa, pedunculo 2–3.3 cm longo, fructibus rubris vel nigris glabris globosis vel suboblongis 8 mm longis, pyrenis in dorso costatis, endospermio ruminato.

Small tree, 3–5 m high, young stems glabrous. Stipules valvate, triangular, 0.4–0.5 cm long, glabrous; apex acute to shortly acuminate. Leaves: petioles 1.2–2.5 cm long, glabrous; leaf blades elliptic to subobovate, often assymetrical in outline, 12.4–17.5 × 4.9–7 cm, subcoriaceous, glabrous above, glabrous

FIG. 77. *Psychotria panayensis* (*Ramos & Edaño BS 31198*) (K, lectotype).

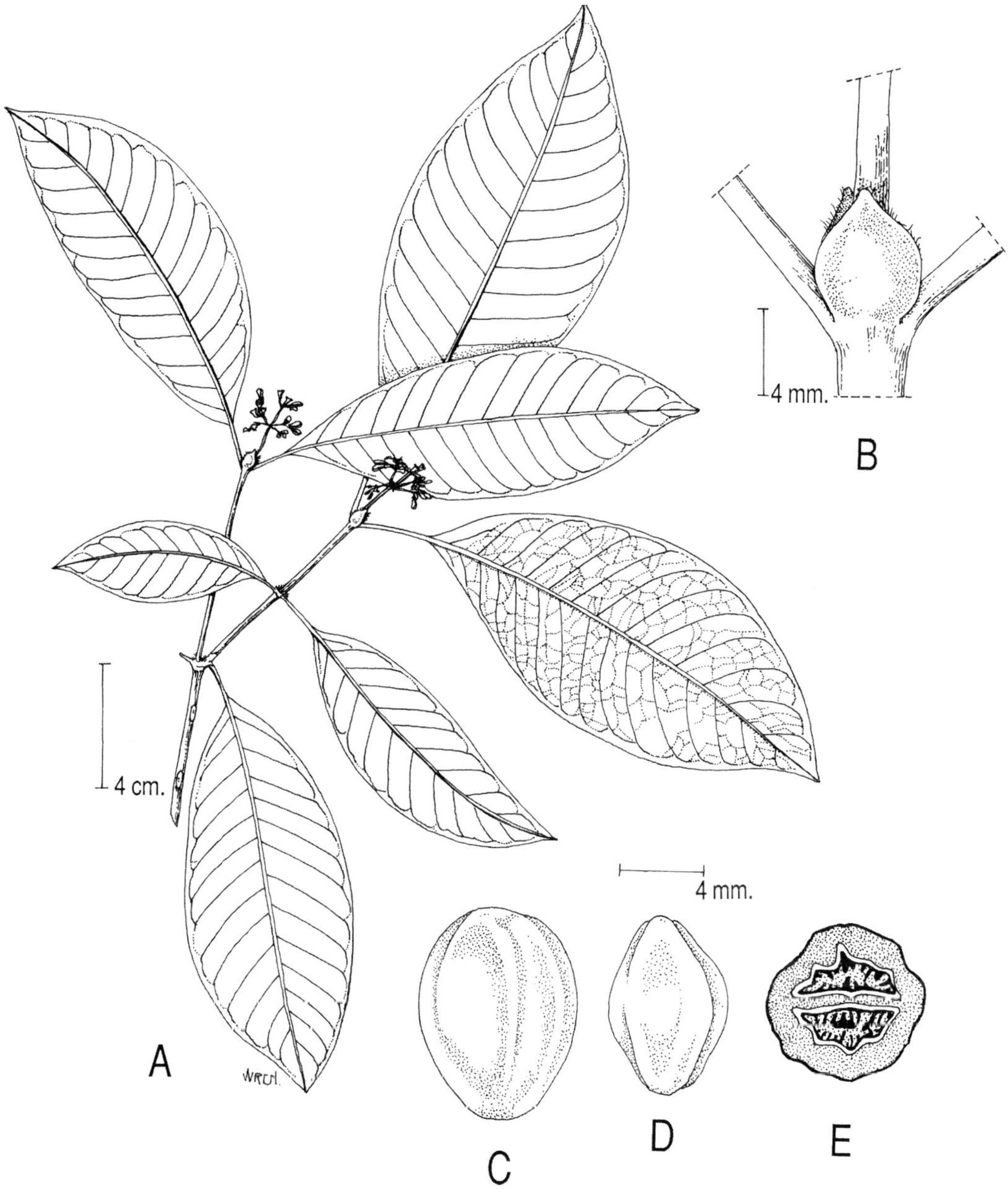

Fig. 78. *Psychotria frakei* (*Frake PNH 38038*) (A, holotype). **A,** habit of branch; **B,** stipule; **C,** fruit; **D,** pyrene; **E,** fruit in cross section.

to sparsely pubescent on costa below, otherwise glabrous; secondary veins (13–)16–17 pairs; tertiary venation manifest to somewhat prominent; base obtuse to cuneate; apex abruptly acute to acuminate. Inflorescence/infructescence monochotomous, (3–)3.6–5.4 cm long, glabrous, principal axis (peduncle) (1.5–)2–3.3 cm long, with 1–2 primary nodes and opposite branching at each node, secondary branches terminating in few-flowered cymules; bracts absent (?caducous); bracteoles unknown (?caducous). Flowers 5-merous, pedicels absent; calyx (incl. hypanthium) ca. 2 mm long, glabrous; calyx lobes shallowly triangular, ca. 0.3 mm long, or obscure; corolla: probably white; corolla tube ca. 2 mm long, external surface glabrous, villous at throat within; corolla lobes ca. 3 mm long; stamens: filaments ca. 1 mm long; anthers ca. 1 mm long (long styled); style ca. 3.5 mm long (long styled), lobes ca. 1 mm long. Fruit globose to somewhat ellipsoid, ca. 8 mm long, ribbed, red or black, glabrous; calyx limb and lobes persistent; pedicels absent; pedicels absent; pedicels absent; pyrenes ribbed on outer surface; endosperm ruminate.

PARATYPES: **Mindanao. Zamboanga del Norte Prov.:** Tulaln, *Frake PNH 38077* (A, L); Dikus, *Frake PNH 38177* (A, K, L).

Distribution.—Only from Zamboanga del Norte Province, Mindanao.

Ecology.—Moist primary or secondary forest; 300–600 m.

Phenology records.—Flowering: January; fruiting: November to January.

Preliminary conservation assessment.—Endangered (**EN B1 ab(iii)**). Last collected in 1958.

Distinguishing characteristics.—Secondary leaf venation prominent; inflorescences monochotomous, peduncles relatively short, ca. 1.5 cm long; fruits ca. 8 mm long.

Remarks.—This taxon probably has very prominent venation in the field, a characteristic that distinguishs this from *P. panayensis* Merr.

68. Psychotria mindanaensis Merr., Philipp. J. Sci. 8:54. 1913; Merr., Enum. Philipp. Fl. Pl. 3:559. 1923. (**Fig. 79**). *Grumilea mindanaensis* (Merr.) Merr., Philipp. J. Sci. 10:130. 1915. TYPE: PHILIPPINES. MINDANAO: Basilan Prov.: Basilan Island, Sax River Mountains back of San Ramon, between 800–900 m, Nov 1911, *Merrill BS 8096* (LECTOTYPE: P, designated here).

[*Psychotria malayana* Jack, sensu auct.]

Shrub or small tree, 3–5 m high; young stems glabrous. Stipules valvate, lanceolate-triangular, 0.7–1 cm long, glabrous; apex acute to obtuse. Leaves: petioles 0.7–1.8 cm long, glabrous; leaf blades narrowly elliptic to elliptic-oblanceolate, 9.8–15 × 2–3.9 cm, subcoriaceous, glabrous above and below; secondary veins 9–11 pairs; tertiary veins obscure to manifest; base acuminate; apex acuminate. Inflorescence/infructescence monochotomous, 3.6–9 cm long, glabrous to puberulous; principal axis (peduncle) 2.4–5.7(–7.4) cm long, with 1–2 primary nodes and opposite or verticellate branching at each node, secondary branches terminating in lax, few-flowered cymules; bracts unknown (caducous); bracteoles broadly ovate, ca. 0.5 mm long. Flowers 5-merous, pedicels 1–2 mm long; calyx (incl. hypanthium) ca. 2.5 mm long, minutely puberulent; calyx lobes obscure; corolla: probably white; corolla tube ca. 1.5 mm long, external surface puberulent, villous at throat within; corolla lobes ca. 2 mm long; stamens: filaments ca. 1 mm long; anthers 0.75–1 mm long; style ca. 1.5 mm long, lobes ca. 0.5 mm long. Fruit obovoid or turbinate, 6–10 mm long, ribbed (when dry), reported to be yellow but probably red at maturity, glabrous; calyx limb and lobes persistent but not prominent; pedicels 2–4.5 mm long; pyrenes ribbed on outer surface; endosperm ruminate.

Other specimens examined: **Luzon. Rizal Prov.:** Mt. Lumutan, *Ramos BS 42250* (MO). **Mindanao. Surigao del Norte Prov.:** Dinagat Island, without further data, *Ramos & Convocar BS 83831* (A), *Ramos & Pascasio BS 35225* (NY). **Davao Prov.:** Mati, *Ondrada FB 31355* (NY). **Zamboanga Prov.:** without further data: *Merrill 8084* (A, BM, K, L, P, US). **Palawan. Palawan Prov.:** Mt. Victoria, Trident Mine, *Dransfield SMHI 1273* (L), *SMHI 1279* (L).

Distribution.—Luzon, Mindanao (including Dinagat Island), and Palawan.

Ecology.—Primary forest; 400–900 m.

Phenology records.—Flowering: November; fruiting: November, December.

Preliminary conservation assessment.—Near Threatened (**NT**). Last collected in 1984.

Distinguishing characteristics.—Leaves narrow, usually less than 3 cm wide; stipules valvate; inflorescences monochotomous.

Remarks.—The lectotype and syntype collections (Merrill 1913) seem distinct, but the variation evident in the other collections indicates that this taxon is not well-defined. There seems little to distinguish this taxon from *P. gitingensis* Elmer, except for the relatively longer leaves of *P. mindanaensis*.

69. Psychotria pyramidata Elmer, Leafl. Philipp. Bot. 4:1350. 1912; Merr., Enum. Philipp. Fl. Pl. 3:562. 1923. (**Fig. 80**). *Grumilea pyramidata* (Elmer) Merr., Philipp. J. Sci. 10:134. 1915. TYPE: PHILIPPINES. PALAWAN: Palawan Prov.: Mt. Pulgar, shaded forest near watercourses from about 200–800 m, May 1911, *Elmer 13183* (LECTOTYPE: NY, designated here; ISOLECTOTYPES: BISH, BM, L).

Small tree or shrub, 1–4 m high, ± glabrous throughout. Stipules valvate, ± triangular, 0.3–0.6 cm long, glabrous; apex acute or narrowly obtuse. Leaves: petioles 1.4–3.8 cm long, glabrous or rarely

Fig. 79. *Psychotria mindanaensis* (*Merrill, BS 8096*) (P, lectotype).

FIG. 80. *Psychotria pyramidata* (*Elmer 13183*) (NY, lectotype).

puberulous; leaf blades oblong-elliptic to oblanceolate, 10.8–19.7(–22.2) × (4–)5.2–8.5(–9.2) cm, subcoriaceous, drying grayish green or brownish, glabrous above, glabrous or puberulous below on midrib and veins; secondary veins 11–16 pairs; tertiary venation manifest to prominent; base acute, narrowly acute, or cuneate; apex abruptly acute or abruptly obtuse. Inflorescence/infructescence monochotomous, (9.6–)13.5–17.5 cm long, puberulous to glabrous; principal axis (peduncle) (3–)6.2–9.1 cm long, with 2–4 primary nodes and verticillate branching at each node, the secondary branches often reflexed, branches themselves and terminating in cymules; bracts usually fused and sheathing the inflorescence axes, broadly ovate to truncate, ca. 0.7 mm long; bracteoles similar to the bracts, ca. 0.4 mm long. Flowers 5-merous, pedicels 0.1–0.3 mm long or absent; calyx (incl. hypanthium) ca. 2 mm long, glabrous to finely puberulous; calyx lobes obscure; corolla: color unknown; corolla tube ca. 2 mm long, external surface puberulous, villous at throat within; corolla lobes ca. 1 mm long. stamens unknown; style unknown. Fruits obovate-turbinate, 7–9 mm long, obscurely ribbed to ribbed, orange, glabrous; calyx limb and lobes not persistent; pedicels 0.4–1.4 mm long; pyrenes ribbed or ridged on outer surface; endosperm ruminate.

Other specimens examined: **Palawan. Palawan Prov.:** Puerto Princesa (Mt. Pulgar), *Elmer 13093* (A, K, US, W); Irawan River Valley, *Ridsdale SMHI 10* (L), *SMHI 139* (L); Mt. Mantalingahan, Brooke's Point, *Edaño PNH 6* (A), *PNH 48* (A, L); Mt. Capoas, *Merrill 9518* (K); Busuanga Island, *Ramos BS 41276* (A, P); Sumindap River, Brgy Bundog, Municipality of Quezon, *Pipoly & Romero et al. PPI 38126* (BRIT); Rizal, Ransang, Aga, *Gaerlan, Romero & Sagcal PPI 13225* (BRIT); Taytay, Bambanan, Mt. Capoas, *Reynoso, Alvarez & Fuentes PPI 11160* (BRIT). **Palawan. Without further data:** *Foxworthy BS 900* (US).

Distribution.—Palawan.

Ecology.—Moist forest (undisturbed or disturbed), including mixed dipterocarp forest and dipterocarp-swamp forest; 150–1200 m.

Phenology records.—Flowering: January, April, May; fruiting: April to June, August.

Preliminary conservation assessment.—Endangered (**EN B1ab(iii)**). Last collected in 1996.

Distinguishing characteristics.—Leaves relatively large, 17–22.2 cm long; stipules valvate; inflorescences monochotomous, 13.5–17.5 cm long, elongate, secondary branches verticillate, widely spreading and often reflexed.

Remarks.—This is a well-circumscribed species, which is apparently restricted to Palawan. There are, however, morphological intergradations to *P. gitingensis.*

70. Psychotria sorsogonensis Elmer, Leafl. Philipp. Bot. 9:3264. 1934. (**Fig. 81**). Type: PHILIPPINES. Luzon: Sorsogon Prov.: Irosin (Mt. Bulusan), deeply shaded forest at 606 m, Jul 1916, *Elmer 16772* (lectotype: A, designated here; isolectotypes: BISH, BM, GH, K, L, MO, P, UC, US).

Shrub or small tree, 1–4 m high, ± glabrous throughout. Stipules valvate, triangular to ovate-triangular, 0.6–0.8 cm long, glabrous; apex acute. Leaves: petioles 1–3 cm long; glabrous; leaf blades broadly obovate, broadly obovate-elliptic, or broadly elliptic, 11–23.4 × 6.2–11 cm, subcoriaceous, glabrous above and below; secondary veins 10–15 pairs; tertiary venation weak to obscure; base cuneate or acute; apex abruptly acute or cuneate. Inflorescence/infructescence monochotomous, 3–9 cm long, glabrous; principal axis (peduncle) 2–4.5 cm long, with 2–4 primary nodes and verticillate branching at each node, the branches terminating in lax cymules; bracts present, insufficiently known; bracteoles broadly ovate, ca. 0.4 mm long. Flowers unknown. Fruit obovoid-globose, 5–7 mm long, ribbed, reported to be yellow but probably red or black at maturity, glabrous; calyx limb and lobes persistent but obscure; pedicels 2–3.7 mm long; pyrenes ribbed on outer surface; endosperm ruminate.

Other specimens examined: Known only from the type collection.

Distribution.—Found only at Irosin on Mt. Bulusan.

Ecology.—Humid forest; at mid altitude (e.g. ca. 600 m).

Phenology records.—Flowering: unknown; fruiting: July.

Preliminary conservation assessment.—Critically Endangered (**CR B1ab(iii)**) or possibly Extinct (**EX**). Only collected once in 1916.

Distinguishing characteristics.—Leaf-blades broadly obovate to broadly obovate-elliptic, 6.2–11 cm wide; stipules valvate, 6–8 mm long; inflorescences monochotomous, peduncle usually more than 3 cm long.

FIG. 81. *Psychotria sorsogonensis* (*Elmer 16772*) (A, lectotype).

Remarks.—We are uncertain of the status of this taxon. It apparently differs from *P. gitingensis* by its broadly obovate leaf blades. It is also very similar in appearance to *P. panayensis* Merr. and is separated from that taxon primarily on the basis that the peduncle of *P. sorsogonensis* is longer than in *P. panayensis.* The endosperm of *P. payanensis* is said to be non ruminate (ruminate in *P. sorsogonensis*) but we have not been able to confirm this. There is so little evidence from which to draw conclusions that we have maintained both species. We have tentatively referred one other collection to *P. sorsogonensis* (*Edaño BS 76458* (NY) from Camarines Sur Province) but have not cited it in this treatment.

71. Psychotria crassifolia Miq., Annales Musei Botanici: Lugduno-Batavi IV: 205. 1868. (**Fig. 82**). TYPE: BORNEO: "M. Pamatton," without precise locality, *Korthals s.n.* (LECTOTYPE: L, designated here).

[*Psychoria malayana* Jack, sensu auct.]

Shrub or small tree, 1–5 m high, glabrous throughout. Stipules valvate, ovate to ovate-triangular, 0.5–0.9 cm long, glabrous; apex narrowly obtuse. Leaves: petioles 0.8–2 cm long, glabrous; leaf blades oblanceolate, elliptic-obovate to broadly obovate, or ± elliptic, (9–)12–16 × 3.5–4 6 cm, subcoriaceous to ± coriaceous, often shiny, dark reddish brown when dry, glabrous above and below; secondary veins 9–16 pairs; tertiary venation manifest to obscure; base cuneate, narrowly acute or acute; apex obtuse or abruptly cuneate to round or abruptly short-obtuse. Inflorescence/infructescence monochotomous, 6–19 cm long, glabrous, principal axis (peduncle) 2.5–10 cm long, with 2–5 primary nodes and verticillate branching at each node, the branches terminating in cymules; bracts ovate to deltate, 1.5–2 mm long, margins usually conspicuously cilate (?colleters); bracteoles (pairs) fused and sheathing the infloresсnece axes, broadly ovate to truncate, ca. 1.2 mm long. Flowers 5-merous, pedicels 0.1–0.7 mm long or absent; calyx (incl. hypanthium) 2–3.5(–5.5) mm long, glabrous; calyx lobes obscure (calyx limb truncate), or sometimes very broadly ovate to deltate, 0.1–0.4 mm long; corolla: white; corolla tube 8–9 mm long, external surface glabrous, villous at throat within; corolla lobes 2–2.5 mm long; stamens: filaments 1.5–2 mm long; anthers 1–1.5 mm long (short styled); style ca. 3–5 mm long (short styled), lobes 1–2 mm long. Fruit globular to subobovoid, (6–)8–10 mm long, ribbed-ridged, or ridged to ± sharply angular-ridged, sometimes appaering ± smooth, orange to red, glabrous; calyx limb and lobes persistent, very obvious, (1–)2–3 mm long; pedicels 0.1–0.8 mm long; pyrenes ribbed-ridged, or ridged to ± sharply angular-ridged (apical ridge the most prominent) on outer surface; endosperm strongly ruminate.

Distribution.—Borneo and the Philippines (Mindanao and Palawan).

Ecology.—Forest (primary and secondary); sea level to ca. 1000 m.

Phenology records.—Flowering: February to August, October; fruiting: January, March, May to August.

Preliminary conservation assessment.—Least Concern (**LC**). Last collected in 1996.

Distinguishing characteristics.—Leaf blades subcoriaceous to more or less coriaceous, drying shiny reddish brown; stipules valvate, less than 1 cm long; inflorescences monochotomous; corolla tube 8–9 mm long (short styled flowers); pyrenes smooth on outer (dorsal) surface, endosperm strongly ruminate.

Remarks.—This taxon is definitely that which Miquel described from Borneo. The normally coriaceous leaves are characteristic. The line separating this species from *P. gitingensis* Elmer, however, is not strongly defined, and the relationship of these two taxa will be brought into better perspective with further study.

Two distinct varieties that can be separated by the following key.

KEY TO THE VARIETIES OF *PSYCHOTRIA CRASSIFOLIA*

1. Leaf blades oblanceolate; inflorescence peduncle 2.5–4.5 cm long _____ **71a.** var. **crassifolia**
1. Leaf blades obovate; inflorescence peduncle 9–10 cm long _____ **71b.** var. **culionensis**

71a. Psychotria crassifolia var. **crassifolia**

Leaves: petioles 1–2 cm long; leaf blades oblanceolate; secondry veins 12–16 per side; base narrowly acute to acute; apex obtuse or abruptly cuneate. Inflorescence 6–9 cm long, main axis (peduncle) 2.5–4.5 cm long, with 2–4 primary nodes.

FIG. 82. *Psychotria crassifolia* var. *crassifolia* (*Korthals s.n.*) (L, lectotype).

Other specimens examined: **Palawan. Palawan Prov.:** Tabon, *Reynoso PNH 87734* (BISH, L, PNH); Puerto Princesa, Mt. Beaufort, *Podzorski SMHI 558* (K), *Podzorski SMHI 570* (K), *Ridsdale SMHI 275* (L), *Ridsdale SMHI 1879* (K); Puerto Princesa, *Cenabre FB 29168* (A, K, US), *Kondo & Edaño PNH 36560* (BISH, K, L, PNH), *McGregor BS 45913* (UC); Puerto Princesa, Bacungan, *Edaño PNH 276* (A, PNH, UC); Puerto Princesa, Malinão Mountain, *Ebalo 619* (A, NY, PNH, RSA, UC); Puerto Princesa, Irawan, Tatanarum, *Stone PPI 50* (K); *Stone PPI 55* (BRIT); *Stone & Madulid et al. PPI 27* (BRIT, K); Puerto Princesa, Irawan, Venture Area, *Madulid 1032* (K); Puerto Princesa to Roxas, km 36, *Dransfield SMHI 1240* (K); Puerto Princesa, Irawan River Valley, *Stone & Madulid PPI 64* (BRIT); Mt. Pulgar, *Elmer 12981* (A, BM, GH, K, L, MO, NY, P, US, W); Lapulapu, Iwahig, *Edaño PNH 144* (A), *Edaño PNH 227* (A, L, UC); Victoria Mountains, Aborlan, *Sulit PNH 12302* (A, L, SING, US); Mt. Iraan, *Sulit PNH 12545* (A, L); Mt. Gantung, *Edaño BS 77607* (A); NNW slopes at headwaters of Tagwaya River, Mt Gantung, Mantalingajan Mt. Range, District of Tindogan, *Pipoly & Romero et al. PPI 37851* (BRIT); Mt. Mantalingajan, *Edaño BS 77545* (A), *PNH 389* (A). **Palawan Prov.:** Island of Paragua, E-wi-ig River[*sic* = Iwahig River], *Merrill 742*, (K, NY). **Palawan Prov**.: without further data: *Danao FB 21589* (BM, K, US), *Escritor BS 21534* (A, BM, GH, K, L, MO, NY, P, US), *McGregor BS 45925* (UC), *Curran FB 3540* (K). **Mindanao. Bukidnon Prov.:** vicinity of Tanculan, *Fenix BS 24976* (A, K). **Davao Prov.:** Mt. Galintan, *Ramos & Edaño BS 48895* (BM, NY, UC).

Distribution.—In the Philippines known principally from Palawan and Davao.

Ecology.—Forest, sometimes in secondary forest; often recorded on rocks and on steep slopes; (10–)250–650 m.

Phenology records.—Flowering: February to June; fruiting: January, March, May to August.

Preliminary conservation assessment.—Least Concern (**LC**). Last collected in 1996.

Distinguishing features.—Leaf blades oblanceolate, lateral veins 12–16 per side, apex obtuse or cuneate, petioles 1–2 cm long; inflorescence peduncle 2.5–4.5 cm long.

Remarks.—This variety is distinguished from var. *culionensis* by its oblanceolate (as opposed to obovate) leaf blades and much shorter peduncles.

71b. Psychotria crassifolia var. **culionensis** Sohmer & A.P. Davis, var. nov. (**Fig. 83**). TYPE: PHILIPPINES. PALAWAN: Palawan Prov.: Culion Island, no further data, Oct 1922, *Ramos BS 41319* (HOLOTYPE: L; ISOTYPES: K, PNH).

Frutices sunt (?) stipulis minus que 5 mm longis crassis lucidis coriaceis glabris, laminis 3.5 × 9–8.5 × 16 cm latis et longis quando siccis lucidis rubri-badiis obovatis breve obtusis basi cuneata ad acuta nervis latoralibus 9–14 in dimidio quoque, petiolis 0.8–2 cm longis glabris, inflorescentia 18–19 cm longa monochotoma, pedunculo 9–10 cm longo, nodis primariis 3–5 ramis verticillatis vel supra oppositis longis et eis inferis 5–6 cm longis 1–2-ramificantis, cymulis remotis, bracteis et bracteolis praesentibus conspicuis 1–1.5 mm, calycilus et hypanthiis 2–2.5 mm.

Leaves: petioles 0.8–2 cm long; leaf blades obovate, 9–16 × 3.5–8.5 cm; secondary veins 9–14 per side; base cuneate to acute; apex round to abruptly short obtuse. Inflorescence 18–19 cm long, main axis (peduncle) 9–10 cm long, with 3–5 primary nodes. Fruits unknown.

PARATYPES: None—known only from the type collection.

Distribution.—Apparently restricted to Culion Island.

Ecology.—Unknown

Phenology records.—Flowering: October; fruiting: unknown.

Preliminary conservation assessment.—Critically Endangered (**CR B1ab(iii)**), or possibly Extinct (**EX**). Only collected once in 1922.

Distinguishing features.—Leaf blades obovate, lateral veins 9–14 per side, apex round to abruptly short obtuse, petioles 0.8–2 cm long; inflorescence peduncle 9–10 cm long.

Remarks.—This variety is distinguished from the nominate variety by its obovate leaf blades and longer peduncles.

72. Psychotria gitingensis Elmer, Leafl. Philipp. Bot. 3:1024. 1911; Merr., Enum. Philipp. Fl. Pl. 3:556. 1923. (**Figs. 84, 85**). TYPE: PHILIPPINES. SIBUYAN: Romblon Prov.: Mt. Giting-giting [on forested ridge], May 1910, *Elmer 12431* (LECTOTYPE: BM, designated here; ISOLECTOTYPES: A, BISH, GH, K, L, MO, P, US, W).

Psychotria similis Elmer, Leafl. Philipp. Bot. 3:1025. 1911. *Grumilea similis* (Elmer) Merr., Philipp. J. Sci. 10:134. 1915. TYPE: PHILIPPINES. SIBUYAN: Romblon Prov.: Mt. Giting-giting, on a wind-swept forested ridge, May 1910, *Elmer 12439* (LECTOTYPE: GH, designated here; ISOLECTOTYPES: A, BISH, BM, K, L, MO, US, W).

Psychotria versicolor Elmer, Leafl. Philipp. Bot. 4:1352. 1912. *Grumilea versicolor* (Elmer) Merr., Philipp. J. Sci. 10:130. 1915. TYPE: PHILIPPINES. PALAWAN: Palawan Prov.: Puerto Princessa (Mt. Pulgar), banks of Iwahig River, Mar 1911, *Elmer 12763* (LECTOTYPE: A, designated here; ISOLECTOTYPES: BISH, BM, GH, K, L, MO, NY, P, US, W).

Psychotria lucida Merr., Philipp. J. Sci. 17:452. 1921. TYPE: PHILIPPINES. LUZON: Rizal Prov.: Mt. Lumutan, Jul 1917, *Ramos & Edaño BS 29716* (LECTOTYPE: US, designated here; ISOLECTOTYPES: K, NY, P).

[*Psychoria malayana* Jack, sensu auct.]

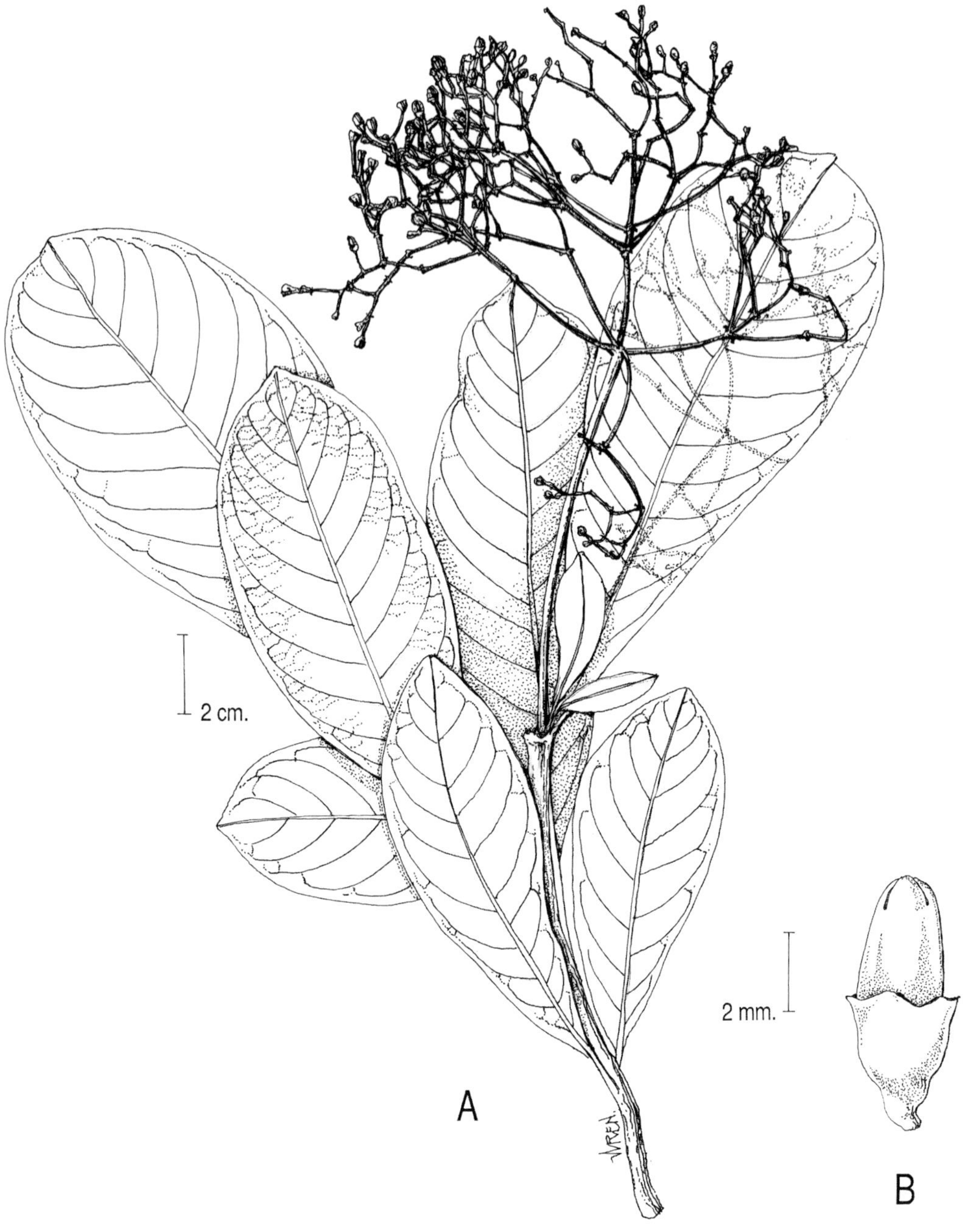

Fig. 83. *Psychotria crassifolia* var. *culionensis* (*Ramos BS 41319*) (L, holotype). **A,** habit of flowering branch; **B,** flower in bud.

Shrub or small tree, (1–)3–5(–8) m high, ± glabrous throughout. Stipules valvate, ovate-deltate to ovate-triangular, 0.2–0.5(–0.8) cm long, glabrous; apex obtuse or acute. Leaves: petioles (0.7–)1–3.1 cm long, glabrous; leaf blades elliptic, obovate, oblong-elliptic to oblanceolate, (6.5–)9.5–18.5 × (2.4–)4–5.6 cm, chartaceous to subcoriaceous, drying greenish brown to brown, glabrous above and below; secondary veins 9–14(–16) pairs; tertiary venation weak to obscure; base acute, narrowly acute to subattenuate; apex narrowly acute to abruptly acuminate. Inflorescence/Infructescence monochotomous, or rarely trichotomous, (4.7–)7–15.5 cm long, glabrous, or puberulous to sparsely pubescent; principal axis (peduncle) 2.5–8.5 cm long, with 2–4 primary nodes and verticillate, or

Fig. 84. *Psychotria gitingensis* (*Elmer 12431*) (BM, lectotype).

FIG. 85. *Psychotria gitingensis* (type of *P. similis* Elmer) *Elmer 12439*) (GH, lectotype).

infrequently opposite, branching at each node, the secondary branches terminating in lax cymules; bracts ovate to deltate, 1–1.5 mm long, margins usually conspicuously cilate (?colleters); bracteoles (pairs) pairs fused and ± sheathing the inflorescnece axes, broadly ovate, 0.5–1 mm long. Flowers 4- or 5-merous, pedicels 0.1–0.5 mm long or absent; calyx (incl. hypanthium) 2–2.5 mm long, glabrous; calyx lobes obscure (calyx limb truncate), or sometimes deltate, 0.3–1 mm long; corolla: white to yellowish white; corolla tube 3.5–9 mm long, glabrous, villous at throat within; corolla lobes ca. 3.5 mm long; stamens: filaments 0.5–1 mm (long styled), or 2–3 mm long (short styled); anthers ca. 1 mm long (long styled), or 1–2 mm long (short styled); style 4.5–6 mm long (long styled) or 2–5 mm long (short styled), lobes ca. 2 mm long (long styled), or 0.7–2 mm long (short styled). Fruit subobovoid, obovoid or turbinate, 5–8.5 mm long, ribbed, reported to be yellow, orange and black, glabrous; calyx limb and lobes absent or persistent, obvious to obscure, 0.2–0.5(–0.9) mm long; fruits sessile, or pedicels 0.2–0.6 mm long; pyrenes ± smooth or ribbed on outer surface; endosperm ruminate.

Other specimens examined: **Luzon. Cagayan Prov.:** Calayan Island, East of St. Vincente Bo, *Quisumbing & del Rosario PNH 79811* (PNH); Bagio [Baguio (?)] Cove, *Allen 44-81* (BISH). **Ilocos Norte Prov.:** Bangui to Claveria, *Ramos BS 33054* (A,BM,PNH). **Ifugao Prov.:** Lepanto, Vidal 1488 (K). **Ilocos Norte Prov.:** without further data: *Adduru FB 22980* (A), *vel aff.* **Benguet Prov.:** Twin Peaks, *Elmer 6341* (K, NY, P). **Nueva Ecija Prov.:** Caraballo Mountains, *Vidal 3051* (K). **Quezon Prov.:** Quezon National Forest Park, Pinagbanderahan Trail, *Sohmer 12102* (BISH); Baler Municipality, Barangay Cemento, on road ca. 8 km from Baler, *Soejarto, Fernando & Fernando PPI 7944* (BRIT); Baler, *Escritor BS 21196* (K); Baler, Pricipe, *Elmer 1103* (K); Cemento River, *Quisumbing PNH 2446* (L), *PNH 2481* (L); Polillo Island, without further data, *Robinson BS 9295* (BM, US). **Rizal Prov.:** Mt. Angilog, *Ramos BS 40756* (A, K, P, US); Mt. Irid, *Ramos & Edaño BS 48417* (K, NY, UC), *BS 48390* (NY, UC); Monte Batay, *Loher 6443* (K); Montalban, *Loher 6434* (K), *s.n.* (UC); Mt. Lumutan, *Ramos BS 29632* (K, P), *BS 42175* (SING). **Rizal Prov.:** without further data: *Loher 12246* (A, UC), *14201* (UC, W), *s.n.* (UC). **Camarines Norte Prov.:** Paracale, *Ramos & Edaño BS 33699* (A,UC); Paracale, *Ramos & Edaño BS 33708* (L). **Batangas Prov.:** without further data: *Cuming 1526* (BM, K, W). **Albay Prov.:** without further data: *Vidal 3054* (K). **Mindoro. Mindoro Occidental Prov.:** Mt. Iglit-Baco National Park, Sitio Sapaton, Barangay Poypoy, Calintaan, *Barbon & Garcia PPI 22685* (BRIT, K); *Barbon & Garcia PPI 22656* (BRIT). **Mindoro Oriental Prov.:** Puerto Galera, *Alegaen et al. 207* (L); *Keinholz BS 15191* (K, NY, UC); on hill top, *Santos 5305* (L); Mt. Halcon, *Ramos & Edaño BS 40736* (A, W); *Ramos & Edaño BS 40566* (K); Pinamalayan, *Ramos BS 41304* (K); Bongabon and Pinamalayan, *Maliwanag 330* (BRIT); Mt. Yagaw (eastern slope), *Sulit & Conklin PNH 17603* (K). **Tablas. Romblon Prov.:** So. Barikwang, Barangay Marigondon Sur, San Andres, *Reynoso & Fuentes PPI 24575* (BRIT, gy. Marigondon Sur, San Andres, *Reynoso & Fuentes PPI 24705* (BRIT, K). **Sibuyan. Romblon Prov.:** Magallanes (Mt. Giting-giting), *Elmer 12125* (A); Sibuyan Island, Punong Mt., near trail to Barrio Jaosan, *Sohmer 12466* (K); *Sohmer 12487*(K); *Sohmer 12490* (K); Sibuyan Island, Punong Mt., on Jaosan-Punong River trial, *Sohmer 12393* (K); *Sohmer 12394* (K); Sibuyan Island, along Punong River, *Sohmer 12414* (K); Mt. Guiting-guiting [Giting-giting], *Reynoso, Fuentes & Rivers PPI 14304* (BRIT); Mt. Guiting-guiting, *Reynoso, Fuentes & Rivers PPI 14054* (BRIT); Mt. Guiting-guiting, 6 km east of Magdiwang, NE slopes, Tampayan, *Reynoso et al. PPI 7608* (BRIT); Magdiwang, SE of Binalabag Valley, *Stone, Reynoso & Fernando PPI 6876* (BRIT, K). **Masbate. Masbate Prov.:** 7-R Ranch, Sitio Kalunakan, Barangay Matipuron, Milagros, *Barbon, Garcia & Fernando PPI 12732* (BRIT, K). **Samar. Eastern Samar Prov.:** Sitio Cantinoc Barangay 1, Balangiga, *Gaerlan & Chavez PPI 26352* (K). **Western Samar Prov.:** Pinabacdao, near lakes, vicinity of km 56, *Conese PNH 13789* (A); Barangay Cambayan, Basey, *Barbon, Garcia & Sagcal PPI 6035* (K); Barangay Hinamok, Basey, *Barbon, Garcia & Sagcal PPI 6067* (BRIT, K); Sitio Kapinyahan, Barangay Lokilokon, Paranas, *Reynoso, Sagcal & Garcia PPI 7450* (BRIT). **Samar.** prov unkown: Quinay, *Vidal 1481c* (A); San Juanico, Barbon, *Garcia & Sagcal PPI 6010* (BRIT). **Panay. Antique Prov.:** San Remigio, *Taleon PNH 34017* (PNH); Libo/Sigayan/Balibago, Barangay Semirara Island, Caluya, *Romero & Majaducon PPI 29604* (BRIT). **Antique Prov.:** without further data: *McGregor BS 32511* (UC). **Iloilo Prov.:** Mt. Bulilão, *Martelino & Edaño BS 35741* (A, BM, L, P); Mt. Salibongbong, *Martelino & Edaño BS 35635* (BM); Miagao, *Vidal 3079* (A, K); Guimaras Island, Buenavista-Bo. Tubod, *Sulit PNH 11727* (A, K, L). **Capiz Prov.:** Libacao, *Martelino BS 35467* (BM, L), *Martelino & Edaño BS 35540* (A, BM, K, P); Mt. Timbaban, *Edaño BS 42442* (A), *BS 42491* (MO); Mt. Madiaas, *Ramos & Edaño BS 30669* (K, L), *BS 30708* (K, UC); Mt. Macosolon, *Ramos & Edaño BS 30743* (BM). **Leyte. Leyte Prov.:** Palo, *Elmer 7048* (A, K, NY), *7055* (A, K, NY). **Leyte. Prov. unkown:** Mt. Abucayan, *Edaño BS 41652* (A, K, P, US). **Leyte. Without further data:** *Wenzel 1297* (A, BM, MO), *Wenzel 1361* (A, MO), *Wenzel 1376* (A, MO), *Wenzel 1416* (A, MO, NY). **Cebu. Cebu Prov.:** Alcoy, Manguera Forest, *Madulid & Reynoso PPI 27378* (BRIT, K); Mantalongon, *Madulid & Reynoso et al. PPI 27487* (BRIT); Malac, Corte, Carmen, *Gaerlan & Reynoso PPI 20288* (BRIT). **Cebu. Without further data:** *Ramos BS 11072* (BM, K), *BS 11076* (BM, K). **Negros. Negros Occidental Prov.:** Danjugan Island, Barangay Bulata, Cauayan, *Madulid & Majaducon PPI 36075* (BRIT). **Palawan. Palawan Prov.:** Puerto Princesa, Irawan River Valley, *Ridsdale SMHI 102* (L); Irawan River Valley, base of Mt. Beaufort, Tatanarom, *Soejarto, Reynoso, Sagcal & Rutz 6430* (K); Mt. Gantung, Mantalingajan Mt. Range, District of Tindogan, *Pipoly & Romero et al. PPI 37983* (K); Barangay Guadalupe, *Madulid, Reynoso & Agoo PPI 7954* (BRIT);Victoria Mountains, Aborlan, *Sulit PNH 12356* (A, L), *PNH 12432* (A, L); Narra, Mt. Victoria, Trident Mining area, *Podzorski SMHI 2142* (L); Brookes Point Municipality, Pulot, roadside Aboabo-Brookes Point [sic.], *Soejarto, Reynoso, Sagcal & Edrada 6507* (K). **Palawan Prov.:** without further data: *Curran FB 3876* (K), *Fernandez FB 21496* (US), *Merrill BS 1229* (A, BM, L, MO, NY, P), *Moseley s.n.* (K), *Vidal 3049* (K). **Mindanao. Surigao del Norte Prov.:** Loreto (Dinagat Island), Cambinliw, Santiago, *Gaerlan, Sagcal & Fernando PPI 4538* (BRIT, K); .Siargao Island, Poneas Islet, Sitio Dabo, Sunkoi Hill, *Stone, Barbon & Garcia PPI 9462* (BRIT); *Stone, Barbon & Garcia PPI 9463* (BRIT); Bucas Grande Island, N. end, Barangay Salug, Sitio Lawigan, *Stone, Barbon & Garcia PPI 12077* (BRIT); Bucas Grande Island, without further data, *Ramos & Pascasio BS 35053* (A, BM). **Surigao del Sur Prov.:** Adlay, *Anonuevo PNH 13451* (A, L). **Surigao Prov.:** without further data: *Ramos & Pascasio BS 34616* (A, UC), *BS 36416* (US). **Agusan**

del Norte Prov.: Butuan subprov., without further data, *Rafael & Ponce FB 20734* (K). **Misamis Prov.:** without further data: *Miranda FB 17986* (K). **Basilan Prov.:** Basilan, *Reillo BS 16084* (K, P). **Zamboanga Prov. [sic]:** without further data: *Merrill 11621* (A, K, L, P). **JOLO. Sulu Prov.:** Jolo Island, Mt. Daho, *Ramos & Edaño BS 43903* (A, NY, UC, US). **Tawi Tawi. Tawi Tawi Prov.:** Birad-dali (Angel), Languyan, *Gaerlan & Sagcal PPI 10087* (K).

Distribution.—Widely distributed throughout the Philippines and possibly also in Borneo.

Ecology.—Moist forest, including beach and mangrove forest, and secondary forest, sometimes on cultivated land; on well-drained soils and limestone; 0–1000 m.

Phenology records.—Flowering: January to September; fruiting: all year round.

Preliminary conservation assessment.—Least Concern (**LC**). Last collected in 1996.

Distinguishing characteristics.—Leaf blades chartaceous to subcoriaceous, (6.5–)9.5–18.5cm long; stipules valvate, as broad or broader than long, usually less than 5 mm long; inflorescences monochotomous.

Remarks.—This is one of the most widely distributed *Psychotria* species in the Philippines and may well extend into Borneo. It may be conspecific with *P. valetonii* Hochr., originally called *P. aurantiaca* Blume. Variation within *P. gitingensis* is considerable. On Palawan, many of the species assigned to *P. gitingensis* have thick, coriaceous leaves that are reddish brown and shiny when dry and look very much like representatives of *P. crassifolia* Miq., which is found in Borneo. If *P. gitingensis*, as here conceptualized, and *P. crassifolia* are conspecific, then this taxon would be known by the latter epithet. If *P. crassifolia*, *P. gitingensis*, and *P. valetonii* turn out to be conspecific, then the earliest name for this assemblage would be *P. valetonii*. Slight differences in leaf size and shape, and inflorescence structure are not sufficient to warrant the separation of *P. gitingenis* and *P. similis* and thus we have here treated the latter as a synonym of the former.

VIII.—Luzoniensis Species Group

Shrubs, treelets or small trees. Stipules imbricate or valvate. Leaf blades ***usually more than 3 cm wide;*** very rarely more than 30 cm long; chartaceous to coriaceous, rarely almost membranaceous. ***Inflorescences with axes and branches, trichotomous or rarely monochotmous, rarely with more than three main axes (e.g. pentachotomous),*** very rarely monochotomous. Flowers pedicellate. Fruits usually red, sometimes yellow, brown or black; pyrene inner (adaxial) surface flat; seed producing a red color when placed in 70% ethanol, due to a soluble seed coat pigment (SCP positive), rarely SCP negative; endosperm ruminate or entire, usually dark. Comprising 30 species.

Note.—Text in ***bold italics*** denotes features useful for the placement of taxa within the respective species group.

KEY TO THE SPECIES OF THE LUZONIENSIS SPECIES GROUP

1. Leaf base cordate.
   2. Leaf blades with 15–19 lateral veins per side; fruit obovoid to turbinate; pyrene ribbed on outer (dorsal) surface; endosperm entire (non ruminate) ______ **73. P. cardiophylla**
   2. Leaf blades with 16–22 lateral veins per side; fruit globose to obovoid; pyrene smooth on outer (dorsal) surface; endosperm ruminate ______ **74. P. cordatula**
1. Leaf base acuminate, acute, cuneate, attenuate, obtuse or round, never cordate.
   3. Inflorescences nearly as long or longer than the subtending leaves.
      4. Leaf blades > 10 cm long, lateral veins 14–17 per side ______ **75. P. bontocensis**
      4. Leaf blades 8 cm long , lateral veins < 10 per side ______ **64. P. merrittioides**
   3. Inflorescences usually much shorter than subtending leaves.
      5. Inflorescences lax, rather fragile (inflorescence axes and pedicels 0.4–0.6 mm in diam.) ______ **76. P. merrittii**
      5. Inflorescences never lax, ± robust (inflorescence axes and/or pedicels > 1 mm in diam.).
         6. Calyx tubes 4–6 mm in diam. ______ **91a. P. paloensis** var. **bukidnonensis**
         6. Calyx tubes usually < 2 mm in diam.
            7. Leaf blades pubescent or tomentose below, the petioles, inflorescence axes and other plant parts also often pubescent to lanate; leaf blades often > 12 cm long.
               8. Calyx (incl. hypanthium) with erect to spreading, wooly hairs; leaf blades drying light grayish green ______ **77. P. ramosii**
               8. Calyx (incl. hypanthium) glabrous or lacking wooloy hairs; leaf blades drying dark brown, orange-brown, or grayish brown.
                  9. Endosperm entire (non ruminate); leaf blades usually < 12 cm long, lateral veins 15–20 per side ______ **78. P. capizensis**

9. Endosperm ruminate; leaf blades > 12 cm long, lateral veins 9–16 per side.
10. Leaf blades drying dark orange-brown, pubescence dark, rusty brown _____ **79. P. dingalanaensis**
10. Leaf blades drying olivaceous green, or olivaceous brown to grayish brown, as well as reddish brown, the pubescence chocolate-brown or whitish.
11. Pubescence brownish _____ **80. P. lagunensis**
11. Pubescence whitish or yellowish _____ **81. P. velutina**
7. Leaf blades glabrous or with scattered puberulence; leaf blade (if pubescent) usually < 12 cm long.
12. Endosperm entire (non ruminate) _____ **66. P. panayensis**
12. Endosperm ruminate.
13. Dry fruits usually with sharp, prominent ridges; pyrenes prominently ridged on outer (dorsal) surface.
14. Leaf blades with 18–24 lateral veins per side; petioles 4.8–6.7 cm long _____ **83. P. longissima**
14. Leaf blades usually with 10–18 lateral veins per side; petioles usually < 5 cm long.
15. Leaf blades usually < 5 cm wide (often < 4 cm wide), often assymetric, apex long-acuminate.
16. Inflorescences and floral parts glabrous _____ **84. P. pinnatinervia**
16. Inflorescences and floral parts puberulous, often densely so _____ **85. P. pubilimba**
15. Leaf blades usually > 5 cm wide, if less, apex never long-acuminate.
17. Leaf blade undersurface lacking a prominent network of tertiary veins (use 10× lens).
18. Leaf blades ovate to oblong-ovate, coriaceous, base round to obtuse _____ **86. P. catanduaniensis**
18. Leaf blades obovate-elliptic to oblong-elliptic, subcoriaceous, base acute to narrowly acute, rarely attenuate _____ **87. P. banahaensis**
17. Leaf blade undersurface with a prominent network of tertiary veins (use 10× lens).
19. Leaf blades obovate, drying reddish brown, apex abruptly acute _____ **88. P. galintanensis**
19. Leaf blades of various shapes but not obovate, usually drying khaki-green or grayish green, apex acute to acuminate.
20. Fruits globose to ellipsoid, 8–11 mm long; pyrenes distinctly ridged on outer (dorsal) surface, endosperm strongly ruminate; principal axis of inflorescence 2.3–5.1 cm long _____ **90. P. reynosoi**
20. Fruits obovoid to turbinate, 5–7 mm long; pyrenes ridged on outer (dorsal) surface, endosperm ruminate to weakly ruminate; principal axis of inflorescence 1.5–3.5 cm long _____ **89. P. pallidifolia**
13. Dry fruits usually smooth, ribbed or wrinkled, without sharp, prominent ridges; pyrenes smooth on on outer (dorsal) surface or with irregular, shallow ribbing.
21. Leaf blades usually < 4.5 cm wide.
22. Inflorescence irregularly branched, compact, flat-topped; fruits obovoid _____ **91. P. paloensis**
22. Inflorescence regularly branched, spreading, not flat-topped; fruits turbinate or sometimes globose, often ± bilobed _____ **92. P. luzoniensis**
21. Leaf blades usually > 4.5 cm wide.
23. Flowers and fruits sessile or subsessile _____ **42. P. wenzelii/35. P. subselliflora**
23. Flowers and fruits usually pedicellate.
24. Leaf blades distinctly coriaceous; petioles 3–5 mm wide at mid-point _____ **93. P. bayombongensis**
24. Leaf blades chartaceous to ± coriaceous, petioles 1–3 mm wide at mid-point.
25. Inflorescences condensed; leaf undersurface often somewhat warty _____ **82. P. fenicis**
25. Inflorescences not condensed; leaf undursurface smooth.
26. Inflorescences lax at anthesis _____ **94. P. negrosensis**
26. Inflorescences rather robust at anthesis.
27. Leaf blades conspicuously brochidodromous below (lateral veins prominently looped to each other near the margin) _____ **95. P. patulinervia**
27. Leaf blades not conspicuously brochidodromous (lateral veins not prominently looped to each other).
28. Fruits often appearing bilobed towards the apex, and often with one major longitudinal sulcus _____ **92. P. luzoniensis**
28. Fruits entire, lacking a major longitudinal sulcus.
29. Petioles usually > 4 cm long.
30. Fruits 9–11 mm long _____ **96. P. casiguranensis**
30. Fruits 5–7 mm long _____ **97. P. mariguidonensis**

29. Petioles usually < 4 cm long, usually < 3 cm long.
  31. Leaf blades broadly obovate to subovate, leaf base attenuate, leaf apex abruptly acute to abruptly acuminate ______ **98. P. plumeriaefolia**
  31. Leaf blades obovate, oblanceolate to elliptic or ovate, leaf base never attenuate, leaf apex acuminate, acute or obtuse.
    32. Inflorescences/infructescences 3.8–9.5 cm long (at least the main axis), not compact and often spreading.
      33. Leaf blades ovate to oblanceolate, often drying shiny yellow-brown or orange or reddish brown, ______ **99. P. nitens**
      33. Leaf blades broadly elliptic, often drying dark olive-green, or grayish, greenish brown ______ **100. P. mendozii**
    32. Inflorescences/infructescences 1–6.5 cm long, usually compact and not spreading.
      34. Short styled flowers with corolla tube 2.5–3 mm long; fruits smooth or sometimes wrinkled when dry; pyrenes smooth to very obscurely ribbed-striated on outer (dorsal) face ______ **101. P. yatesii**
      34. Short styled flowers with corolla tube 1–2 mm long; fruits smooth or ribbed when dry; pyrenes ribbed on outer (dorsal) face ______ **102. P. weberi**

73. Psychotria cardiophylla Merr., Philipp. J. Sci. 17:446. 1921; Merr., Enum. Philipp. Fl. Pl. 3:553. 1923. (**Fig. 86**). TYPE: PHILIPPINES. LUZON: Quezon Prov.: Infanta-Siniloan Trail, "along small streams in forests," 12 Jun 1917, *Ramos & Edaño BS 29166* (LECTOTYPE: A, designated here; ISOLECTOTYPES: K, P).

Shrub, height unknown, young stems puberulous to pubescent Stipules unknown. Leaves: petioles 2.1–3.3 cm long, pubescent; leaf blades ovate to oblong-elliptic, 16.9–23.2 × 6.4–12 cm, subcoriaceous, drying rusty-brown, glabrous above, pubescent below, particularly on midrib and lateral veins; secondary veins 15–19 pairs; tertiary venation rather prominent; base cordate with rounded or blunt lobes; apex obtuse, acute to abruptly acuminate. Inflorescence/infructescence trichotomous, 3.4–4.8 cm long, puberulous to ± pubescent; principal axis 2.2–4.5 cm long, each axis with 2–3 primary nodes and verticellate branching at each node, the branches terminating in 1–several cymules; bracts and bracteols unknown (?absent). Flowers unknown. Fruit obovoid to turbinate, 6–7 mm long, smooth to slightly ribbed, reported to be brown but probably dark red to black, sparsely puberulous to sparsely pubescent; calyx limb and lobes persistent but inconspicuous; pedicels 3–6 mm long; pyrenes ribbed on outer surface; endosperm entire.

Other specimens examined: Known only from the type collection.

Distribution.—Known only from the Infanta-Siniloan area of Quezon province.

Ecology.—Moist forest, including forest along streams; lower elevations.

Phenology records.—Flowering: unknown; fruiting: June.

Preliminary conservation assessment.—Extinct (**EX**). Only collected once in 1917.

Distinguishing characteristics.—Leaf blades large, lateral veins 15–19 per side, base cordate; inflorescences trichotomous; fruit turbinate; pyrenes ribbed on outer (dorsal) surface, endosperm entire (non ruminate).

Remarks.—Merrill (1921) described this taxon at the same time he described *P. cordatula*. In his treatment of the latter taxon, he stated that these two taxa are closely related and distinct on the basis of the number of lateral veins per leaf. In his discussion for *P. cardiophylla*, however, he states that 'the alliance of this species is manifestly with *P. alvarezii* Merr. from which it differs in its broader, differently shaped leaves ....' Actually, this taxon is quite distinct from *P. alvarezii*, which we have synonymized with *P. bataanensis* Elmer. The latter taxon has sessile flowers and fruit and belongs in another species group.

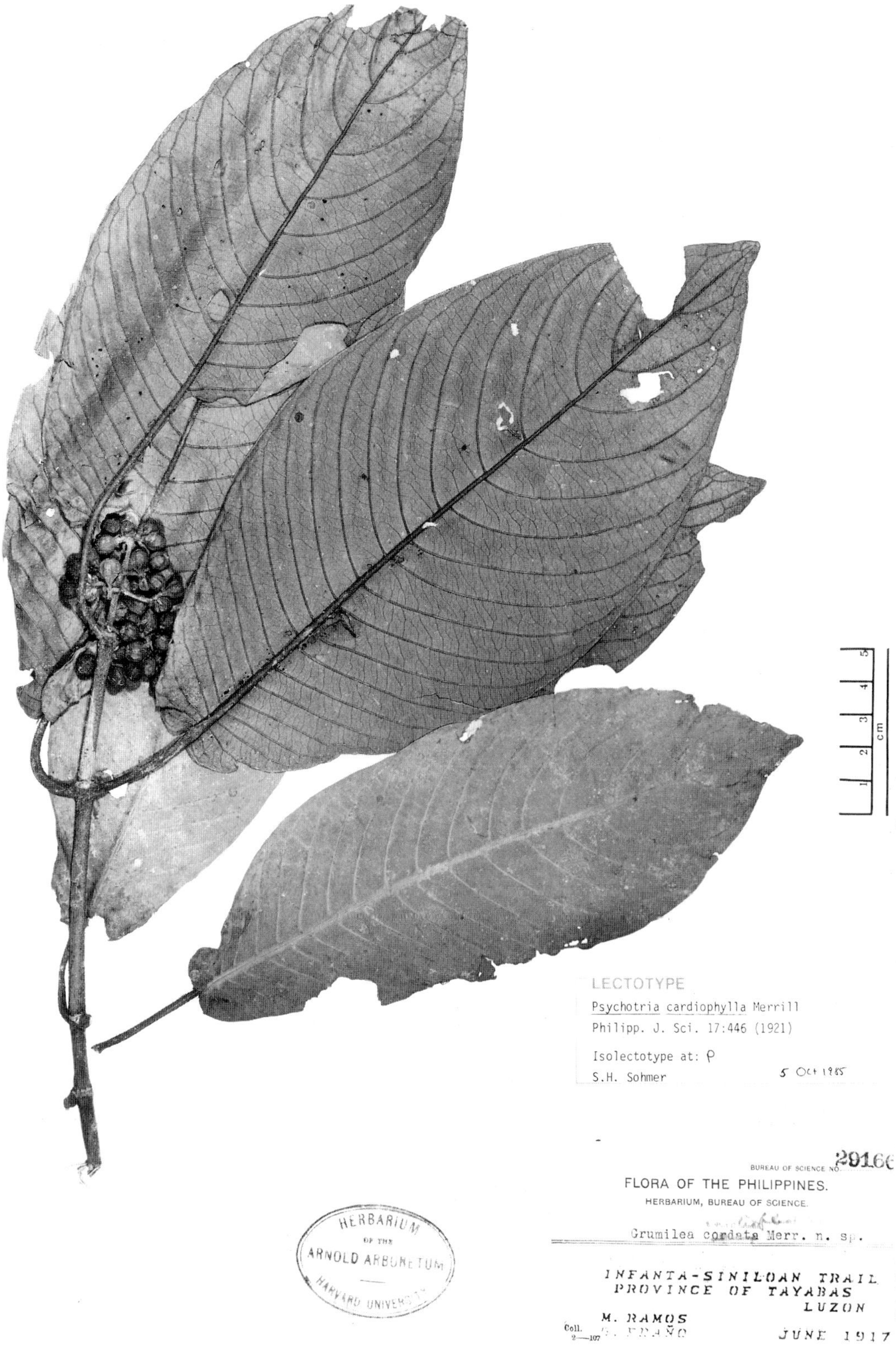

Fig. 86. *Psychotria cardiophylla* (*Ramos & Edaño BS 29166*) (A, lectotype).

74. Psychotria cordatula Merr., Philipp. J. Sci. 17:447. 1921; Merr., Enum. Philipp. Fl. Pl. 3:554. 1923. (**Fig. 87**). TYPE: PHILIPPINES. LUZON: Mountain Prov.: Mt. Masapilid, mossy forests at about 1250 m, 15 Mar 1920, *Ramos & Edaño BS 37854* (LECTOTYPE: A, designated here; ISOLECTOTYPES: K, US).

Shrub, ca. 8 m high, young stems glabrous to pubescent. Stipules valvate or imbricate, lanceolate-triangular, 1.5–1.7 cm long, glabrous; apex acute. Leaves: petioles 1.5–3.5 cm long, puberulous to pubescent, or glabrous; leaf blades lanceolate, slightly assymetrical, (9.5–)12.5–18.5 × 4–6.3 cm, ± coriaceous, drying reddish or chocolate-brown, glabrous above, pubescent on veins and midrib below; secondary veins 16–22 pairs; tertiary venation manifest to rather prominent; base cordate with rounded lobes; apex acuminate. Inflorescence/infructescence trichotomous, 6–8.8 cm long, pubescent to puberulous, or rarely glabrous; principal axis 2.7–7 cm long, with 2–3 primary nodes and (if branched) verticellate branching, the branches terminating in small cymules, or nodes with clusters of flowers/fruits; bracts and bracteoles unknown (?caducous or absent). Flowers largely unknown (immature); calyx (incl. hypanthium) ca. 1 mm long, puberulous to pubescent; calyx lobes obscure; corolla and other flower parts unknown. Fruit globose to obovoid, 5–7 mm long, smooth or somewhat wrinkled (when dry) but not ribbed, yellow but possibly red at maturity, puberulous or glabrous; calyx limb and lobes persistent but not obvious; pedicels 2–5 mm long; pyrenes smooth on outer surface; endosperm ruminate.

Other specimens examined: **Luzon. Ifugao Prov.:** Bayninan, Banaue, *Conklin & Buwaya PNH 79575* (K, PNH); Barangay Pula, Mt. Hagada, Banaue, *Reynoso, Garcia & Sagcal PPI 14492* (BRIT, K); Mt Bukung, Barangay Dalikan, Banaue, *Reynoso, Sagcal & Garcia PPI 7030* (BRIT).

Distribution.—In north central Luzon, in the provinces of Ifugao and Mountain.

Ecology.—Mid-montane forest, including secondary forest; 1250–1600 m.

Phenology records.—Flowering: unknown; fruiting: March, July, August.

Preliminary conservation assessment.—Endangered (**EN B1ab(iii)**). Last collected in 1992.

Distinguishing characteristics.—Leaf blades lanceolate, slightly assymetrical, more or less coriaceous, secondary veins 16–22 pairs, base cordate with rounded lobes; inflorescences trichotomous; fruit globose to obovoid, pyrenes smooth on outer (dorsal) surface, endosperm ruminate.

Remarks.—Merrill (1921) states that *P. cordatlula*: '... is allied to *P. cardiophylla* Merr. but has differently shaped, more-numerously nerved leaves.' In addition, the fruit shape is different between these otherwise morphologically very similar taxa, being globose to obovoid in *P. cordatlula* and turbinate in *P. cardiophylla*; the endosperm is ruminate in the former and homogeneous (non ruminate) in the latter.

75. Psychotria bontocensis Merr., Philipp. J. Sci. 9:456. 1915; Merr., Enum. Philipp. Fl. Pl. 3:553. 1923. TYPE: PHILIPPINES. LUZON: Mountain Prov.: Bontoc subprov., Bauco, along streams at about 1500 m, 9 Mar 1913, *Vanoverbergh 2610* (not seen).

Small tree, 3–4 m high, young stems glabrous. Stipules imbricate, ± triangular, 1.5–2.3 cm long, glabrous; apex acute or acuminate. Leaves: petioles 1.5–5 cm long, glabrous; leaf blades elliptic-obovate to elliptic-lanceolate to obovate, 12.2–21 × 4.4–6.3(–7.4) cm, subcoraiceous, drying olivaceous brown, glabrous above, usually glabrous below but sometimes pubescent on midrib; secondary veins 14–17 pairs; tertiary venation manifest, ± scalariform; base acute to narrowly acute; apex obtuse, or abruptly acute to narrowly acute. Inflorescence/infructescence trichotomous, 4.8–14.4 cm long, glabrous or sometimes puberulous; principal axis 8.4–20.5 cm long, with 1 node and veticillate branching, or 1–3 nodes with verticillate (bottom node) or opposite branching (upper node(s)), terminating is cymules; bracts ovate to ± triangular, 1.4–3 mm long, but often absent (?caducous); bracteoles triangular to very narrowly triangular, ca. 0.7–2 mm long. Flowers 4-merous, pedicels 0.7–2 mm long; calyx (incl. hypanthium) ca. 2 mm long, glabrous; calyx lobes obscure to ovate-deltate, ca. 0.7 mm long; corolla: white; corolla tube ca. 2 mm long, external surface glabrous, villous at throat within; corolla lobes ca. 1.5 mm long; stamens: filaments ca. 2 mm long (short styled); anthers ca. 0.7 mm long; style ca. 2 mm long, lobes ca. 0.5 mm long. Fruit unknown.

Other specimens examined: **Luzon. Kalinga-Apayao Prov.:** Mt. Masingit, Lubuagan, *Ramos & Edaño BS 37550* (A). **Mountain Prov.:** Bontoc, *Vanoverbergh 4366* (A, L, MO, S, W); Mt. Masapilid, *Ramos & Edaño BS 37856* (K). **Ifugao Prov.:** Batad, Banaue, *Mendoza PNH 97485* (L, PNH). **Benguet Prov.:** Pauai, *Santos BS 31737* (A, BM, K, P, US). **Rizal Prov.:** Balacbac, *Loher 13017* (UC). **Romblon.**

Fig. 87. *Psychotria cordatula* (*Ramos & Edaño BS 37854*) (A, lectotype).

Romblon Prov.: no further data, *Elmer 12165* (BISH, BM, GH, K, L, MO, NY, US, W). **Mindanao. Agusan Prov.:** without further data: *Miras & Oliveros FB 24524* (US), *vel aff.*

Distribution.—In Ifugao, Rizal, Benquet, and Mountain provinces of Luzon, as well as one uncertain report from Mindanao.

Ecology.—High elevation forest; in ravines and by streams; 1300–1800 m.

Phenology records.—Flowering: March to June; fruiting: unknown.

Preliminary conservation assessment.—Near Threatened (**NT**). Last collected in 1967.

Distinguishing characteristics.—Leaves coriaceous, prominently veined; inflorescences trichotomous as large or larger than the subtending leaves.

Remarks.—Merrill (1915a) stated that this species was 'in a group with *P. banahaensis* Elmer, but not closely allied to that species.' It is clearly separated from *P. banahaensis* by the nature of its inflorescence, this being the only species in the Luzoniensis Species Group in which the inflorescence is relatively robust and as large as or larger than the subtending leaves. *Psychotria bontocensis* stands apart on this characteristic alone within the species group, except for *P. merrittioides* Sohmer & A.P. Davis, which differs from the former by its relatively small leaves (which are less than 10 cm long) and generally monochotomous inflorescences.

76. Psychotria merrittii Merr., Philipp. J. Sci. 26:493. 1925. (**Fig. 88**). TYPE: PHILIPPINES. MINDORO: Mindoro Oriental Prov.: Mt. Halcon, on forested slopes at about 1400 m, 16 Mar 1922, *Ramos & Edaño BS 40599* (LECTOTYPE: P, designated here; ISOLECTOTYPES: A, K, US).

Treelet or shrub, 2–4 m high, glabrous throughout. Stipules imbricate, narrowly triangular-lanceolate to ± ligulate, 0.9–2.5 cm long, glabrous; apex acuminate to acute. Leaves: petioles 1.5–2.5 cm long, glabrous; leaf blades elliptic to elliptic-ovate, rarely elliptic-obovate, (4.9–)9.2–16 × (1.9–)2.6–5.2(–6.1) cm, semimembranceous to chartaceous, drying olivaceous black to brownish black, glabrous above and below; secondary veins 10–13 pairs; tertiary venation rather prominent; base acute to narrowly acute; apex abruptly acuminate to subcaudate. Inflorescence/infructescence trichotomous, 3.8–7.5 cm long, glabrous; principal axis (1.3–)2.5–4 cm long, all axes very slender (0.4–0.6 mm in diam. when dry), each axis with primary 1 node, terminating in 1–3 lax cymules, inflorescence parts widely separated; bracts and bracteoles apparently absent. Flowers 6-merous, pedicels 0.9-3.5 mm long; calyx (incl. hypanthium) 1.5–2.5(–3) mm long, glabrous; calyx lobes obscure or absent (apex truncate); corolla: white; corolla tube distinctly broadening towards the apex, 1.2–3 mm long, external surface glabrous, densely villous at throat within (short styled); corolla lobes (1.1–)2.1–4 mm long; stamens: filaments ca. 2 mm long; anthers ca. 1 mm long (short styled); style ca. 1 mm long, lobes ca. 0.5 mm long. Fruit ellipsoid, ca. 8 mm long, smooth, color unknown, glabrous; calyx limb and lobes persistent; pedicels 1–5 mm long; pyrenes smooth or slightly ribbed on outer surface; endosperm ruminate.

Other specimens examined: **Mindoro. Mindoro Oriental Prov.:** Mt. Halcon, *Ridsdale 1733* (K, L), *1738* (K, L), *1738A* (K, L); north face of Mt. Halcon, Lantuyan, *Stone, Reynoso & Sagcal PPI 513* (BRIT, K); Subaan River, inland from San Teodoro, *Coode 5334* (K); Inland from San Teodoro, at and above Subaan River, *Ridsdale 1150* (K).

Distribution.—Restricted to Mt. Halcon and the Subaan River, northern Mindoro.

Ecology.—Forest, including stunted, mossy forest, often on slopes; usually at relatively high elevations (e.g. 1350 m, and 1450 m), but also at low altitudes (e.g. 150–200 m).

Phenology records.—Flowering: March to May; fruiting: April.

Preliminary conservation assessment.—Endangered (**EN B1ab(iii)**). Last collected in 1991.

Distinguishing characteristics.—Leaves glabrous, apex abruptly acuminate to subcaudate; inflorescences lax, few-flowered, peduncles slender; pedicels 0.4–0.6 mm in diam.; corolla tubes 3 mm long, widely dilated towards the apex.

Remarks.—This species is easily recognizable and apparently rather rare. It apparently still exists on Mt. Halcon, as three collections were made by Ridsdale in 1986, and one by Stone, Reynoso & Sagcal in 1991 (see above). Merrill (1925) believed that there was a resemblance to *P. pinnatinervia* Elmer, but we are cimcumspect about agreeing with his observation. Two collections (*Coode 5334*, *Ridsdale 1150*) made at low attitude (150 m and 200 m, respectively) differ slightly from the collections made on Mt. Halcon. The leaves of the former tend to be slightly larger are often elliptic-obovate rather than elliptic to elliptic-ovate, otherwise they fall within the variation observed for this species.

Fig. 88. *Psychotria merrittii* (*Ramos & Edaño BS 40599*) (P, lectotype).

77. Psychotria ramosii Merr., Philipp. J. Sci. 5:244. 1910; Merr., Enum. Philipp. Fl. Pl. 3:562. 1923. (**Fig. 89**). TYPE: PHILIPPINES. LUZON: Cagayan Prov.: near Pamplona, 17 Mar 1909, *Ramos BS 7499* (LECTOTYPE: US, designated here; ISOLECTOTYPE: K).

Shrub, ca. 2 m high, young stems glabrous. Stipules valvate or imbricate, triangular, ca. 0.7 cm long, pubescent (covered with wooly, white hairs); apex acute. Leaves: petioles 1.8–2.3 cm long, glabrous to sparsely pubescent; leaf blades obovate-oblanceolate, 9–12 × 4–5.5 cm, chartaceous, drying grayish green, glabrous above, pubescent below mainly on midrib and lateral veins (hairs white, crisped, appressed); secondary veins 10–14(–16) pairs; tertiary venation manifest to quite prominent; base attenuate to acute; apex obtuse to abrubtly acute. Inflorescence/infructescence with 3–several main axes, 3–3.5 cm long, pubescent (hairs erect to spreading, wooly, white); principal axes 1.5–2 cm long, each axis with 1 primary node, terminating in small cymules; bracts and bracteoles apparently absent. Flowers 5-merous, pedicels 1–2 mm long; calyx (incl. hypanthium) ca. 3 mm long, pubescent (as inflorescence); calyx lobes deltate, ca. 1 mm long; corolla: white; corolla tube distinctly broadening towards the apex, 2.5–3 mm long, external surface glabrous to sparsely pubescent), sparsely pubescent at throat within; corolla lobes ca. 3 mm long; stamens: filaments ca. 0.5 mm long; anthers ca. 0.5 mm long (short styled); style ca. 2.5 mm long, lobes ca. 0.8 mm long. Fruit unknown.

Other specimens examined: Known only from the type collection.

Distribution.—Known only from Pamplona, Luzon.

Ecology.—Forest; low elevations.

Phenology records.—Flowering: March; fruiting: unknown.

Preliminary conservation assessment.—Extinct (**EX**). Only collected once in 1909.

Distinguishing characteristics.—Nearly all the plant parts covered with wooly hairs, including the calyx. inflorescences trichotomous or with several main axes arising from the shoot apex.

Remarks.—This is a rather unique-looking species and is unlike anything else in the Philippines. As it is known from only one flowering specimen we are not able to make any assumptions about its relationships with other *Psychotria* species. There is a strong possibility that *P. ramosii* is extinct.

78. Psychotria capizensis Merr., Philipp. J. Sci. 17:445. 1921. Merr., Enum. Philipp. Fl. Pl. 3:553. 1923. (**Fig. 90**). TYPE: PHILIPPINES. PANAY: Capiz Prov.: Mt. Madigas, forests, without further data, 13 May 1918, *Ramos & Edaño BS 30657* (LECTOTYPE: A, designated here; ISOLECTOTYPES: K, US).

Shrub, height unknown, young stems puberulous to pubescent. Stipules imbricate, ± elliptic to lanceolate, ca. 1 cm long, glabrous; apex acute. Leaves: petioles 1.4–2.2 cm long, pubescent; leaf blades elliptic-obovate, or ± elliptic, 9–12.2 × 3.8–5.3 cm, subcoriaceous to coriaceous, glabrous above, puberulous below on midrib and sometimes lateral veins; secondary veins 15–20 pairs; tertiary venation rather prominent; base narrowly obtuse to nearly round; apex abruptly acute to acuminate. Inflorescence/infructescence with 2–4 axes, 2.6–4.6 cm long, pubescent to densely pubescent; principal axis 2–3.1 cm long, each axis with 1 primary node, terminating in small lax clusters of 1–several flowers/fruits; bracts and bracteoles unknown (?caducous or absent). Flowers unknown. Fruit obovoid-turbinate, 5–6 mm long, somewhat wrinkled, probably red, usually glabrous; calyx limb and lobes persistent but not obvious; pedicels 1–3 mm long; pyrenes smooth on outer surface; endosperm entire.

Other specimens examined: Known only from the type collection.

Distribution.—Known only from Mt. Madigas, Panay.

Ecology.—Forest, probably at mid altitudes, otherwise unknown.

Phenology records.—Flowering: unknown; fruiting: April, May.

Preliminary conservation assessment.—Extinct (**EX**). Only collected once in 1918.

Distinguishing characteristics.—Leaves relatively small, lateral veins 15–20 per side, densely pubescence; inflorescences with 2–4 main axes arising from the shoot apex; endosperm non ruminate.

Remarks.—This species resemblers members of the Linearis Species Group, in which the inflorescences are very much reduced (lacking axes and branches) and the flowers are pedicellate. If *P. capizensis* had reduced inflorescnences, it would appear very similar to species in the Linearis Species Group. Otherwise *P. capizensis* resembles *P. fenicis* Merr., due to its non ruminate endosperm and leaf color (when dry).

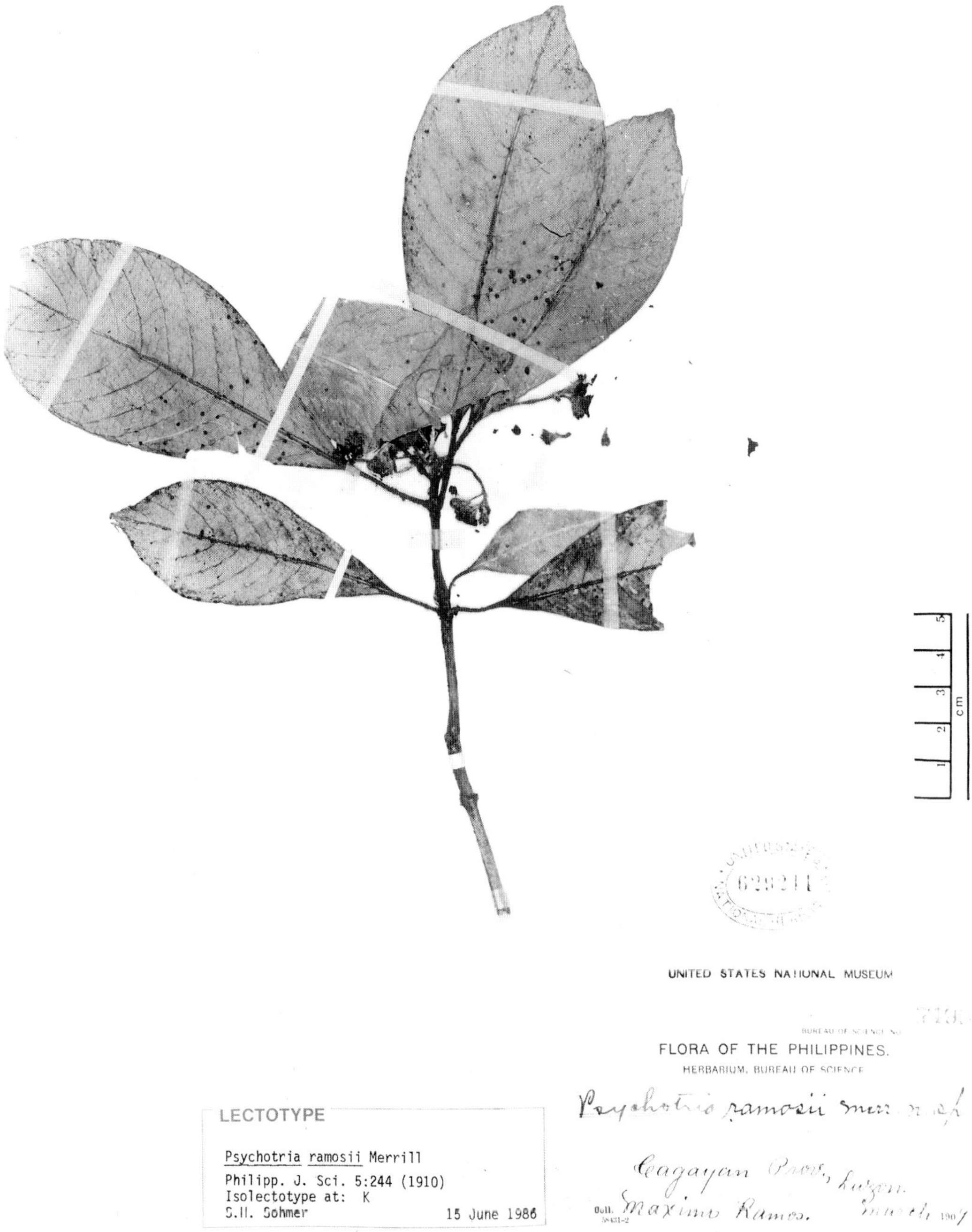

FIG. 89. *Psychotria ramosii* (*Ramos BS 7499*) (US, lectotype).

79. Psychotria dingalanensis Sohmer & A.P. Davis, sp. nov. (**Fig. 91**). TYPE: PHILIPPINES. LUZON: Aurora Prov.: Sierra Madre Mountains north-northeast of Dingalan, primary dipterocarp forest at 300–400 m, 12 Mar 1968, *Jacobs 7654* (HOLOTYPE: K; ISOTYPE: L).

Arbor parva est, foliis cum laminis 4.2 × 12–8 × 21 cm latis et longis oblanceolati-obovatis apice obtuso vel acuto basi acuta vel cuneata subcoriaceis supra glabris infra ferruginose badie pubescentibus praecipue in nervis, nervis lateralibus 9–12 in dimidio quoque in sicco aurantiaci-badiis, petiolis 1.2–4.5 cm longis ferruginose badie pubescentibus, inflorescentia trichotoma axibus 2.5 cm longis dense pubescentibus, pedunculis 0.5–1.5 cm longis, nodis 1–2 et irregulariter ramosis, ramulis ultimis cum fructibus

Fig. 90. *Psychotria capizensis* (*Ramos & Edaño BS 30657*) (A, lectotype).

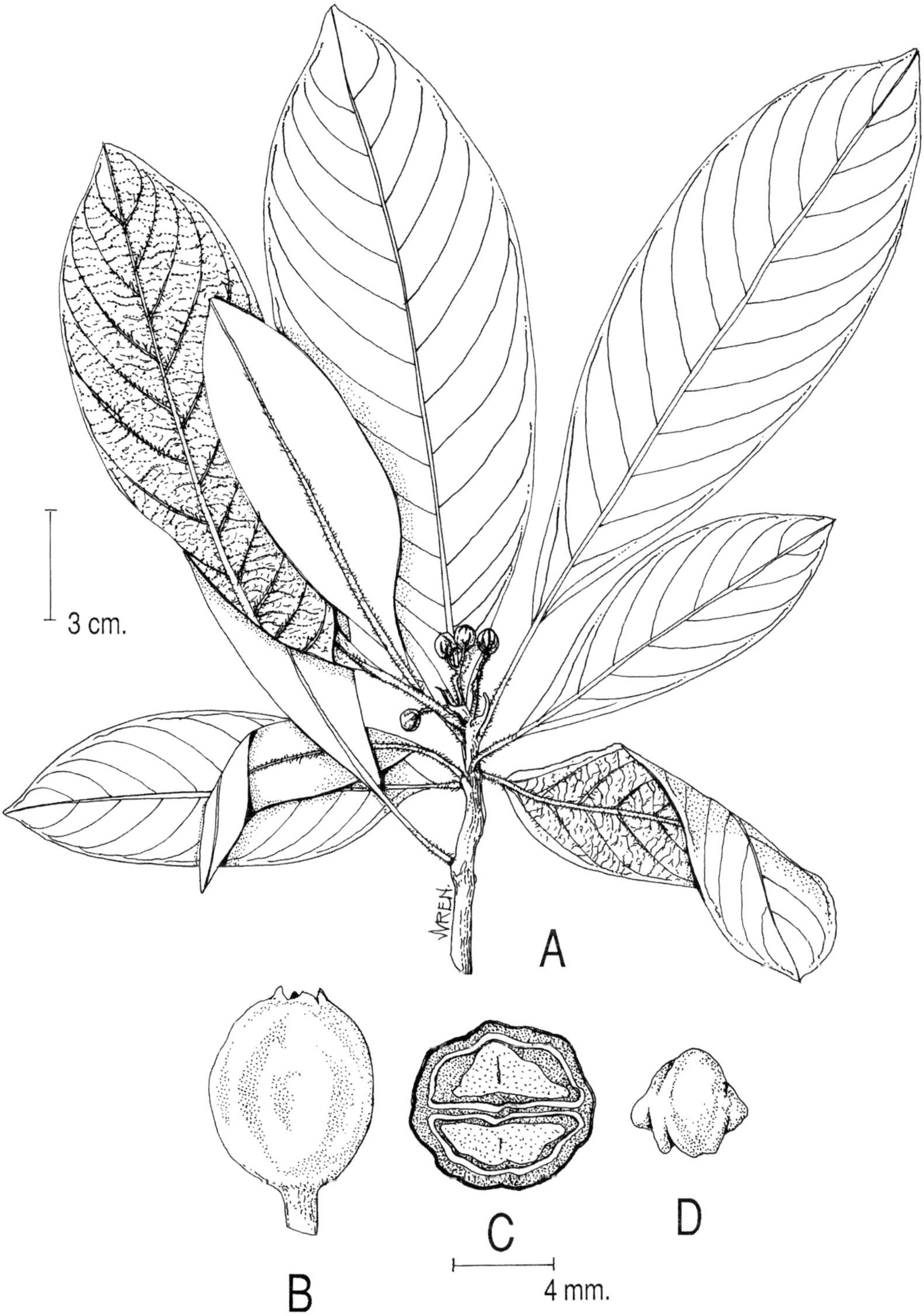

Fig. 91. *Psychotria dingalanensis* (*Jacobs 7654*) (K, holotype). **A,** habit of fruiting branch; **B,** fruit; **C,** fruit in cross section; **D,** pyrene.

paucis, pedicellis 2-5 mm longis, fructibus maturis 9-10 mm longis rubris obovoideis vel turbinatis laevibus sed cum rugis paucis calycibus persistentibus, pyrenis in dorso laevibus, endospermo ruminato.

Small tree, height unknown, young stems pubescent. Stipules unknown. Leaves: petioles 1.2-4.5 cm long, pubescent (hairs rusty-brown); leaf blades oblanceolate-obovate to oblanceolate-elliptic, 12-21 × 4.2-8 cm, subcoriaceous to ± coriaceous, drying dark orange-brown, glabrous above, pubescent below, particularly on midrib and lateral veins (hairs rusty-brown); secondary veins 9-12 pairs; tertiary venation quite prominent; base acute to cuneate; apex obtuse to broadly acute. Inflorescence/infructescence trichotomous, ca. 2.5 cm long, rather densely pubescent; principal axis 0.5-1.5 cm long, each axis with 1-2 nodes, terminaing in small rather irregularly branched cymules; bracts and bracteoles unknown (?caducous or absent). Flowers unknown. Fruit obovoid to weakly turbinate, 9-10 mm long, mostly smooth with some wrinkles (when dry), red, glabrous; calyx limb and lobes persistent but not conspicuous; pedicels 2-5 mm long; pyrenes smooth on outer surface; endosperm ruminate.

PARATYPES: None—known only from the type collection.

Distribution.—Known only from one location in the Sierra Madre Mountains, Luzon.

Ecology.—Primary dipterocarp forest; 300-400 m.

Phenology records.—Flowering: March; fruiting: March.

Preliminary conservation assessment.—Critically Endangered (**CR B1ab(iii)**), or possibly Extinct (**EX**). Only collected once in 1968.

Distinguishing characteristics.—Leaves 12-21 cm long, pubescent below, and drying orange-brown; inflorescences trichotomous; fruit 9-10 mm long, smooth, pyrenes smooth on outer (dorsal) surface.

Remarks.—This taxon stands out from all others in its group in the herbarium. In overall morphology, coloring, and pubescence it resembles *P. rubiginosa*, but the latter has monochotomous inflorescences that are many times longer than that produced by *P. dingalanensis*. It is doubtful whether the original habitat is as undisturbed now as it was when Jacobs collected there.

80. Psychotria lagunensis (Merr.) Merr., Enum. Philipp. Fl. Pl. 3:556. 1923. (**Fig. 92**). *Grumilea lagunensis* Merr., Philipp. J. Sci. 122:169. 1917. TYPE: PHILIPPINES. LUZON: Laguna Prov.: San Antonio, in damp forest, 19 Oct 1915, *Ramos BS 23826* (LECTOTYPE: K, designated here).

Shrub, 1-2 m high, young stems puberulous to pubescent. Stipules imbricate, ± lanceolate, 0.8-1.3 cm long, pubescent; apex acute. Leaves: petioles 1.1-5.6 cm long, puberulous to pubescent; leaf blades ± elliptic to subobovate, 13.7-20 × 5.8-9.4(-11.2) cm, chartaceous to subcoriaceous, drying dark olivaceous green; glabrous above, pubescence below on midrib and lateral venis and puberulous to pubesecent on venation and leaf surface; secondary veins (10-)12-17 pairs, sharply ascending; tertiary venation manifest to rather prominent; base acute to round; apex obtuse to acute or abruptly acute. Inflorescence/infructescence trichotomous, (2.6-)3.2-7.2 cm long, puberulous to pubescent; principal axis (1.3-)3.8-5.7 cm long, each axis with 1-2 nodes and opposite branching at each node, branches terminating in cymules; bracts apparently absent; bracteoles ovate, ca. 1 mm long, or absent (caducous). Flowers very poorly known; calyx (incl. hypanthium) ca. 1.2 mm long, puberulous; calyx lobes very shallowly triangular, ca. 0.2 mm long, or obscure; corolla and other flower parts unknown. Fruit obovoid, 6-10 mm long, sulcate to smooth, probably maturing red or black, reported yellow, glabrous or puberulent; calyx limb and lobes not persistent; pedicels 1-4 mm long; pyrenes ribbed on outer surface; endosperm ruminate.

Other specimens examined: **Luzon. Quezon Prov.:** Tagcauayan [Tagkawayan], *Ramos BS 13362* (K); Mt. Binuang, *Ramos & Edaño BS 28680* (NY); Polillo Island, Anibawan, *Castro & Anonuevo PNH 6525* (A), *McGregor BS 10212* (US); Alabat Island, *Ramos & Edaño BS 48008* (A, BM, K, P, SING, UC, US, W), *BS 48167* (UC). **Nueva Vizcaya Prov.:** Mt. Alzapan, *Ramos & Edaño BS 45583* (BM, UC). **Rizal Prov.:** Mt. Irid, *Ramos & Edaño BS 48461* (UC); Montalban, *Loher 12774* (UC); Mt. Susun Dalaga, *Ramos & Edaño BS 29335* (A); Mt. Lumutan, *Ramos & Edaño BS 29738* (A, K). **Laguna Prov.:** Majayjay, *Rivera & Duyag BS 49953* (UC). **Camarines Sur Prov.:** Sarapan, *Edaño BS 76147* (NY); Mt. Isarog, *Ramos BS 22098* (BM, K). **Camarines Norte Prov.:** Parcale, *Ramos & Edaño BS 33637* (A, UC, US), *BS 33746* (SING). **Catanduanes. Catanduanes Prov.:** without further data: *Ramos & Edaño BS 75251* (SING), *BS 75288* (UC), *BS 75323* (SING).

Distribution.—Eastern provinces of Luzon, Polillo, Alabat, and Catanduanes Islands.

Ecology.—Moist forest and forested slopes; 70-1532 m.

FIG. 92. *Psychotria lagunensis* (*Ramos BS 23826*) (K, lectotype).

Phenology records.—Flowering: March; fruiting: June to December.

Preliminary conservation assessment.—Vulnerable (**VU B1ab(iii)**). Last collected in 1928.

Distinguishing characteristics.—Leaf blades relatively large, covered with a brownish or reddish brown pubescence when dry, lateral veins 12–17 per side (steeply ascending); inflorescences trichotomous.

Remarks.—An easily recognizable species which has (or had) a wide distribution in the northern part of the archipelago. Two of the collections tentatively assigned here are the only two collections made since 1928 (*Edaño PNH 40174* (BM), *PNH 40195*) and both fit the concept of this species in nearly every way but they are both nearly glabrous.

81. Psychotria velutina Elmer, Leafl. Philipp. Bot. 5:1894. 1913; Merr., Enum. Philipp. Fl. Pl. 3:564. 1923. (**Fig. 93**). *Grumilea velutina* (Elmer) Merr., Philipp. J. Sci. 10:33. 1915. TYPE: PHILIPPINES. MINDANAO: Agusan Prov.: Cabadbaran (Mt. Urdaneta), in moist fertile soil of dense forests at 1060 m, Aug 1912, *Elmer 13615* (LECTOTYPE: UC, designated here; ISOLECTOTYPES: A, BISH, BM, GH, K, L, MO, NY, P, US, W).

Shrub, sparingly branched, 1–2 m high, young stems glabrous to puberulous. Stipules imbricate, narrowly lanceolate, 0.8–1.2(–1.6) cm long, densely pubescent (hairs whitish or yellowish); apex acute. Leaves: petioles 2–4.8 cm long, pubescent, sometimes glabrous; leaf blades broadly elliptic to broadly obovate, 16–23 × 7.8–11.5 cm, subcoriaceous, drying light olivaceous silvery green to olivaceous green, glabrous above, pubscent below mainly on midrib and lateral venis (hairs erect, whitish or yellowish); secondary veins 13–16 pairs, steeply ascending; tertiary venation rather prominent; base acute to obtuse; apex abruptly acute or accuminate. Inflorescence/infructescence trichotomous, total length unknown, probably puberulous to pubescent; principal axis 2–6(–8.1) cm long, each axis with 2–3 nodes, terminating with 2–4 flowers/fruit per node; bracts and bracteoles unknown. Flowers 5-merous, pedicels 1–2 mm long, pubescent; calyx (incl. hypanthium) ca. 2 mm long, glabrous; calyx lobes obscure; corolla: white; corolla tube 0.5–3 mm long, external surface glabrous, villous at throat within; corolla lobes ca. 2.5 mm long; stamens: filaments 1–2 mm long; anthers ca. 1 mm long (short styled); style ca. 1.5 mm long, lobes 0.5–0.8 mm long. Fruit obovoid, 7–10 mm long, ribbed, red, sparsely pubescent; calyx limb and lobes ± persistent but not obvious; pedicels unknown; pyrenes ribbed or ridged on outer surface; endosperm ruminate.

Other specimens examined: **Luzon. Quezon Prov.:** Malbog, *Paduada FB 29511* (A, BM, UC).

Distribution.—Luzon and Mindanao.

Ecology.—Humid forest; ca. 1060 m.

Phenology records.—Flowering: unknown; fruiting: August.

Preliminary conservation assessment.—Critically Endangered (**CR B1ab(iii)**). Last collected in 1923.

Distinguishing characteristics.—Leaf blades (and inflorescence axes) often densely covered with a whitish or yellowish pubescence, lateral veins 13–16 per side, ascending steeply and joining each other; inflorescences trichotomous.

Remarks.—This species is morphologically nearly indistinguishable from *P. lagunensis* except for the color of the dry pubescence. In *P. velutina* the pubescence is whitish or yellowish, whereas in *P. lagunensis* it is brownish or red-brown. There have been few new collections of this species since the early part of this century (the only recent collection available to us is from 1923), and with this lack of additional material it is probably wiser to leave the status of this species as is. What makes us suspect that it has (or had) a much wider distribution is the fact that the only two collections available come from widely differing parts of the Archipelago.

82. Psychotria fenicis Merr., Philipp. J. Sci. 17:449. 1921; Merr., Enum. Philipp. Fl. Pl. 3:555. 1923. (**Fig. 94**). TYPE: PHILIPPINES. LUZON: Pangasinan Prov.: Mt. San Isidro, Nov 1917, *Fenix BS 29914* (LECTOTYPE: US, designated here; ISOLECTOTYPE: L).

Shrub or treelet, 1–3 m high, glabrous throughout. Stipules imbricate, deltate to triangular, 0.2–0.4 cm long, glabrous; apex acute. Leaves: petioles 0.8–1.8 cm long, glabrous; leaf blades obovate to obovate-elliptic, 8–16.1 × 2.7–5.6 cm, ± coriaceous, often rather warty, drying reddish brown, glabrous above and below; secondary veins 10–14 pairs; tertiary venation prominent; base acute to attenuate; apex abruptly acuminate to subacute. Inflorescence/infructescence trichotomous, (0.7–)1–1.5(–2.8)

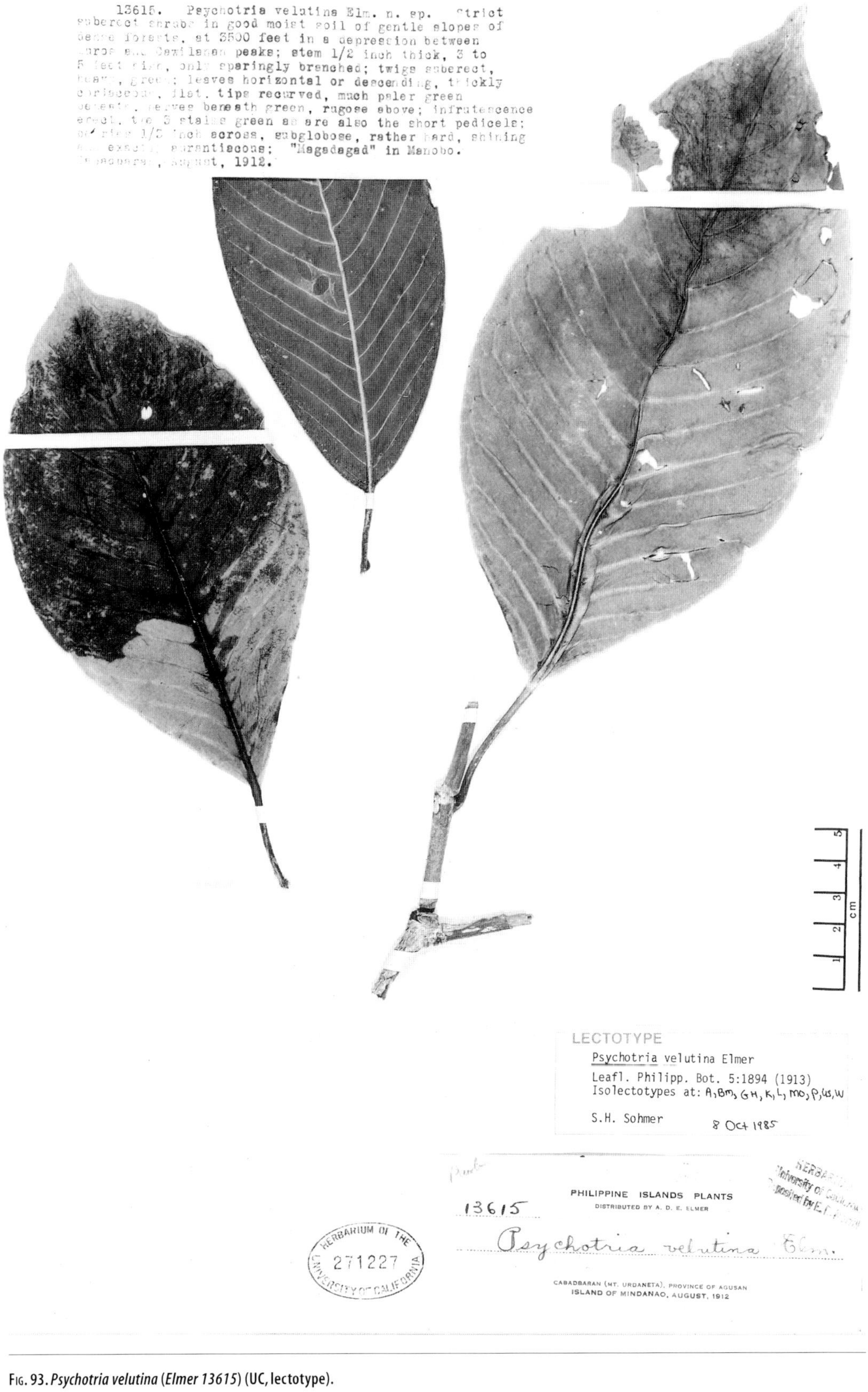

FIG. 93. *Psychotria velutina* (*Elmer 13615*) (UC, lectotype).

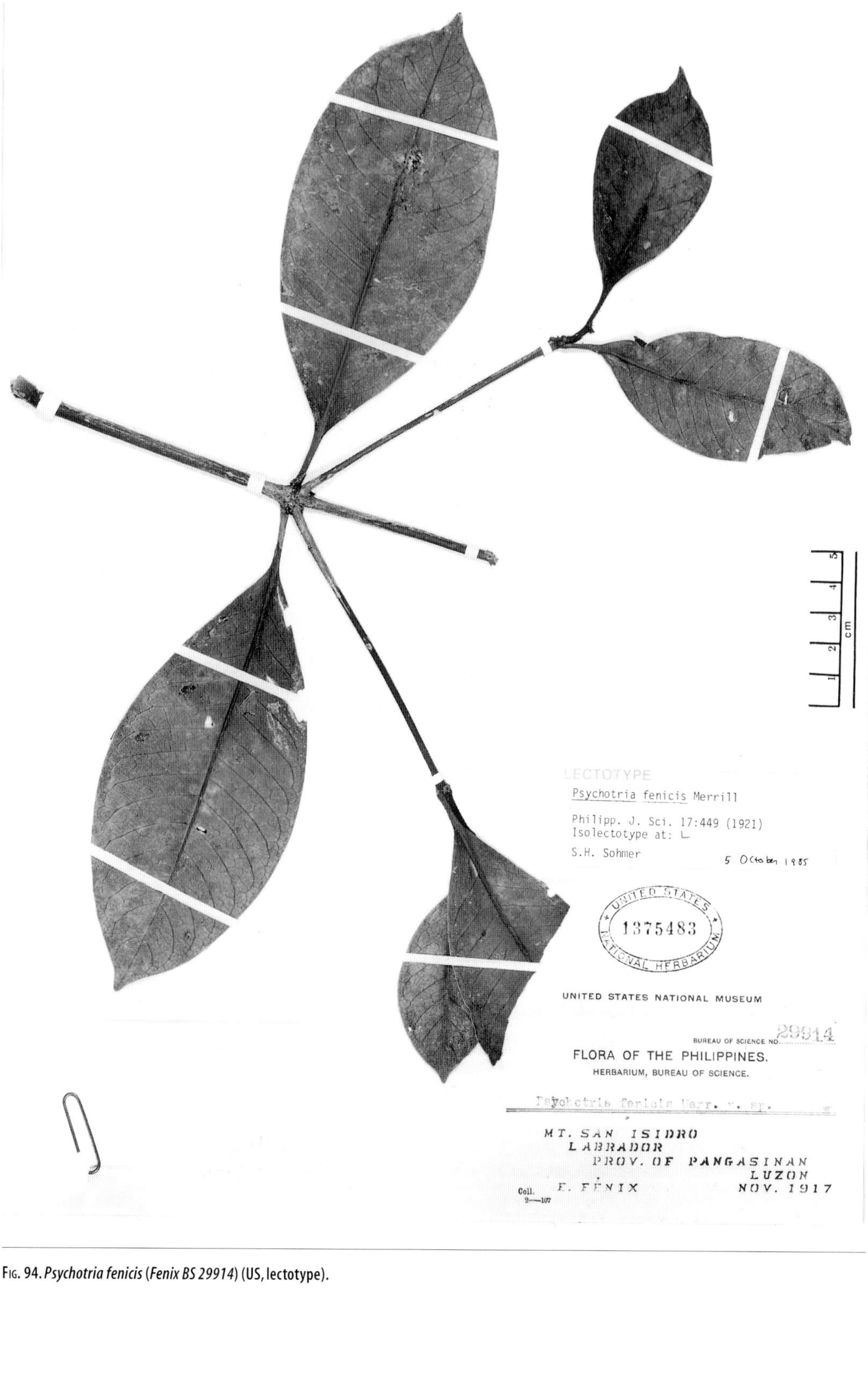

FIG. 94. *Psychotria fenicis* (*Fenix BS 29914*) (US, lectotype).

cm long, glabrous; principal axis (0.2–)0.5–1(–1.6) cm long, each axis with 1(–2) primary nodes and opposite branching, the branches probably terminating in 1 or 2 cymules, each cymule 3-flowered; bracts and bracteoles unknown (?caducous or absent). Flowers unknown. Fruit obovoid to somewhat turbinate, 6–8 mm long, wrinkled or somewhat ribbed, orange to red or black, glabrous, calyx limb and lobes not persistent; pedicels 2–3 mm long; pyrenes ribbed on outer surface; endosperm ruminate.

Other specimens examined: **Luzon. Pangasinan Prov.:** Mt. San Isidro, Labrador, *Fenix BS 29967* (BM, K, P). **Zambales Prov.:** Mt. Aglao, San Marcelino, *Edaño BS 26803* (UC); Mt. Pinatubo, *Fox PNH 4600* (A); Subig, *Hallier 4237* (L); Mt. Tapolao, *Ramos & Edaño BS 44726* (L,S,UC), *BS 44772* (S,UC); Masinloc, Coto Mines, *Reynoso, Alvarez & Fernando PPI 4144* (BRIT, K), *PPI 4163* (K). **Bataan Prov.:** Tawawa, slope of Mt. Natib, *Reynoso, Sagcal & Fuentes PPI 17236* (BRIT, K); *Reynoso, Sagcal & Fuentes PPI 17253* (BRIT, K); Mt. Natib, So. Ulingan, Barangay Banawang, Municipality of Bagac, *Fuentes & de la Rosa PPI 38708* (BRIT). **Bataan Prov.:** without further data: *Alhambra & Caulas FB 23223* (A).

Distribution.—Restricted to western Luzon, in the provinces of Pangasinan, Zambales and Bataan.

Ecology.—Dry forest or well-drained ridge forest, including secondary forest; recorded on rocky soils; 200–400 m.

Phenology records.—Flowering: May, November; fruiting: April, November to January.

Preliminary conservation assessment.—Endangered (**EN B1ab(iii)**). Last collected in 1992

Distinguishing characteristics.—Leaves coriaceous, venation prominent, undersurface of leaf often somewhat warty (verrucose), drying dark reddish brown; inflorescences trichotomous, short and condensed; endosperm ruminate.

83. Psychotria longissima Quisumb. & Merr., Philipp. J. Sci. 37:206. 1928. (**Fig. 95**). Type: PHILIPPINES. Luzon: Quezon Prov.: Mt. Alzapan, forested slopes, at ca. 1300 m, 7 Jun 1925, *Ramos & Edaño BS 45681* (lectotype: UC, designated here; isolectotypes: A, K, US).

Shrub, ca. 2 m high, ± glabrous throughout. Stipules unknown. Leaves: petioles 4.8–6.7 cm long, glabrous; leaf blades lanceolate, 21–27 × 4.5–6.2 cm, coriaceous, khaki-green to khaki-brown when dry, glabrous above and below; secondary veins 18–24 pairs; tertiary venation prominent; base narrowly acute to attenuate; apex obtuse to acute. Inflorescence/infructescence trichotomous, 6.7–9.5 cm long, minutely puberulous; principal axis 2.7–5.1 cm long, each axis with 2–3 primary nodes and verticellate branching at each node, branches terminating in cymules, these positioned towards the ends of the inflorescence branches; bracts absent; bracteoles very broadly ovate, ca. 0.4 mm long. Flowers 5-merous, pedicels 2–4 mm long; calyx (incl. hypanthium) ca. 3 mm long, glabrous to minutely puberulous; calyx lobes obscure, calyx limb truncate; corolla: white; corolla tube 3–4 mm long, external surface glabrous, villous at throat within; corolla lobes 1.6–2.5(–3) mm long; stamens: filaments 1–2 mm long (long styled); anthers ca. 0.7 mm long; style ca. 3.5 mm long (long styled), lobes ca. 0.7 mm long. Fruit obovoid to subturbinate, 7–8 mm long, ribbed, probably red, glabrous; calyx limb and lobes persistent; pedicels 2–6 mm long; pyrenes ribbed to ridged on outer surface; endosperm ruminate.

Other specimens examined: Known only from the type collection.

Distribution.—Mt. Alzapan, Quezon Province, Luzon.

Ecology.—Forest; ca. 1300 m.

Phenology records.—Flowering: May, June; fruiting: unknown.

Preliminary conservation assessment.—Critically Endangered (**CR B1ab(iii)**) or possibly Extinct (**EX**). Last collected in 1925.

Distinguishing characteristics.—Leaf blades large, lanceolate, coriaceous, lateral veins 18–24 per side, tertiary venation pattern prominent; inflorescences trichotomous; pyrenes ridged on outer (dorsal) surface.

Remarks.—Though known only from the type collection, this taxon is extremely distinct, and we have no hesitation whatever in maintaining it.

84. Psychotria pinnatinervia Elmer, Leafl. Philipp. Bot. 1:26. 1906; Merr., Enum. Philipp. Fl. Pl. 3:561. 1923. (**Fig. 96**). Type: PHILIPPINES. Luzon: Quezon Prov.: Mt. Banahão, Oct 1904, *Whitford 932* (lectotype: US, designated here; isolectotypes: K, P).

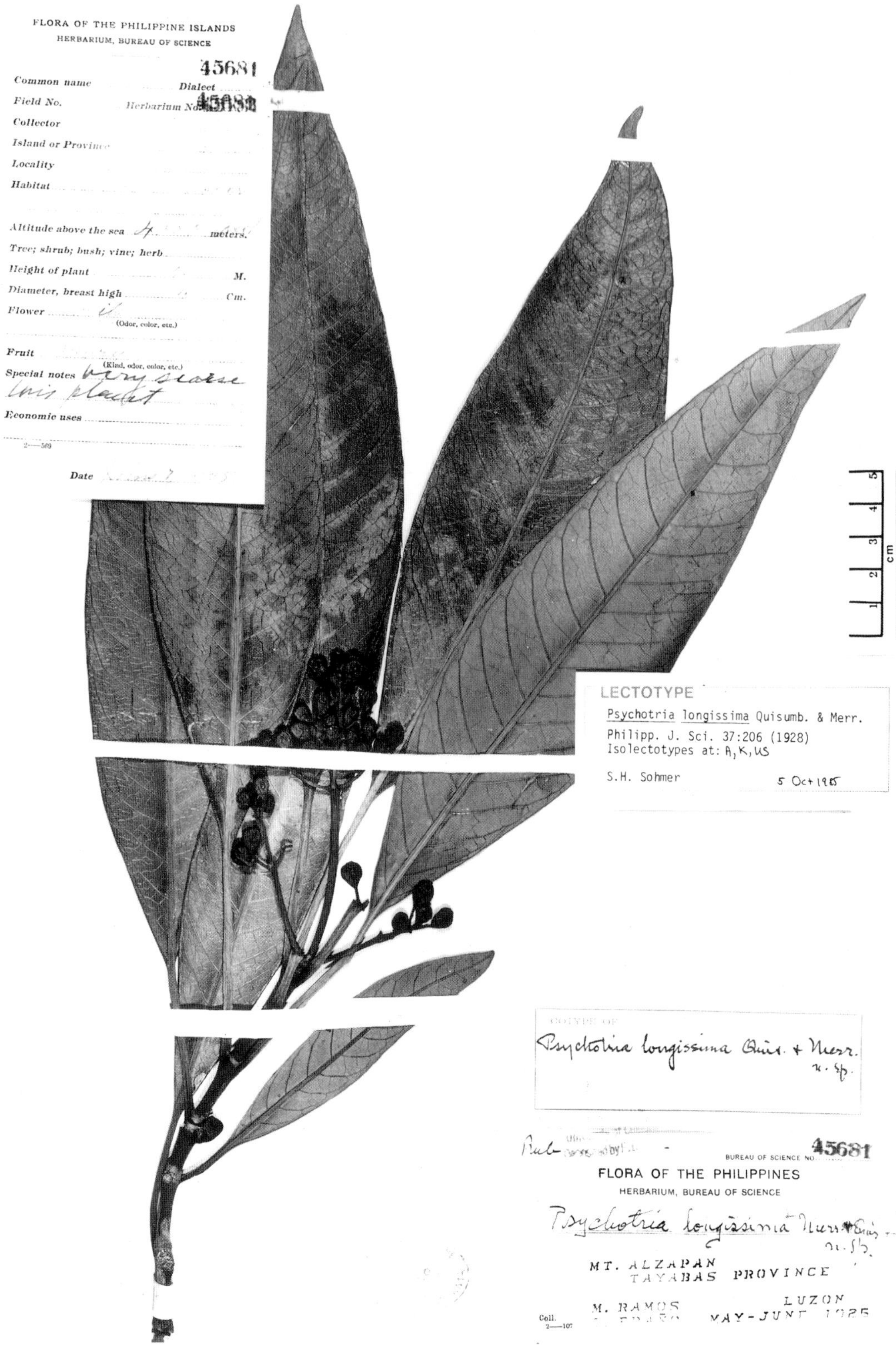

Fig. 95. *Psychotria longissima* (*Ramos & Edaño BS 45681*) (UC, lectotype).

Shrub, reported (1–)3–5(–7) m high, ± glabrous throughout. Stipules imbricate, ± narrowly elliptic to vary narrowly elliptic-lanceolate, 0.8–2.4 cm long, glabrous; apex acuminate or acute. Leaves: petioles 1–3.7 cm long, glabrous; leaf blades elliptic to elliptic lanceolate, somewhat asymmetrical, 7–17.5(–20.8) × 2.5–5(–5.4) cm, chartaceous to subcoriaceous, drying grayish green to grayish brown, glabrous above and below; secondary veins 10–18 pairs; tertiary venation rather prominent to obscure; base acute to attenuate; apex usually long-acuminate. Inflorescence/infructescence trichotomous, (1.4–)3–5.5(–9) cm long, glabrous; principal axis 1–4.3(–6) cm long, each axis with 1–3(–5) primary nodes and opposite branching at each node, terminating in rather lax cymules, these occurring towards the ends of the inflorescence branches; bracts ovate to ± deltate, ca. 0. 6 mm long, but often absent (?caducous); bracteoles similar to the bracts, 0.3 mm long, often absent (?caducous). Flowers 5-merous, pedicels (1–)2–4 mm long; calyx (incl. hypanthium) 1–2 mm long, glabrous; calyx lobes obscure; corolla: white; corolla tube ca. 3 mm long (long and short styled), external surface glabrous, glabrous within; corolla lobes 2–3 mm long; stamens: filaments ca. 2 mm (short styled) or ca. 0.5 mm long (long styled); anthers ca. 1 mm long (short and long styled); style ca. 2 mm (short styled) or ca. 4.5 mm long (long styled), lobes ca. 0.5 mm long (short and long styled). Fruit obovoid, 5–6(–7) mm long, ribbed, color unknown, glabrous; calyx limb and lobes ± persistent; pedicels 2–5(–6) mm long; pyrenes ridged on outer surface; endosperm ruminate.

Other specimens examined: **Luzon. Cagayan Prov.:** Mt. Cagua, *Edaño BS 78435* (A); Mt. Dos Cuernos, *Ramos BS 77037* (A); Mt. Balatongan, *Edaño BS 78521* (A). **Isabela Prov.:** Mt. Moises, *Clemens 17022* (UC). **Mountain Prov.:** Lugu, Banaue, *Banlugan et al. PNH 72467* (K, L). **Quezon Prov.:** Lucban, *Elmer 7894* (A, K, L, MO, US, W), *7905* (A, K, L, MO, W), *9084* (A, K, L); Mt. Banahāo, *Foxworthy BS 2405* (NY), *Iwatsuki et al. P-1161* (L), *P-1188* (L), *P-1191* (L), *P-1196* (A, L), *Loher 6334* (K), *Ramos BS 19555* (K, US), *Vidal 3043* (K); Dolores, *Sulit FB 30064* (NY, UC); Mt. Alzapan, *Ramos & Edaño BS 45666* (A, K, UC); Polillo Island, Bo. Apad., *Salvoza FB 29634* (UC). **Rizal Prov.:** Mt. Irid, *Ramos BS 41956* (A, K, L, W), *Ramos & Edaño BS 48432* (UC), *BS 48486* (NY, UC); Montalban, *Loher 12751* (W), *s.n.* (UC); Balacbac, *Loher 12991* (UC); Antipolo, *Ramos BS 22255* (BISH, US); Bosoboso, *Ramos BS 2099* (GH, K); Mt. Lumutan, *Ramos & Edaño BS 29767* (A, P), *BS 29693* (L), *BS 29793* (BM, P); Paningtingan, *Loher 13513* (UC), *13515* (A), *s.n.* (UC); San Isidro, *Ramos BS 337* (US); Mt. Angilog, *Loher 6327* (K). **Rizal Prov.:** without precise locality: *Loher 14324* (UC); *Ramos BS 2634* (K). **Laguna Prov.:** Mt. Maquiling, *Brown BS 17907* (BM, US), *Elmer 18109* (GH, MO), *Sulit PNH 8491* (A); Ranahao[?], *Loher 6372* (K). **Camarines Sur Prov.:** Ocampo, Mount Isarog National Park, *Delprete 6455* (BRIT). **Mindoro. Mindoro Occidental Prov.:** Paluan, *Ramos BS 39618* (A, L). **Leyte. Leyte Prov.:** Mt. Abucayan, *Edaño BS 41810* (A, K). **PHILIPPINES. Prov. unknown:** Mt. Banahas, *Baker 2457* (K). **PHILIPPINES.** without precise locality: *Vidal 1455e* (A).

Distribution.—Principally on Luzon, with reports from Mindoro and Leyte.

Ecology.—Drier parts of montane forest; 600–1600 m.

Phenology records.—Flowering: February to May; fruiting: January, February, May to August, October to December.

Preliminary conservation assessment.—Near Threatened (**NT**). Last collected in 1994.

Distinguishing characteristics.—Leaf blades generally narrow elliptic to elliptic-lanceolate, often asymmetrical, apex long-acuminate; inflorescences trichotomous, glabrous; pyrenes ridged on outer (dorsal) surface.

Remarks.—The leaves of this species are a very attractive, glossy green in the field and the fruits ripen through yellow and orange to deep red. *Psychotria pinnatinervia* is a slender shrub, found in the drier, better drained sides of mountains on Luzon. It appears to very closely related to *P. pubilima* Quisumb., from which it is distinguished by glabrous inflorescence and flower parts (densely puberulent in *P. pubilima*).

85. Psychotria pubilimba Quisumb. in Philipp. J. Sci. 41:366. 1930. (**Fig. 97**). TYPE: PHILIPPINES. LUZON: Isabela Prov.: San Mariano, along streams in damp forests at ca. 700 m, 1 Mar 1926. *Ramos & Edaño BS 47221* (LECTOTYPE: A, designated here; ISOLECTOTYPES: BM, NY).

Small tree or shrub, 1–2 m high, young stems puberulous. Stipules imbricate, narrowly triangular-lanceolate, 0.3–0.8(–2) cm long, glabrous to puberulous; apex acute. Leaves: petioles 1–2 cm long, glabrous to puberulous; leaf blades lanceolate to elliptic-lanceolate, 9–17.5 × 3.5–4.4 cm, subcoriaceous to coriaceous, grayish brown to dark reddish brown, glabrous above, glabrous to puberulous on midrib and lateral veins below; secondary veins 18–24 pairs; tertiary venation manifest to rather prominent, rarely obscure to invisible; base acute to obtuse; apex long-acuminate. Inflorescence/infructescence trichotomous, 1.9–4(–8) cm long, puberulous; principal axis 0.9–4.5 cm long, each

Fig. 96. *Psychotria pinnatinervia* (*Whitford 932*) (US, lectotype).

Fig. 97. *Psychotria pubilimba* (*Ramos & Edaño BS 47221*) (A, lectotype).

axis with 1–3 primary nodes and opposite branching at each node, branches terminating in cymules, these occurring towards the ends of the inflorescence branches; bracts apparently absent; bracteoles ± deltate, ca. 0.3 mm long, often absent (rare or caducous). Flowers 5-merous, pedicels 1–2 mm long; calyx (incl. hypanthium) ca. 3 mm long, minutely puberulous; calyx lobes shallowly triangular to deltate, ca. 0.3 mm long, sometimes obscure; corolla: white to yellowish white; corolla tube ca. 1.5 mm long, external surface minutely puberulous, densely villous within; corolla lobes ca. 2 mm long; stamens unknown; style unknown. Fruit obovoid, ca. 6 mm long, weakly ridged, probably red, glabrous; calyx limb and lobes persistent; pedicels 1.2–3.7 mm long; pyrenes weakly ridged on outer surface; endosperm ruminate.

Other specimens examined: **Luzon. Cagayan Prov.:** Ilang River, *Allen 238-81* (BISH). **Isabela Prov.:** Sierra Madre Mountains, Palanan, *Gutierrez PNH 78259* (A, K, L, PNH); Digallorin, Divinisa camp site, *Ridsdale & Baquiran et al. ISU 470* (K); Mt. Moises, *Ramos & Edaño BS 47273* (UC), *vel aff.*, *BS 47321* (A, K, NY).

Distribution.—Found only in Cagayan and Isabela provinces, on Luzon.

Ecology.—Streamside forest (along water courses), or on forested slopes; (50–)610–1128 m.

Phenology records.—Flowering: April, May; fruiting: March to May.

Preliminary conservation assessment.—Endangered (**EN B1ab(iii)**). Last collected in 1992.

Distinguishing characteristics.—Leaves: lateral veins relatively numerous, apex long-acuminate; inflorescences trichotomous; flowers puberulent; pyrenes ridged on the outer (dorsal) surface.

Remarks.—This is a highly distinctive taxon in the herbarium due to the shape and color of the leaves and relatively short and delicate inflorescences. We believe that it is closely related to the strictly glabrous species *P. pinnatinervia* Elmer, from which it appears to be distinguished principally by its puberulence, particularly on the floral parts. Quisumbing (1930) provided a very good, complete description of his new species and obviously had good flowering material at hand. We did not have flowering material, and the available fruits were immature. Therefore, we cannot verify the fact that Quisumbing reported the pyrenes as being smooth, not ridged, on the outer (dorsal) surface. If this is true and can be confirmed, it would change the position of this taxon in this revision.

86. Psychotria catanduaniensis Sohmer & A.P. Davis, sp. nov. (**Fig. 98**). TYPE: PHILIPPINES. CATANDUANES: Masanganbahi, in forest streams about 160 m, 21 Aug 1928, *Ramos & Edaño BS 75589* (HOLOTYPE: NY).

Frutices vel arbores parvi 1–4 m alti glabri sunt, laminis 5.3 × 15.5–9 × 24.3 cm latis et longis quando siccis griseis ad subrubribadiis crasse coriaceis vel coriaceis glabris ovatis ad oblongo-ovatis apice acuto vel abrupte acuto ad obtuso basi rotundata vel obtusa, petiolis 1.7–3.9(–5.5) cm longis, infructescentia 5.1–7 cm longa trichotoma glabra, pedunculis 3–6 cm longis, nodis primariis 1–2, ramis verticillatis, fructibus 6–7 mm longis turbinatis liratis dicunt subviridi-luteis versimiliter rubris, pedicellis 3–8 mm longis, pyrenis in dorso liratus, endospermo ruminato.

Shrub or small tree, 1–4 m high, glabrous throughout. Stipules unknown. Leaves: petioles 1.7–3.9(–5.5) cm long, glabrous; leaf blades ovate to oblong-ovate, or ± narrowly elliptic, 15.5–24.3 × 5.3–9 cm, coriaceous, grayish to reddish brown when dry, glabrous above and below; secondary veins 14–16 pairs; tertiary venation manifest to obscure; base round to obtuse; apex acute to obtuse to abruptly acute to obtuse. Inflorescence/infructescence trichotomous, 5.1–7 cm long, glabrous; principal axis 3–6 cm long, each axis with 1–2 primary nodes and verticillate branching at each nodes, branches terminating in cymules; bracts and bracteoles unknown (?caducous or absent). Flowers unknown. Fruit turbinate, 6–7 mm long, ridged, probably red but reported greenish yellow, glabrous; calyx limb and lobes ± not persistent; pedicels 3–8 mm long; pyrenes ridged on outer surface; endosperm ruminate.

PARATYPES: **Luzon. Camarines Sur Prov.:** Lagonoy, *Espinas FB 30342* (BM, K, NY, UC). **Panay. Capiz Prov.:** Hamanay [?], *Edaño BS 46160* (BM, UC); Libacao, *Ramos & Edaño BS 31441* (SING)?.

Distribution.—The taxon is known from Catanduanes and Camarines Sur provinces of Luzon and perhaps from Panay.

Ecology.—Forest, including streamside forest; sea level to ca. 160 m.

Phenology records.—Flowering: unknown; fruiting: April, May, August.

Preliminary conservation assessment.—Critically Endangered (**CR B1ab(iii)**). Last collected in 1928.

Distinguishing characteristics.—Leaf blades relatively large, 15.5–24.3 cm long, coriaceous, base

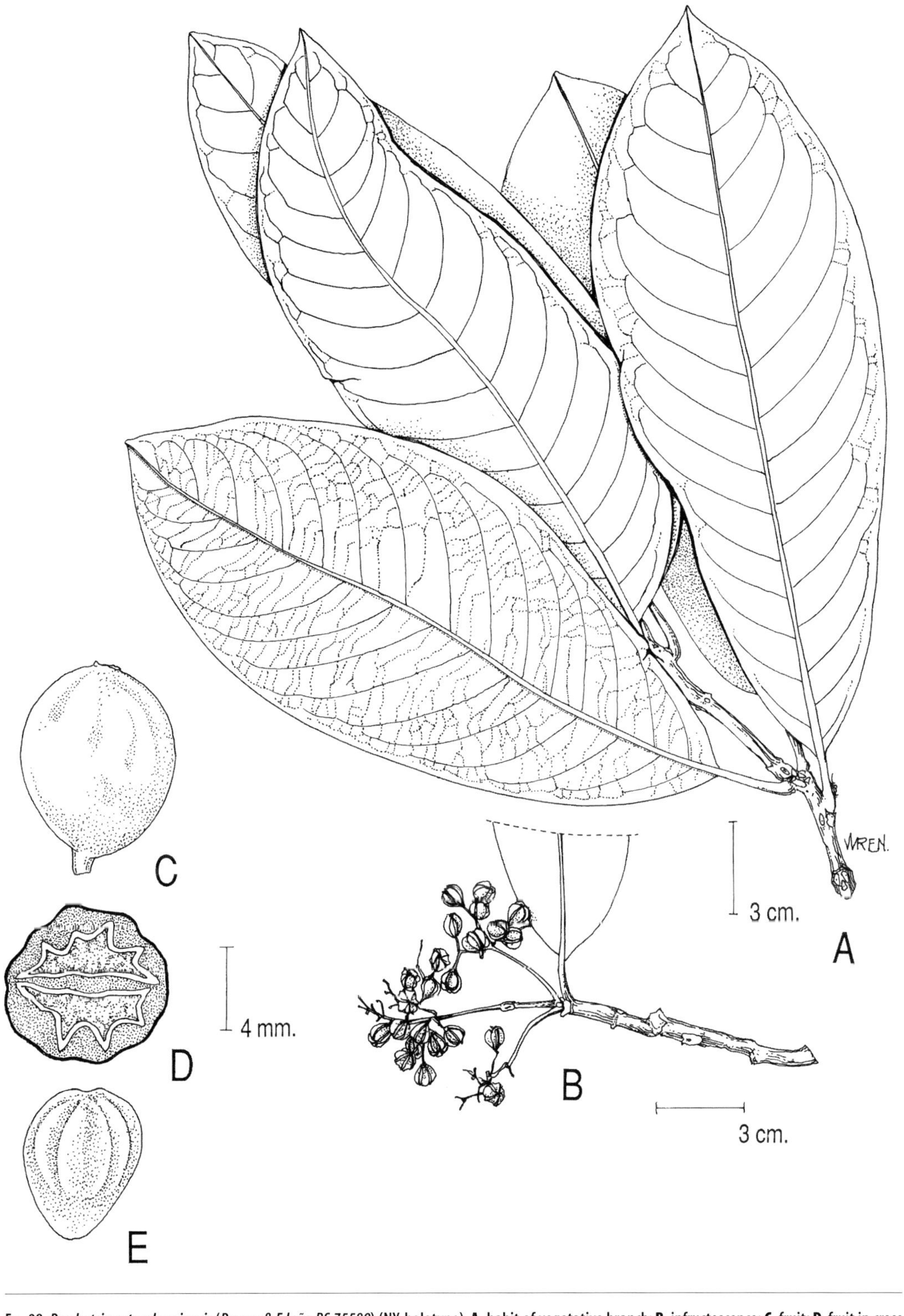

Fig. 98. *Psychotria catanduaniensis* (*Ramos & Edaño BS 75589*) (NY, holotype). **A,** habit of vegetative branch; **B,** infructescence; **C,** fruit; **D,** fruit in cross section; **E,** pyrene.

round to obtuse, tertiary venation obscure; inflorescences trichotomous; pyrenes prominently ridged on the outer (dorsal) surface.

Remarks.—Despite the lack of material, this species is distinct enough to merit recognition. We have tentatively assigned several collections to *P. catanduaniensis* that we suspect may constitute a separate taxon. If these dubious collections are indeed representative of a distinct taxon, it would be very closely related to *P. catanduaniensis*. There are notes on *Edaño BS 46030* and *BS 46160* indicating that someone, probably Quisumbing, was corresponding with Merrill after the latter had gone to Berkeley, and that he (Quisumbing) considered those collections to constitute a new taxon. Merrill, apparently, wrote back and stated: '*46160, 46030 Psychotria ampla*; never published as I could not be sure of its validity, being afraid it was some of Elmer's species of which we had only poor material. I would let it ride under "ampla," and leave *Psychotria* alone, except for very characteristic species, until you orient yourself a little better on the evaluation of characteristics.' This is very sound advice indeed and true even today.

87. Psychotria banahaensis Elmer, Leafl. Philipp. Bot. 1:26. 1906; Merr., Enum. Philipp. Fl. Pl. 3:53. 1923. (**Fig. 99**). *Grumilea banahaensis* (Elmer) Elmer, Philipp. J. Sci. 10:132. 1915. TYPE: PHILIPPINES. LUZON: Quezon Prov.: Mt. Banahão, Oct 1904, *Whitford 983* (LECTOTYPE: US, designated here).

Small tree or shrub, 1–10 (–15) m high, young stems glabrous. Stipules imbricate, narrowly triangular-lanceolate to ± narrowly triangular, (0.5–)1.4–2.6(–4.3) cm long, glabrous; apex acute. Leaves: petioles 1.3–5.5 cm long, glabrous; leaf blades oblong-elliptic to obovate-elliptic, 9.5–12.2 × 3.1–2.6(–3.9) cm, subcoriaceous, usually drying a grayish or khaki-green or green, glabrous above and below; secondary veins 11–16 pairs; tertiary venation manifest to obscure; base acute to narrowly acute, rarely attenuate; apex usually acute to obtuse or acuminate. Inflorescence/infructescence trichotomous, or rarely monochotomous, 3.3–9(–14) cm long, glabrous to puberulous; principal axis (1.4–)2–4.6(–5.5) cm long, each axis with 2–3 primary nodes and verticillate or opposite branching at each node, branches terminating in cymules; bracts absent; bracteoles narrowly triangular, ca. 0.7 mm long. Flowers 5-merous, pedicels 1–4 mm long; calyx (incl. hypanthium) (0.8–)1.1–3 mm long, puberulous; calyx lobes very depressed ovate, or very weakly lobed to obscure and then ± truncate; corolla: white; corolla tube (1.2–)1.8–2.2(–3) mm long, external surface glabrous, villous at throat within; corolla lobes 1.7–2 mm long; stamens: filaments ca. 2 mm long (long styled); anthers ca. 0.8 mm long; style unknown. Fruit obovoid-turbinate, 5–9 mm long, ridged, yellow to orange, then red, glabrous; calyx limb and lobes not persistent; pedicels 2–5 mm long; pyrenes distinctly and sharply ridged on outer surface; endosperm ruminate.

Distribution.—More or less distributed throughout the Philippines.

Ecology.—Forest (primary and secondary); (35–)140–1100 m.

Phenology records.—Flowering: February, March, May; fruiting: February to November.

Preliminary conservation assessment.—Near Threatened (**NT**). Last collection made in 1993.

Distinguishing features.—Leaf blades oblong-elliptic to obovate, usually drying a grayish or khaki-green and with no tertiary venation pattern visible; inflorescences trichotomous; pyrenes sharply ridged on the outer (dorsal) surface.

Remarks.—This taxon has a tendency to produce relatively large stipules. The largest we have seen are from a population on Mantalban near Dingalan. We have recognized these as comprising a variety, which can be separated from the nominate variety as described in the following key.

KEY TO THE VARIETIES OF *PSYCHOTRIA BANAHAENSIS*

1. Stipules < 2.6 cm long ____ **87a.** var. **banahaensis**
1. Stipules > 3 cm long ____ **87b.** var. **montalbanensis**

87a. Psychotria banahaensis var. **banahaensis**

Shrub or small tree, 1–5 (–15?) m high. Stipules (0.5–)1.4–2.6 cm long. Leaf blades 12.2–26 × 3.9–9.5 cm; lateral veins 11–16 per side. Inflorescence 5–9(–14) cm long. Fruits 6–9 mm long.

Other specimens examined: **Luzon. Cagayan Prov.:** north of Bagio Cove, *Allen 393-81* (BISH, PNH). **Isabela Prov.:** San Mariano, *Ramos & Edaño BS 46832* (UC); So. Kapuntian, Barangay San Jose, San Mariano, *Barbon, Romero & Fuentes PPI 13178* (BRIT);

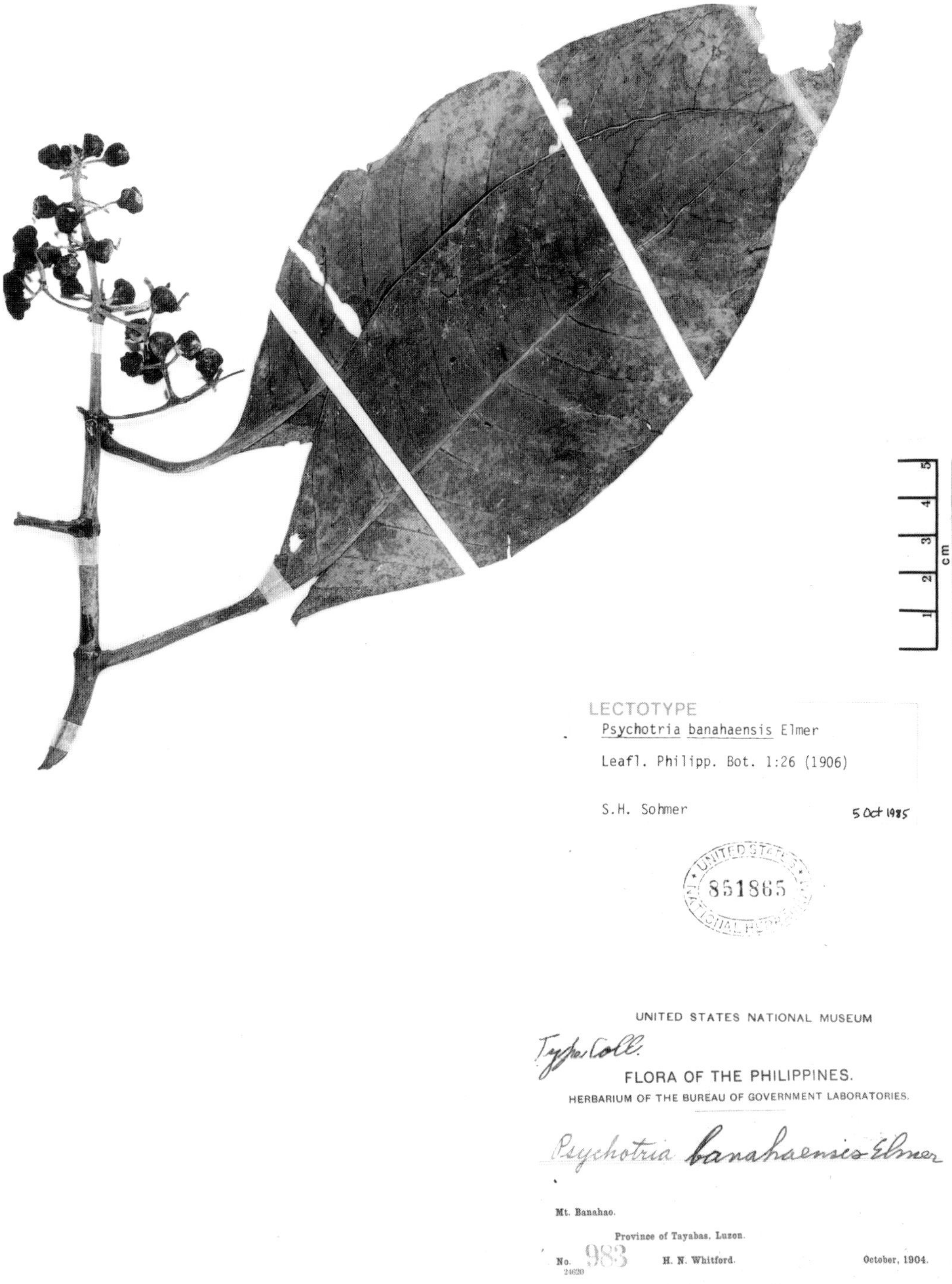

Fig. 99. ***Psychotria banahaensis*** **var.** ***banahaensis*** **(*Whitford 983*) (US, lectotype).**

Palanan, Dikatkotan, *Gaerlan et al. PPI 2953* (BRIT, K); Sierra Madre Mountain Range (E. foothills), Palanan Municipality, Barangay San Isidro, hills N. of Sitio Diago, *Leonardo 3428* (K); Mt. Yalobo region, *Ramos & Edaño BS 47170* (A, NY, P, SING, UC). **Nueva Ecija Prov.:** Mt. Umingan, *Ramos & Edaño BS 26456* (A, US). **Quezon Prov.:** Casiguran, *Ramos & Edaño BS 45386* (UC), *Ramos & Edaño BS 26568* (A, NY); Mt. Binuang, *Ramos & Edaño BS 28524* (A, K, P, SING), *BS 28580* (US); Lucban, *Elmer 7951* (A), *9124* (A, GH, L, MO, US, W); Antimonan, *Steiner PNH 39556* (PNH), *Whitford 625* (NY); Mt. Cadig, *Yates BS 25553* (K); Hinabaan, Guinayangan,

Mendoza PNH 97956 (K, L, PNH); Unisan, *Gojar FB 31466* (NY); Quezon National Forest Park, *Sohmer 12096, 12097, 12098, 12259, 12260, 12261, 12262* (BISH); Barangay Diteki, San Luis, *Fuentes & Fernando PPI 37473* (BRIT); Bischaan, *Logrinan s.n.* (MO); *Reynoso et al. PPI 3882* (BRIT, K); Polillo Island, *McGregor BS 10775* (K); Alabat Island, without further data, *Ramos & Edaño BS 48015* (NY, P, UC), *BS 48219* (NY, UC); Tagkawayan, Barangay Bagong Silang II, *Barbon, Garcia & Fernando PPI 12253* (BRIT, K), *PPI 12311* (BRIT). **Quezon Prov.:** without precise locality: *Edaño BS 26960* (A, P). **Rizal Prov.:** Mt. Angilo, *Loher 6333* (K), *6381* (K); Balacbac, *Loher s.n.* (UC). **Camarines Sur Prov.:** Sarapan, *Edaño BS 76148* (NY), *BS 76151* (NY); Botol River, *Edaño BS 76288* (NY). **Albay Prov.:** Mt. Masaraga, *Sohmer 12236* (BISH). **Sibuyan. Romblon Prov.:** Sibuyan Island, Punong Mt., near trail to Barrio Jaosan, *Sohmer 12472* (K); Sibuyan Island, Magdiwang, Barrio Hawasan, Ating River, *Stone et al. PPI 6773* (BRIT, K); *ibid., Stone et al. PPI 6733* (K); Camp 1 above Magdiwang on rideg leading to Mayos Peak, *Argent & Reynoso 899* (K). **Mindoro. Mindoro Oriental Prov.:** Mt. Halcon, *Ramos & Edaño BS 40576* (A, K). **Masbate. Masbate Prov.:** Masbate Island, 7-R Ranch, Sitio Kalunakan, Barangay Matipuron, Milagros, *Barbon, Garcia & Fernando PPI 12733* (BRIT, K). **Samar. Western Samar Prov.:** Mt. Sohoton, *Gutierrez et al. PNH 117547* (K, L, NY); Sitio Campo-uno, Barangay Tinane, Paranas, *Reynoso, Sagcal & Garcia PPI 7571* (BRIT, K). **Negros. Negros Oriental Prov.:** San Jose, Calo, *Kondo & Edaño PNH 36638* (BISH, K, L, NY, PNH). **Mindanao. Surigao Prov.:** Mt. Kabatuan, *Mendoza & Convocar PNH 10585* (A, L, PNH), *vel aff.* **Palawan. Palawan Prov.:** Mangsee Island, without precise locality, *Wilkes Expedition s.n.* (NY); Quezon, Tabon, *Reynoso PNH 87708* (CANB, K, L, PNH).

Distribution.—More or less distributed throughout the Philippines.

Ecology.—Forest, including lowland dipterocarp forest, primary or secondary (logged-over) forest, sometimes by streamsides; (35–)140–650 m.

Phenology records.—Flowering: February, March; fruiting: February to May, August to November.

Preliminary conservation assessment.—Near Threatened (**NT**). Last collected in 1993.

Distinguishing characteristics.—Shorter stipules distinguish this taxon from var. *montalbanensis*.

87b. Psychotria banahaensis var. **montalbanensis** Sohmer & A.P. Davis, var. nov. (**Fig. 100**). TYPE: PHILIPPINES. LUZON: Aurora Prov.: Sierra Madre Mts. north-northeast of Dingalan at 1000–1100 m, 18 Mar 1968, *Jacobs 7846* (HOLOTYPE: L; ISOTYPE: PNH).

A *Psychotria banahaensis* var. *banahaensis* differt in stipulis longioribus que 3 cm.

Shrub, ca. 2 m high. Stipules 3–4.3 cm long. Leaf blades 9.5–15.6(–18) × 3.1–6.5 cm; lateral veins 11–12(–16) per side. Inflorescence 3.3–4.8 cm long. Fruits 5–6 mm long.

PARATYPE: **Luzon. Rizal Prov.:** Montalban, *Loher 12752* (UC), *12775* (UC).

Distribution.—Known only from Rizal Province.

Ecology.—Forest; 1000–1100 m.

Phenology records.—Flowering: March; flowering: unknown.

Preliminary conservation assessment.—Endangered (**EN B1ab(iii)**). Last collected in 1968.

Distinguishing characteristics.—Stipules more than 3 cm long.

Remarks.—This taxon may well turn out to be superfluous if future collecting shows that stipular sizes as demonstrated here are but a portion of the total variation present in the species. Most of the stipules available were from specimens with young inflorescences, and they are always larger than those associated with leaves and stems.

88. Psychotria galintanensis Sohmer & A.P. Davis, sp. nov. (**Fig. 101**). TYPE: PHILIPPINES. MINDANAO: Davao Prov.: Mt. Galintan, on forest slopes at 1000 m 13 Jun 1927, *Ramos & Edaño BS 48899* (HOLOTYPE: UC).

Arbores parvae vel fructices 2 m altae sunt, stipulis 2.5 cm longis imbricatis glabris foliis 8.5 × 19–10 × 24 cm latis et longis obovatus glabris nervis lateralibus 15 dimidio quoque nervis tertiis prominentibus in sicco subrubri-badiis apice breve acutis basi acutis, petiolis 4–5 cm longis glabris, infructescentia trichotoma axibus minute puberulis eo principali 6 cm longo, pedunculo 2.3 cm longo 3–4 nodis primariis cum ramis oppositis, fructibus 5–6 cm longis obovoidi-turbinatis glabris liratis, pyrenis in dorso liratis, endospermo ruminato.

Small tree or shrub, ca. 2 m high, young stems glabrous. Stipules imbricate, triangular-lanceolate, ca. 2.5 cm long, glabrous; apex acute to slightly acuminate. Leaves: petioles 4–5 cm long, glabrous; leaf blades obovate, 19–24 × 8.5–10 cm, subcoriaceous, drying reddish brown, glabrous above and below; secondary veins ca. 15 pairs; tertiary venation prominent; base acute; apex abruptly acute. Inflorescence/infructescence trichotomous, ca. 6 cm long, minutely puberulous; principal axis ca. 2.3 cm long, each axis with 3–4 primary nodes with opposite branching at each node, branches terminating in cymules; bracts and bracteoles unknown. Flowers unknown. Fruit obovoid-turbinate, 5–6 mm long, ridged, color unknown, glabrous; calyx limb and lobes persistent but obscure; pedicels ca. 1 mm long; pyrenes ridged on outer surface; endosperm ruminate.

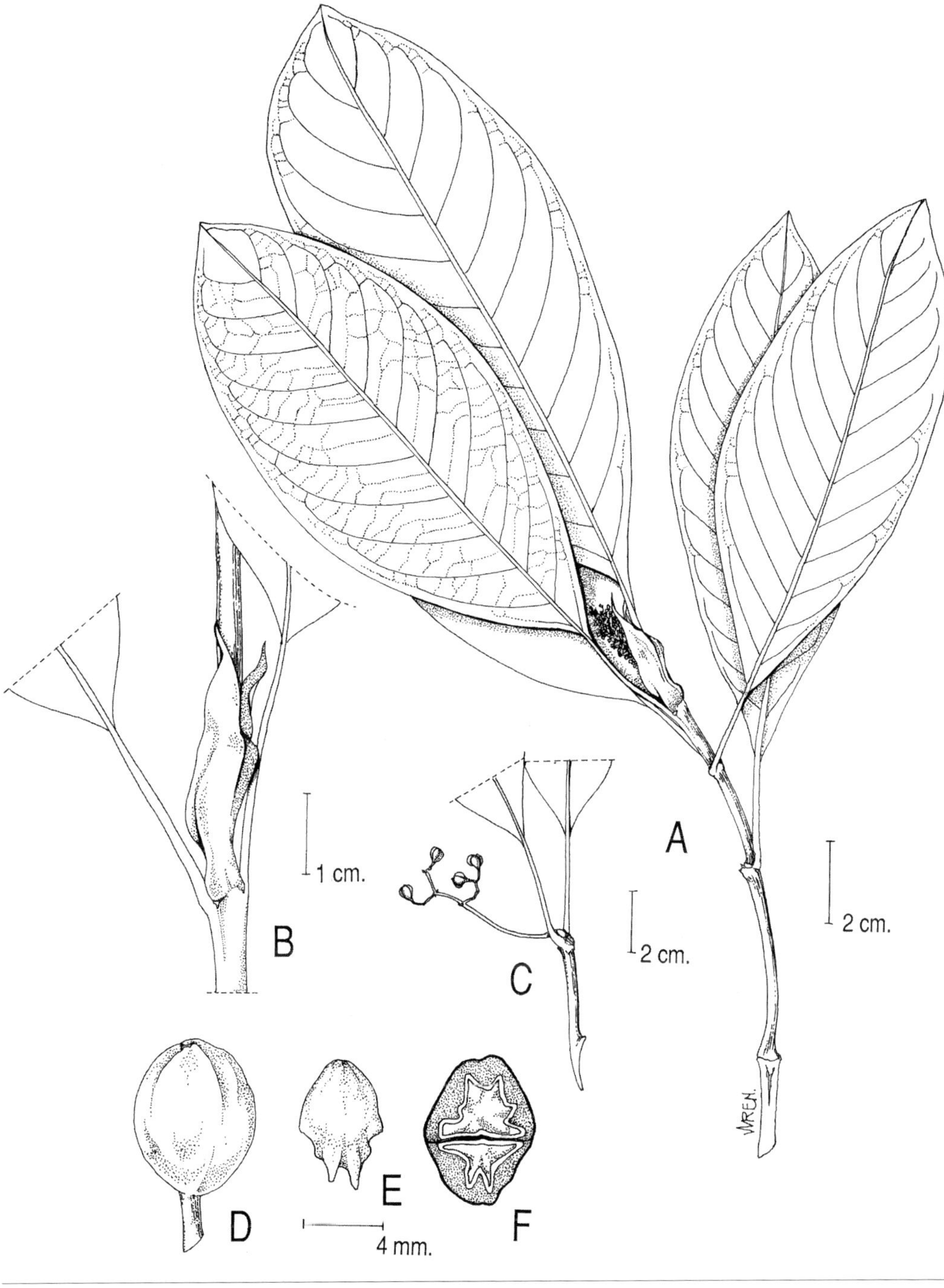

Fig. 100. ***Psychotria banahaensis*** **var.** ***montalbanensis*** **(*Jacobs 7846* (L, holotype). A, habit of flowering branch; B, stipule; C, infructescence; D, fruit; E, pyrene; F, fruit in cross section.**

Paratypes: None—known only from the type collection.

Distribution.—Only known from Mt. Galintan (Mindanao, Davao Province).

Ecology.—Forest; low to mid elevation (e.g. ca. 1000 m).

Phenology records.—Flowering: unknown; fruiting: June.

Preliminary conservation assessment.—Critically Endangered (**CR B1ab(iii)**) or possibly Extinct (**EX**). Only collected once in 1927.

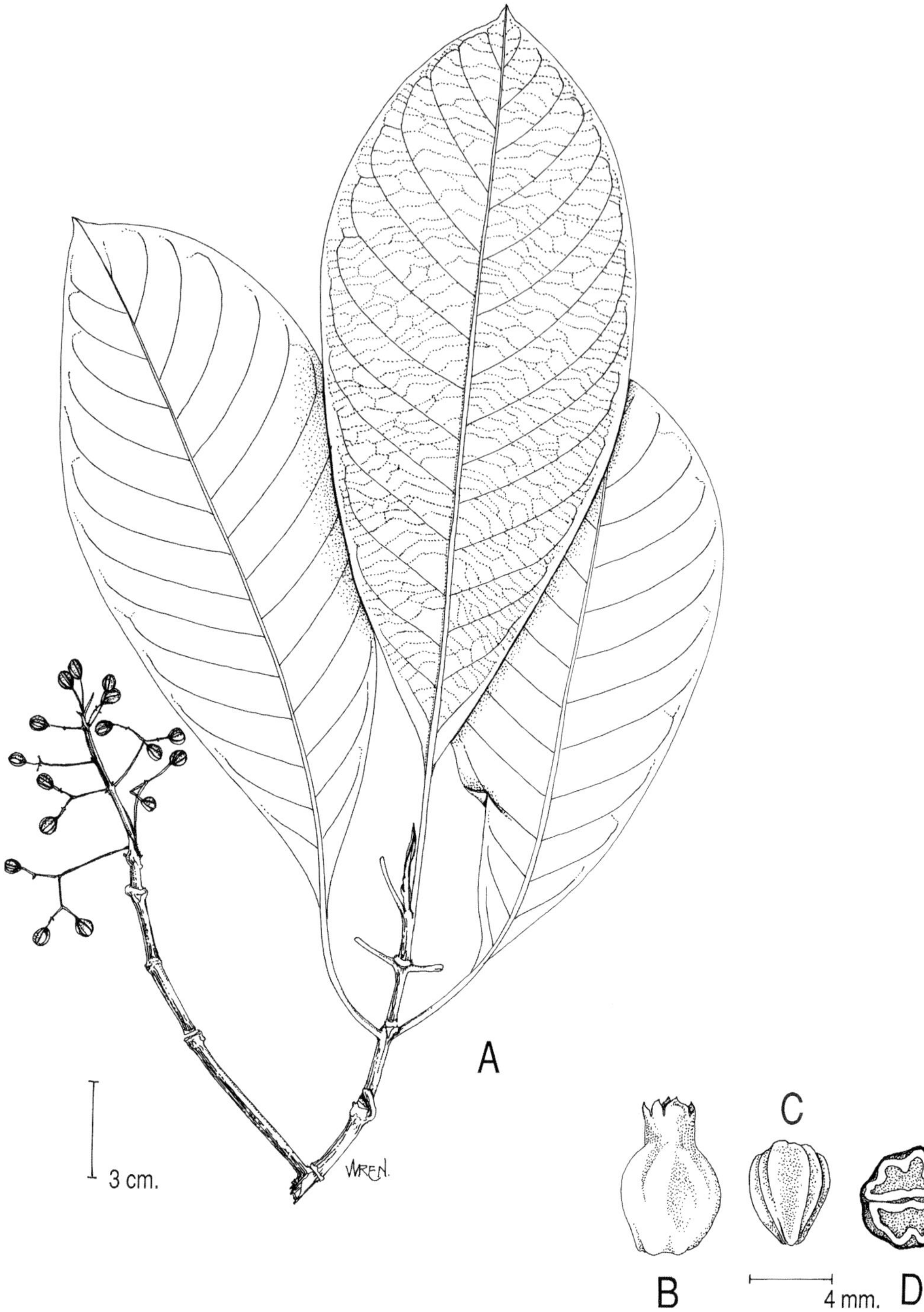

Fig. 101. ***Psychotria galintanensis*** **(*Ramos & Edaño BS 48899*) (UC, holotype). A, habit of fruiting branch; B, fruit; C, pyrene; D, fruit in cross section.**

Distinguishing characteristics.—Leaf blades obovate, 19–24 cm long, glabrous, tertiary venation pattern prominent; inflorescences trichotomous; pyrenes ridged on outer (dorsal) surface.

Remarks.—We would normally not recognize a species on the basis of one specimen, but have done so here because we have not been able to satisfactorily match this specimen with any other taxon within the group to which it belongs. There are typewritten notes attached to the specimen clearly indicating that someone, probably Quisumbing, was going to recognize this as a new species. One note states: 'This species is distinct from any *Psychotria* sp. We have a number of specimens in the herbarium with such name, but could not locate the description. If not described yet do you want me to send to you all specimens here and you describe it or do you want me to go ahead and describe it.' In evident answer to that is another typewritten note, probably from Merrill, that says: 'Same note applied as to *P. ampla*. In both cases, if you are *sure*, descriptions may be written, but you cannot accept all the named specimens in the herbarium as *certainly* representing the species, they need most critical study.'

Obviously, nothing happened in the intervening years, although the only specimen in hand has a handwritten annotation added to the label providing the name *P. obovoidea* Merr. Nothing further appears to have been done with this taxon and, as we now have even less material than originally, the statement which we attribute to Merrill that it requires most critical study is poignant. Indeed, one of the primary reasons we are recognizing this species at this time is the hope that such action will provide the impetus to further such critical study should this taxon still be extant.

89. Psychotria pallidifolia Merr., Philipp. J. Sci. 17:456. 1921; Merr., Enum. Philipp. Fl. Pl. 3:560. 1923. (**Fig. 102**). TYPE: PHILIPPINES. CATANDUANES. Catanduanes Prov.: Mt. Mariguidon and vicinity of Cololbong, on forested slopes, Nov–Dec 1917. *Ramos BS 30510* (LECTOTYPE: US, designated here; ISOLECTOTYPE: K).

Small tree or shrub, 2–3 m high, young stems glabrous to minutely puberulous. Stipules imbricate, triangular-lanceolate, 1.5–2.2 cm long, glabrous; apex acute to acuminate. Leaves: petioles 0.9–3.5 cm long, glabrous; leaf blades elliptic to elliptic-oblanceolate, or elliptic-lanceolate, (7.4–)10.5–22.8 × (1.9–) 3.1–8.4 cm, subcoriaceous or chartaceous, khaki-green to greyish green when dry, glabrous above, glabrous to minutely puberulous below; secondary veins 9–18 pairs; tertiary venation prominent to very prominent; base obtuse to acute; apex acute to acuminate. Inflorescence/infructescence trichotomous, 1.7–6.3 cm long, glabrous to minutely puberulous; principal axis 1.5–3.5 cm long, each axis with 2–3 primary nodes and opposite or verticellate branching at each node; branches terminating in cymules; bracts and bracteoles apparently absent. Flowers 4-merous, pedicels 1–3 mm long; calyx (incl. hypanthium) 2–3 mm long, glabrous; calyx lobes obscure; corolla: white; corolla tube 2–6 mm long, external surface glabrous, villous at throat within; corolla lobes (1–)2–3 mm long; stamens: filaments ca. 2 mm long (short and long styled); anthers ca. 1 mm long; style ca. 1.5 mm long (short styled) to 3.5 (long styled) mm long; lobes ca. 0.5 mm long (short and long styled). Fruit obovoid to turbinate, 5–7 mm long, strongly ribbed or ridged, orange to red or blackish, glabrous; calyx limb and lobes not persistent; pedicels 2–4 mm long; pyrenes ridged on outer surface; endosperm ruminate to weakly ruminate.

Other specimens examined: **Luzon. Cagayan Prov.:** Kilingkiling River, *Edaño BS 79563* (A). **Isabela Prov.:** Palanan, Diguyo, *Ridsdale, Dejan & Baquiran ISU 109* (K). **Zambales Prov.:** Subic, *Edaño BS 38318* (NY). **Rizal Prov.:** Mt. Lumutan, *Ramos & Edaño BS 29715* (A). **Camarines Sur Prov.:** Kamugong River, *Edaño BS 75908* (NY); Mt. Potianay, *Edaño BS 75944* (NY), *BS 76013* (NY), *BS 76018* (NY); Mt. Bagacay, *Ramos & Edaño BS 33823* (K, US); Parcale, *Ramos & Edaño BS 33484* (A, P) (Syntype for *P. longipetiolata* Merr.). **Catanduanes. Catanduanes Prov.:** Mt. Abucay, *Ramos & Edaño BS 75525* (NY). **Catanduanes Prov.:** without precise locality: *Ramos BS 30320* (K, P, US), *Ramos BS 30567* (K, US), *Ramos & Edaño BS 75440* (NY, UC); *Ramos BS 30446* (K).

Distribution. Principally of central and southern Luzon.

Ecology.—Forest, including dipterocarp forest; 50–400 m.

Phenology records.—Flowering April, September to December; fruiting: September to November.

Preliminary conservation assessment.—Vulnerable (**VU B1ab(iii)**). Last collected in 1946.

Distinguishing characteristics.—Leaves with tertiary venation prominent to very prominent, inflorescences trichotomous, rather short (1.7–6.3 cm long), fruits 5–7 mm long.

90. Psychotria reynosoi Sohmer & A.P. Davis, sp. nov. (**Figs. 2e, 103**). TYPE: PHILIPPINES: LUZON: Zambales Prov.: Mt. Marayep, on forested slopes at about 700 m, 9 Dec 1924. *Ramos & Edaño BS 44808* (HOLOTYPE: UC; ISOTYPES: A, K, L, NY, P, US).

Arbores parvae 1–4 m altae glabrae, stipulis imbricatis 0.6–2.7 cm longis, laminis coriaceis 11.5–24.7 × 3.4–11.5 cm latis et longis nervis lateralibus 12–14(–)17 prominentibus. Infructescentia trichotoma 6.5–8.2(–14.5) cm longa, axe principali 2.3–5.1 cm longo, fructibus globosis vel ellipsoideis 8–11 mm longis, pyrenis dorso valde liratis, endospermo valde ruminato.

Fig. 102. *Psychotria pallidifolia* var. *pallidifolia* (*Ramos BS 30510*) (US, lectotype).

Small tree or shrub, 1–4 m high, glabrous throughout. Stipules imbricate, triangular-lanceolate, 0.6–2.7 cm long, glabrous; apex acute to acuminate. Leaves: petioles 1.2–4.5 cm long, glabrous; leaf blades elliptic to broadly elliptic, or elliptic-ovate, 11.5–24.7 × 3.4–11.5 cm, coriaceous, khaki-green to greyish green when dry, glabrous above and below; secondary veins 12–14(–)17 pairs; tertiary venation prominent to very prominent; base obtuse to acute; apex acute. Inflorescence/infructescence tri-

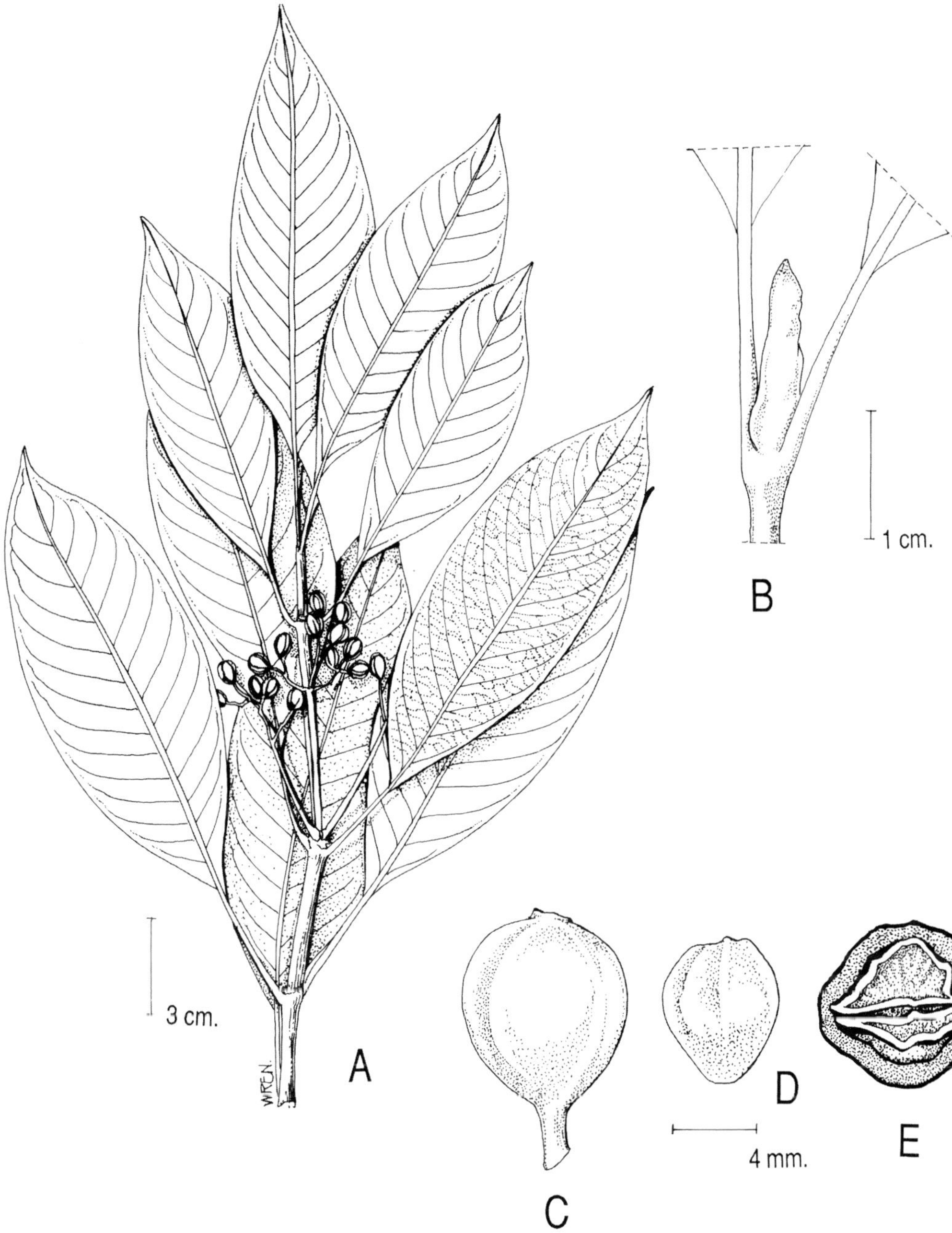

Fig. 103. *Psychotria reynosoi* (*Ramos & Edaño BS 44808*) (UC, holotype). **A,** habit of fruiting branch; **B,** stipule; **C,** fruit; **D,** pyrene [$redraw]; **E,** fruit in cross section.

chotomous, 6.5–8.2(–14.5) cm long, glabrous; principal axis 2.3–5.1 cm long, each axis with 2–3 primary nodes and opposite or verticillate branching at each node; branches termiating in cymules; bracts and bracteoles absent. Flowers 5-merous, pedicels 1.9–6 mm long; calyx (incl. hypanthium) 2–3.2 mm long, glabrous; calyx lobes obscure to shallowly triangular, if present ca. 0.3 mm long; corolla: white; corolla tube 4–6 mm long, external surface glabrous, villous at throat within; corolla lobes 2–3 mm long; stamens: filaments ca. 2.2 mm long (short and long styled); anthers ca. 1 mm long; style ca. 1.5 mm long (short styled) mm long; lobes ca. 0.5 mm long (short styled). Fruit globose to ellipsoid, 8–11 mm long, coarsely ribbed/ridged, probably otrange to red, glabrous; calyx limb and lobes not persistent; pedicels 4–8 mm long; pyrenes distinctly ridged on outer surface; endosperm strongly ruminate.

PARATYPES: **Luzon. Cagayan Prov.:** without further data: *Curran FB 19579* (BM, K, P, US); Mt. Balatongan, *Edaño BS 78518* (A), *BS 78526* (A); Mt. Cagua, *Edaño BS 78413* (A). **Isabela Prov.:** Mt. Moises, *Clemens 16841* (UC), *Ramos & Edaño BS 47288* (K, UC); So. Kapuntian, Barangay San Jose, San Mariano, *Barbon, Romero & Fuentes PPI 13150* (BRIT); So. Agal, Barangay San Jose, San Mariano, *Barbon, Romero & Fuentes PPI 13019* (BRIT). **Quezon Prov.:** Mt. Binuang, *Ramos & Edaño BS 28568* (K, P); Infanta, *Robinson BS 9402* (BM, L); Aurora National Park, So. Dikong, Barangay Diketi, San Luis, *Barbon, Garcia & Fernando PPI 9120* (BRIT). **Camarines Sur Prov.:** Mt. Bagacay, *Ramos & Edaño BS 33865* (L). **Nueva Vizcaya Prov.:** Mt. Alzapan, *Ramos & Edaño BS 45742* (UC). **Negros. Without further data:** *Everett FB 7258* (US).

Distribution.—Principally in northern and central Luzon, with a report from Negros.

Ecology.—Forested slopes, including good condition secondary forest; 400–1300 m.

Phenology records.—Flowering: March, May; fruiting: June, July, December, January.

Preliminary conservation assessment.—Near Threatened (**NT**). Last collected in 1995.

Distinguishing characteristics.—Fruit 8–11 mm long.

Remarks.—Originally we placed this species as a variety of *P. pallidiflora* (as var. *macrocarpa* ined.) but further collections made under the PPI initiative have convinced us that it represents a separate species.

91. Psychotria paloensis Elmer, Leafl. Philipp. Bot. 1:65. 1906, as '*Psychotria paloense*' Elmer; Merr., Enum. Philipp. Fl. Pl. 3:560. 1923. (**Fig. 104**). TYPE: PHILIPPINES. LEYTE: Leyte Prov.: Palo, gently sloping wooded ravine, Jan 1906, *Elmer 7056* (LECTOTYPE: A, designated here; ISOLECTOTYPE: K).

Small tree or shrub, 2–5 m high, young stems pubescent to glabrous. Stipules imbricate, triangular-lanceolate to narrowly triangular lanceolate, 0.2–1(–1.4) cm long, pubescent to glabrous; apex acute. Leaves: petioles 0.5–2.6 cm long, glabrous, or pubescent to puberulous; leaf blades elliptic to elliptic-obovate, or obobovate to oblanceolate, 7.7–13.2(–16.3) × 2.4–5.9(–6.5) cm, chartaceous to subcoriaceous, drying grayish green or oliveaceous, glabrous above, puberulous to pubescent, or glabrous below; secondary veins (9–)10–13(–15) pairs; tertiary venation obscure to rather prominent; base obtuse to acute; apex obtuse to abruptly obtuse, or acute to narrowly acute . Inflorescence/infructescence trichotomous or monochotomous, characteristically compact and flat-topped, (0.9–)1.3–5.6 cm long, puberulous to pubescent, or glabrous; principal axis (0.9–)1.4–5 cm long, each axis with 1–2(–4) nodes and verticillate or opposite branching at each node, branches terminating in cymules; bracts triangular to acuminate, 0.4–0.7 mm long; bracteoles unknown (absent or caducous). Flowers 5-merous, pedicels (0.1–)0.4–2.8 mm long or absent; calyx (incl. hypanthium) 2–4(–6) mm long, glabrous to puberulous; calyx lobes deltate to triangular, 0.3–0.6 mm long; corolla: white or reddish white; corolla tube 2.5–4 mm long (long styled) or 4–9 mm long (short styled), external surface glabrous to puberulous, villous at throat within (long styled) or glabrous (?short styled); corolla lobes 2–3.7 mm long; stamens: filaments ca. 1 mm (long styled) or 2–3 mm long (short styled); anthers ca. 0.4–0.8 mm long; style 4 mm long (long styled), 2–3 mm long (short styled), lobes ca. 1 mm long. Fruit obovoid, 7–10 mm long, weakly ribbed to ± smooth, orange, red or black, glabrous; calyx limb and lobes either not persistent or persistent and inconspicuous; pedicels (0.1–)0.4–3 mm long, or pedicels absent; pyrenes weakly ribbed to almost smooth on outer surface; endosperm strongly ruminate.

Distribution.—Found throughout the Archipelego from Luzon to Mindanao, except the smaller Visayan Islands. Some varieties are restricted to Mindanao.

Ecology.—Forest; low to mid-elevations, to ca. 1500 m.

Phenology records.—Flowering: March to May, August to November; fruiting: January to October.

Preliminary conservation assessment.—Near Threatened (**NT**). Last collected in 1992.

Distinguishing characteristics.—Leaf blades usually relatively short, lateral veins brochidodromous (forming pronounced, closed loops near the margin); inflorescences with a sympoidal branching pattern, generally short, with short axes and branches; fruits relatively large, 8–10 mm long.

Remarks.—*Psychotria paloensis* is widely distributed in the Philippines and is characterized by the nature of the sympodial branching of the inflorescence and the often strongly brochidodromous lateral veins near the margin. However, the variable leaf blade size and extreme differences in

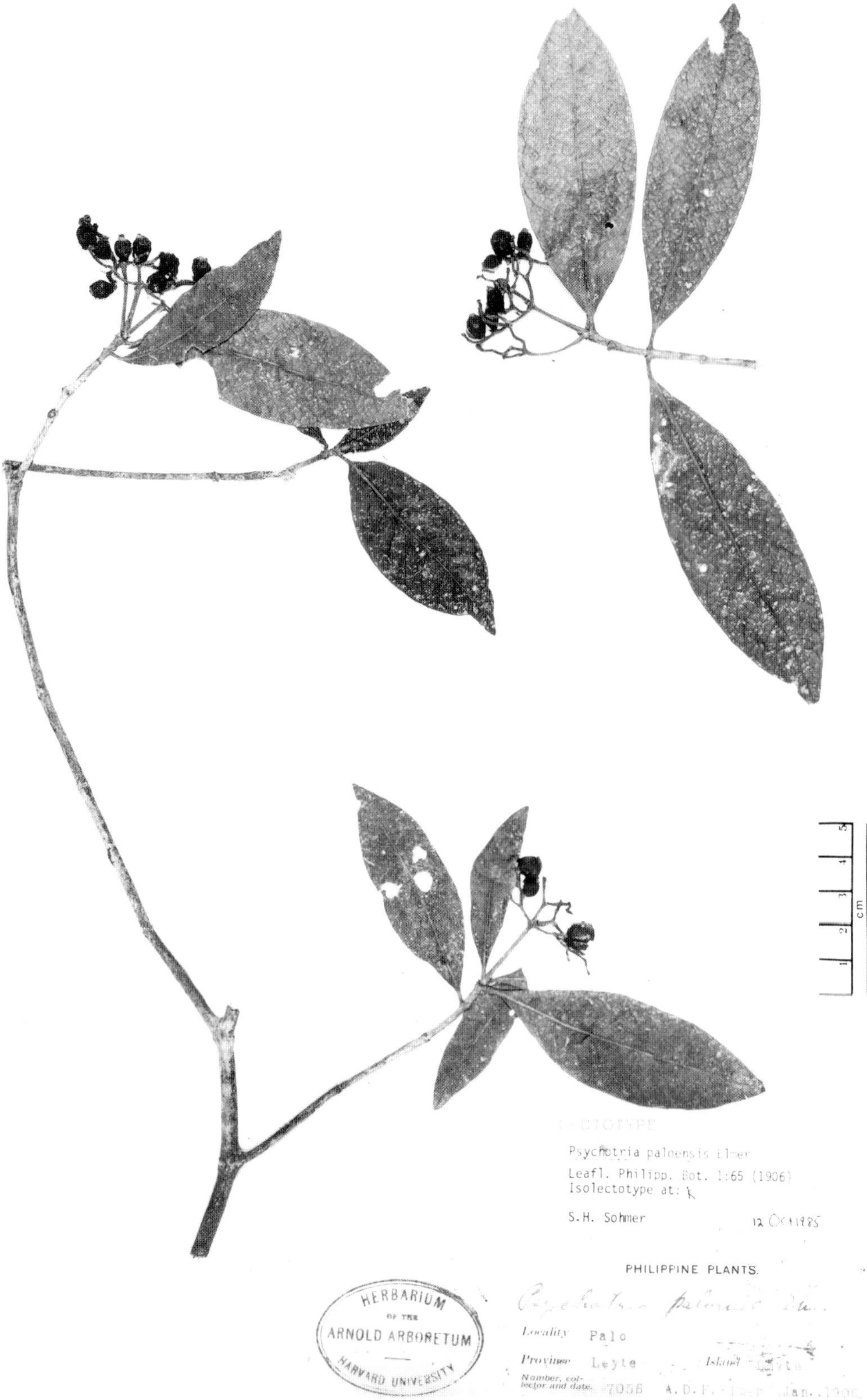

Fig. 104. *Psychotria paloensis* var. *paloensis* (*Elmer 7056*) (A, lectotype).

indumentum from totally glabrous to densely pubescent (only in the type collection) is indicative of what is, or must have been, a widely distributed taxon with a great deal of variation. The extraordinarily large calyx in one variety is remarkable but may represent merely an extreme. More recent collections and field knowledge of *P. paloensis* are required for a better understanding of this species. However, it is possible to recognize a number of varieties, as separated by the following key. With further work it is likely that some of these varieties, and groups within *P. paloensis* var. *paloensis*, may need reconginition as species.

KEY TO THE VARIETIES OF *PSYCHOTRIA PALOENSIS*

1. Calyx limb 4–6 mm wide at anthesis ____ **91b.** var. **bukidnonesis**
1. Calyx limb up to 2.5 mm wide, often < 2 mm wide, at anthesis.
   2. Secondary branches of infructescence sometimes > 3 cm long; the leaf blades drying a greyish green ____ **91e.** var. **wenzeliana**
   2. Secondary branches of infructescence usually much < 3 cm long; the leaf blades grayish green or, more frequently brownish when dry.
      3. Leaf blades broadly obovate to ovate-elliptic, 5.4–6.1 cm wide, totally glabrous ____ **91c.** var. **ovatifolia**
      3. Leaf blades narrowly elliptic, subelliptic, narrowly oblanceolate to obovate, 2.2–4 cm wide, glabrous, pubescent, or puberulent.
         4. Main inflorescence axis 2–4 cm long, puberulent; leaf blades subelliptic to narrowly oblanceolate ____ **91d.** var. **subelliptifolia**
         4. Main inflorescence axis 0.9–4 (usually < 2 cm) long, glabrous to sometimes pubescent or puberulent; leaf blades elliptic to subovate ____ **91a.** var. **paloensis**

91a. Psychotria paloensis var. paloensis

Shrub or small tree, 2–5 m high. Stipules 2–5 mm long, pubescent or glabrous. Leaves: petioles 0.5–1.6 cm long, glabrous to pubescent; leaf blades elliptic to subovate, 7.7–13.2 × 2.4–5.9 cm, drying khaki-green to grayish green, glabrous above, pubescent or glabrous below; secondary veins 10–13 per side; base obtuse to acute; apex obtuse to acute. Inflorescence/infructescence trichotomous or monochotomous, 1.4–4.5(–16) cm long, principal axis 0.9–2(–4) cm long, glabrous or puberulent, secondary branches less than 3 cm long. Flowers 5-merous, sessile or subsessile; calyx (incl. hypanthium and calyx limb) ca. 2.5 mm long, often pubescent or strigose. Fruits obovoid, 8–10 mm long; pyrenes with irregular ribbing on outer (dorsal) surface.

Other specimens examined: **Luzon. Sorsogon Prov.:** Irosin (Mt. Bulusan), *Elmer 14534* (A, BISH, BM, GH, K, L, MO, NY, P, S, UC). **Leyte. Leyte Prov.:** Buenavista, near Jaro, *Wenzel 882* (A, BM, MO, US). **Negros. Negros Occidental Prov.:** Mt Canlaon, *Edaño PNH 21962* (K, L), PNH *21969* (L, PNH); Canlaon National Park, Mt. Canlaon, *Stone et al. PPI 6262* (BRIT, K). **Mindanao.** Butuan subprov., *Weber 1007* (A). **Davao Prov.:** Todaya (Mt. Apo), *Elmer 10587* (A, BISH, BM, K, L, MO, US, W). **Davao Prov.:** without further data: *Elmer FB 28205* (A). **Agusan Prov.:** Cabadbaran, between Duros and Cawilanan, *Elmer 13740* (A, BISH, BM, GH, L, MO, P, UC, US, W). **Misamis Prov.:** without further data: *Ponce & Abejero FB 30446* (UC). **Camaguin Prov.:** Mambajao, *Elmer 14230* (A, BISH, GH, K, L, MO, UC, US, W). **Lanao Prov.:** near Lumbatan & Sawir, *Zwickey 216* (MO). **Palawan. Palawan Prov.:** Mt. Mantalingajan, *Edaño BS 77563* (A); Mt. Mantalingajan, Brooke's Point, *Edaño PNH 46* (A, L, UC), *PNH 91* (A, UC), *PNH 93* (A, L, UC); Tigaplan River, *Edaño BS 77651* (A), *BS 77656* (A), *BS 77670* (A).

Distribution.—Southern Luzon, northern Leyte, Negros Occidental, Mindanao and Palawan.

Ecology.—Forest; 200–1212 m.

Phenology records.—Flowering: April, August to November; fruiting: January, April, May, September, October.

Preliminary conservation assessment.—Near Threatened (**NT**). Last collected in 1992.

Distinguishing characteristics.—Leaf blades elliptic to subovate; inflorescence with peduncle (or main axis) usually less than 2 cm long, secondary branches less than 3 cm long; calyx tube less than 4 mm wide.

91b. Psychotria paloensis var. **bukidnonensis** Sohmer & A.P. Davis, var. nov. (**Fig. 105**). TYPE: PHILIPPINES. MINDANAO: Bukidnon Prov.: Mt. Katanglad, south slope of south peak in dense forest above Barangay Kibangay at 1900 m, 7 Mar 1949, *Sulit PNH 9977* (HOLOTYPE: A; ISOTYPE: PNH).

A *P. paloensi* var *paloensi* differt in tubis calycorum latioribus quam 4 mm.

Shrub, 2–3 m high. Stipules 3–7 mm long, pubescent. Leaves: petioles 0.5–0.8 cm long, glabrous. Leaf blades obovate or elliptic, 6.9–11.6 × 2.6–6 cm, pubescent on veins below; secondary veins 9–12 per

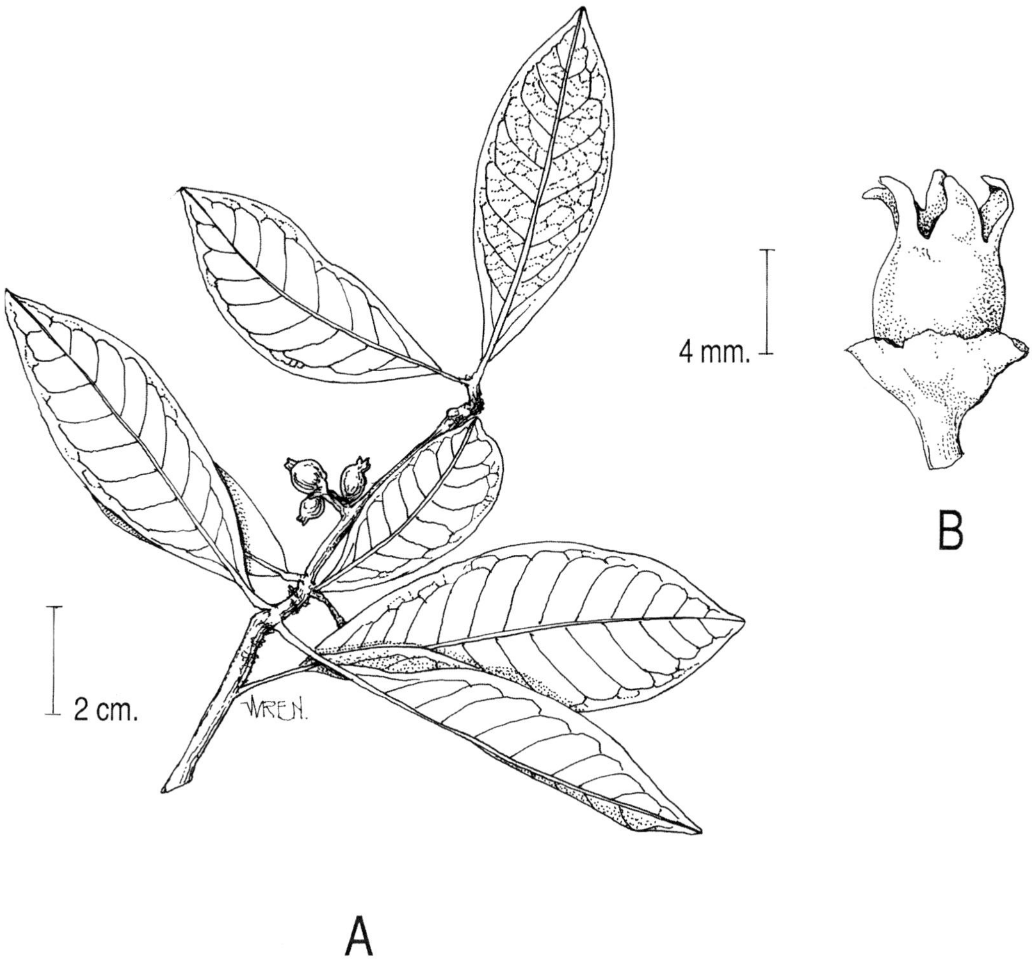

FIG. 105. *Psychotria paloensis* var. *bukidnonesis* (*Sulit PNH 9977*) (A, holotype). **A,** habit fruiting branch; **B,** flower.

side; base acute; apex obtuse to acute. Inflorescence/infructescence trichotomous, 2.5 cm long, principal axis up to 1.2 cm long, glabrous. Flowers 4-merous, sessile or subsessile; calyx (incl. hypanthium and calyx limb) 3–5 × 4–6 mm, glabrous. Fruits ovoid, 7–8 mm long; pyrenes smooth on outer (dorsal) surface.

PARATYPES: **Mindanao. Agusan Prov.:** San Mateo Bo., Butvun, Berdee Line, *Mendoza PNH 42262* (L, PNH). **Bukidnon Prov.:** Mahilucot River, *Ramos & Edaño BS 38614* (A, BM, K); Mt. Candoon, *Ramos & Edaño BS 38884* (A). **Davao Prov.:** Katigan, *Kanehira 2553* (NY), *2564* (NY).

Distribution.—In Bukidnon and Agusan provinces of Mindanao.

Ecology.—Montane forest; moderately high elevations, e.g. ca. 1900 m.

Phenology records.—Flowering: March, April; fruiting: June, July.

Preliminary conservation assessment.—Endangered (**EN B1ab(iii)**). Last collected in 1961.

Distinguishing characteristics.—Calyx 4–6 mm wide.

Remarks.—It would be interesting to be able to study this taxon in the field to see if its unusually large calyx limb is due to an aberration or part of natural variation.

91c. Psychotria paloensis var. **ovatifolia** Sohmer & A.P. Davis, var. nov. (**Fig. 107**). TYPE: PHILIPPINES. MINDORO: Mindoro Occidental Prov.: Paluan, Apr 1921, *Ramos BS 39564* (HOLOTYPE: US; ISOTYPES: A, K, L, P).

A *P. paloensi* varietatis alteris differt in laminis plerumque latioribus que 5 cm late ellipticus ad obovatis et pyrenis in dorso liratis.

Small tree or shrub, height unknown. Stipules ca. 2 mm long, glabrous. Leaves: petioles 1.5–2.6 cm

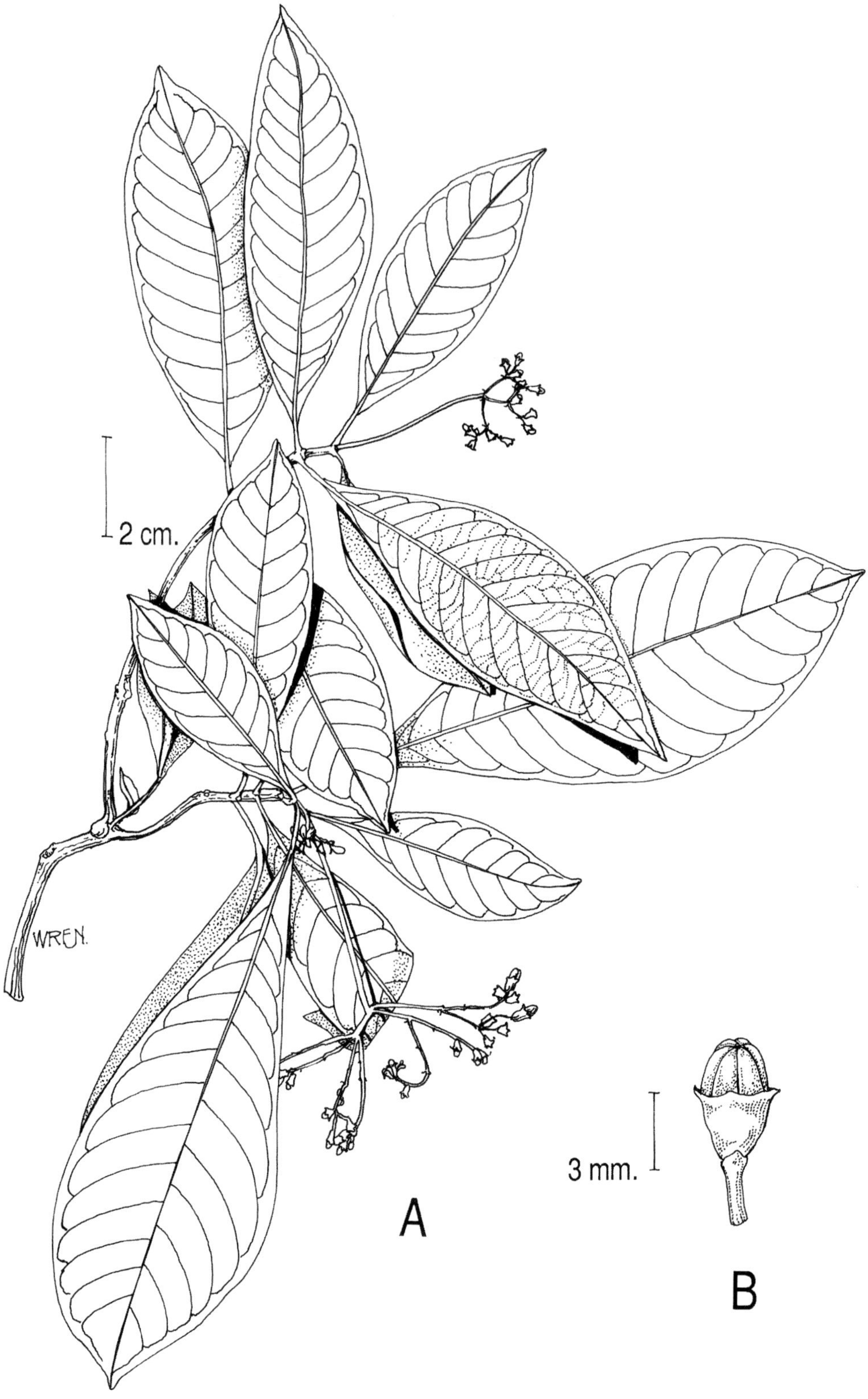

Fig. 106. *Psychotria paloensis* var. *wenzelina* (*Wenzel 3261*) (MO, holotype). **A,** habit of fruiting branch; **B,** flower in bud.

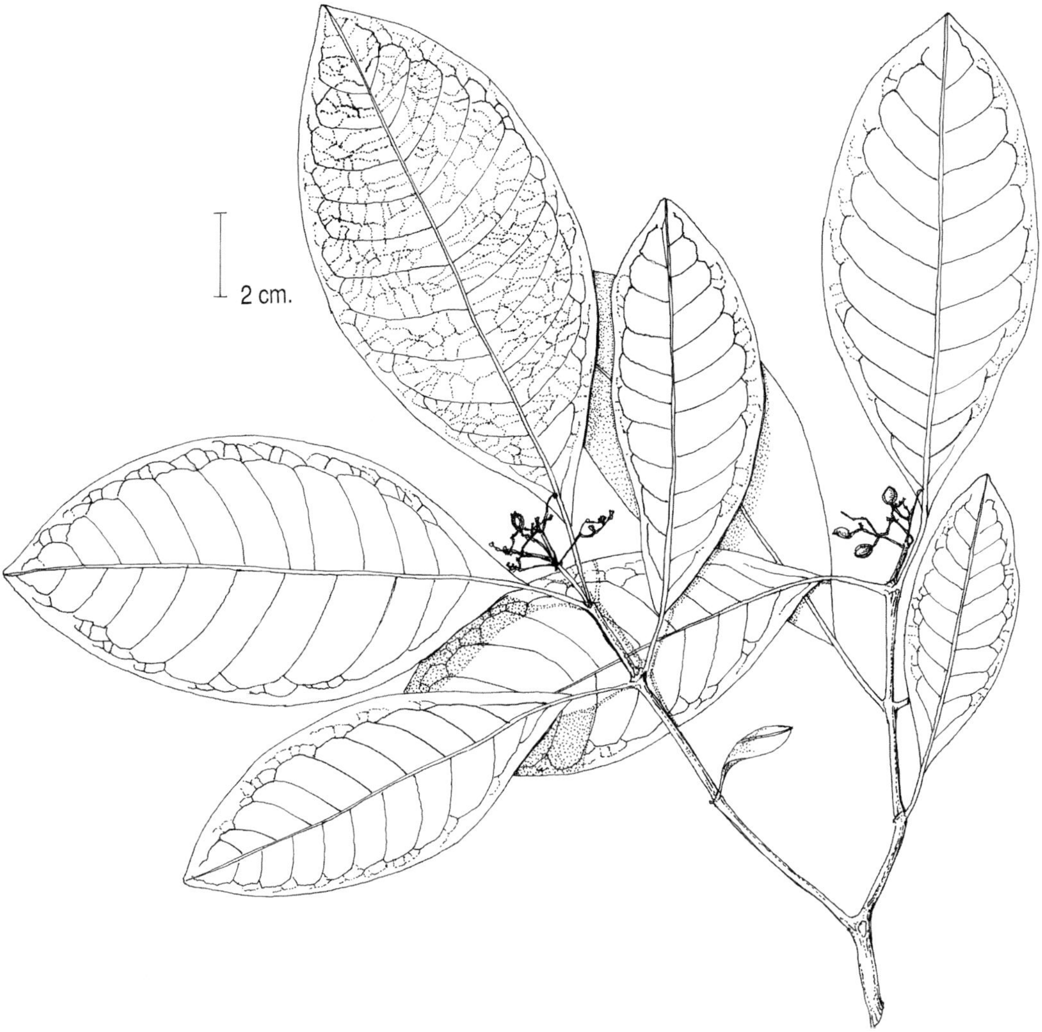

FIG. 107. *Psychotria paloensis* var. *ovatifolia* (*Ramos BS 39564*) (US, holotype). habit of fruiting branch.

long, glabrous; leaf blades broadly obovate to ovate-elliptic, 10–15 × 5.4–6.1 cm, greyish green when dry, glabrous above and below; secondary veins 10–12 per side; base acute to narrowly acute; apex abruptly acute to obtuse. Inflorescence/infructescence monochotomous or trichotomous, 3–5 cm long, the main axis ca. 1 cm long, glabrous, secondary branches less than 3 cm long. Flowers unknown. Fruit ovoid to round, 6–9 mm long; pyrenes ribbed on outer (dorsal) surface.

PARATYPES: **Mindoro. Mindoro Oriental Prov.:** Sibuang River Camp, San Teodoro, *Ridsdale 919* (BISH), *927* (BISH); inland from San Teodoro at and above Subaan River, *Ridsdale 1228* (K); Subaan River, inland from San Teodoro, *Coode 5563* (K).

Distribution.—Apparently restricted to Mindoro.

Ecology.—Forest, often on steep slopes; 300–630 m.

Phenology records.—Flowering: unknown; fruiting: April, May.

Preliminary conservation assessment.—Endangered (**EN B1ab(iii)**). Last collected in 1986.

Distinguishing characteristics.—All plant parts completely glabrous; leaf blades broadly elliptic or ovate; pyrenes ridged on outer (dorsal) surface.

91d. Psychotria paloensis var. **subelliptifolia** Sohmer & A.P. Davis, var. nov. (**Fig. 108**). TYPE: PHILIPPINES. LEYTE: Leyte Prov.: Ormoc, Lake Danao, along forested ridge at 666 m, 21 Mar 1950, *Edaño PNH 11913* (HOLOTYPE: A; ISOTYPES: L, PNH).

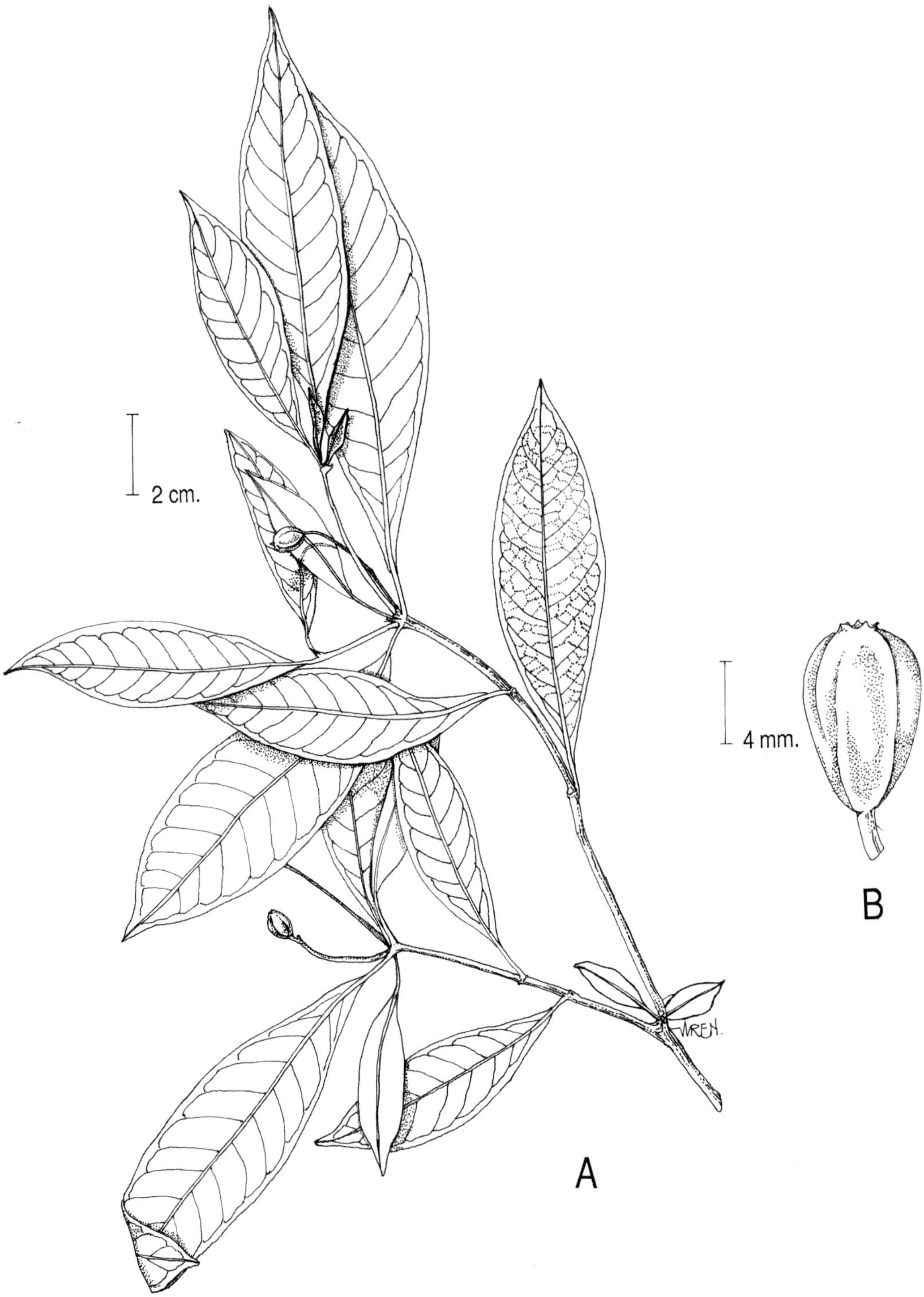

FIG. 108. *Psychotria paloensis* var. *subelliptifolia* (*Edaño PNH 11913*) (A, holotype). **A,** habit of fruiting branch; **B,** fruit.

A *P. paloensi* varietatis alteris differt in laminis plerumque angustioribus que 5 cm anguste ellipticis ad oblanceolatis, et pedunculo 2–4 cm longo.

Shrub, ca. 3 m high. Stipules ca. 5 mm long, puberulent. Leaves: petioles 0.5–2 cm long, glabrous to puberulent; leaf blades subelliptic to narrowly oblanceolate, 7.8–12.5 × 2.2–3.8 cm, pubescent on midrib below; secondary veins 9–10 per side; base acute to narrowly acute; apex acuminate. Inflorescence/infructescence monochotomous or trichotomous, but usually monochotomous, 2.5–3.5 cm long, principal axis 2–4 cm long, puberulent. Flowers unknown. Fruit obovoid-ellipsoid, ca. 10 mm long; pyrenes ribbed on outer (dorsal) surface.

PARATYPES: None—known only from the type collection.

Distribution.—Known only from Lake Danao, Leyte Province, Leyte.

Ecology.—Forested slopes; one record at 666 m.

Phenology records.—Flowering: unknown; fruiting: March.

Preliminary conservation assessment.—Critically Endangered (**CR B1ab(iii)**). Only collected once in 1950.

91e. Psychotria paloensis Elmer var. **wenzeliana** Sohmer & A.P. Davis, var. nov. (**Fig. 106**). TYPE: PHILIPPINES. MINDANAO: Surigao Prov.: "Turugaun Turbid Placer" (?), forest at 150 m, 19 Apr 1928, *Wenzel 3261* (HOLOTYPE: MO; ISOTYPES: A, GH, K, NY, UC).

A *P. paloensi* varietatis aeteris differt in ramulis secundariis inflorescentiae saepe longioribus que 3 cm, et laminis siccis pallide griseo-viridibus.

Small tree, or shrub, 1.8–4.6 m high. Stipules 2–4 mm long, pubescent to glabrous. Leaves: petioles 0.5–1.5 cm long, glabrous to pubescent; leaf blades elliptic to subobovate, 7–13.2 × 2.9–5.9 cm, drying a light khaki-green, glabrous to pubescent along veins below; secondary veins 10–12 per side; base narrowly acute to acute; apex abruptly acute to acuminate. Inflorescence/infructescence trichotomous, 1.6–8 cm long, principal axis 1.1–4 cm long, glabrous, secondary branches often more than 3 cm long. Flowers 5-merous, often sessile; calyx (incl. hypanthium and calyx limb) ca. 2.5 × 2.5 mm, puberulent. Fruit obovoid, 6–8 mm long; pyrenes with irregular ribbing on the outer (dorsal) surface.

PARATYPES: **Mindanao. Surigao Prov.:** without further data: *Wenzel 3120* (A, GH, K, MO), *31209* (A, MO, NY).

Distribution.—Limited to Surigao Province.

Ecology.—Forest; low elevation, e.g. ca. 150 m.

Phenology records.—Flowering: April, August; fruiting: April, August.

Preliminary conservation assessment.—Critically Endangered (**CR B1ab(iii)**). Last collected in 1928.

Distinguishing characteristics.—Leaf blades drying light khaki-green; secondary branches of inflorescence/infructescence often more than 3 cm long;

92. Psychotria luzoniensis (Cham. & Schltdl.) F.-Vill. in F.M. Blanco, Fl. Filip., ed. 3, 4, Nov. App.: 112. 1880. *Grumilea luzoniensis* (Cham. & Schltdl.) Merr., Philip. J. Sci. 10:131. 1915, as '*luçoniensis*.' *Coffea luzoniensis* Cham. & Schltdl. in Linnaea 4:32. 1829, as '*luçoniensis*.' TYPE: PHILIPPINES. Luçonia, *Chamisso s.n.* (HOLOTYPE: LE; ISOTYPE: K).

Psychotria tacpo (Blanco) Rolfe in J. Linn. Soc. (Bot) 21:312. 1884. *Paederia tacpo* Blanco, Fl. Filip. 160. 1837. TYPE: unknown.

Small tree or shrub, 1 7 m high, young stems glabrous. Stipules imbricate, lanceolate to narrowly lanceolate, (0.4–)0.6–1.8(–2.2) cm long, glabrous; apex acute. Leaves: petioles 0.7–2.5(–3.5) cm long, glabrous; leaf blades narrowly elliptic, ± elliptic, or linear-elliptic to subobovate, (4.7–)5.5–16(–18.7) × 2–5.3–4.7 cm, subcoriaceous, usually drying green or brownish green, glabrous above and below, rarely pubescent on veins below; secondary veins (6–)9–17 pairs; tertiary venation usually weak to rather prominent; base acute to narrowly acute; apex abruptly acuminate to acute. Inflorescence/infructescence trichotomous, (1.2–)2–4.5(–9) cm long, glabrous; principal axis 1–5.6 cm long, each axis with 2–3 primary nodes and opposite branching at each node, branches terminating in cymules; bracts triangular, ca. 0.3 mm long, often absent (?caducous); bracteols not seen (?absent). Flowers 4- or 5-merous, pedicels 0.5–3 mm long; calyx (incl. hypanthium) 1.3–2.1 mm long, glabrous; calyx lobes shallowly triangular, ca. 0.2 mm long, or obscure; corolla: white, greenish white to greenish yellow; corolla tube 2–2.5 mm long, external surface glabrous, villous at throat within; corolla lobes 1.6–2 mm long; stamens: filaments ca. 0.5 mm long (long styled) or ca. 2 mm long (short styled); anthers ca. 0.5 mm long (long styled) or ca. 1 mm long (short styled); style 4 mm long (long styled) to 1 mm long (short styled), lobes ca. 0.5 mm long (long styled) to ca. 1 mm long (short styled). Fruit turbinate or sometimes globose, often ± bilobed, 5–9 mm long, smooth or rarely very weakly ribbed, orange or red, glabrous; calyx limb and lobes persistent; pedicels 1–3.8 mm long; pyrenes smooth on outer surface; endosperm ruminate.

Other specimens examined: **Luzon. Cagayan Prov.:** Mt. Bawa, *Edaño BS 78475* (A); Mt. Caboloan, *Edaño BS 78588* (A), *BS 78593* (A), *BS 78596* (A); Peñablanca, Bagaba, Sitio Mororan, *Reynoso, Fuentes & Majaducon PPI 11736* (BRIT); Pena Blanca [Peñablanca], *Bolster 180* (UC), *Ramos BS 76949* (A), *Ramos & Edaño BS 46621* (UC). **Ilocos Norte Prov.:** Sitio Nalamus, Barangay Dampig, Pagudpud, *Garcia & Fernando PPI 25221* (BRIT, K); Mt. Pico de Loro, *Edaño PNH 18032* (A, K, PNH), *PNH 17918* (A, K, L), *PNH 17985* (K, L, PNH), *PNH 17990* (A, K, L, PNH), PNH *17992* (K, L, PNH); Nueva Era, *Iwatsuki, Murata & Gutierrez P-82* (L); Mt. Burnay, south of Mt. Sicapoo, *Iwatsuki, Murata & Gutierrez P-860* (L); Mt. Bubungkayo, en route from Santo Nino to Miner's Camp, *Iwatsuki, Murata & Gutierrez P-380* (A, L); enroute from Nueva Era to Mt. Semenublan, *Iwatsuki, Murata & Gutierrez P-110* (L), *P-111* (L), *P-169* (L); Bangui, *Ramos BS 27459* (BM, P); Mt. Palimlim, *Ramos BS 33260* (A, K, UC), *vel aff.*; *ibid.*, *Ramos BS 33376* (A, UC). **Ilocos Norte Prov.:** without further data: *Merritt & Darling FB 13972* (L), *Vidal 1481* (A, K, L). **Ilocos Sur Prov.:** *Rivera BS 46512* (UC). **Isabela Prov.:** Ilagan to Echague, *Clemens 7552* (UC). **Abra Prov.:** Poblacion Cangal, Municipality of Sallapadan, *Fuentes PPI 38502* (BRIT); Massisiat, *Alcasid et al. PNH 1615* (A). **Mountain Prov.:** Mabuntat Lubuagan, *Celestino PNH 7843* (A); District of Bontoc, *Curran FB 17023* (NY); Benguet, *Costales FB 29787* (SING, UC). **Nueva Vizcaya Prov.:** Carasacas, *Martin PNH 33497* (L, US); vicinity of Dupax, *McGregor BS 11437* (K). **Nueva Vizcaya Prov.:** without further data: *Alvarez FB 18410* (NY). **Benguet Prov.:** Sablang, *Fenix BS 12701* (MO, S); Trinidad to Kapungun, *Clemens 17246* (UC); Baguio, *Basfeinas 76* (L), *Elmer 8953* (K, NY), between Baguio and Sablon, *Williams 1474* (K, NY); Tonglon, *Loher 5092* (US). **Union Prov.:** Santa Rita, *Barbon et al. PPI 8993* (BRIT); Bauang, *Elmer 5545* (K), *Fenix 144* (A, K, MO), *BS 12950* (BM, K, US). **Pangasinan Prov.:** *McGregor BS 41437* (A, W). **Pampanga Prov.:** Arayat, *Merrill 1364* (K, US); Stotsenburg, *Clemens 17652* (UC). **Zambales Prov.:** Mt. Pinatubo, *Clemens 17404* (UC), *Fox PNH 4911* (PNH); Villar, *Fox PNH 4599* (A, L); Hermana Mayer Island, without further data, *Santos 6194* (L, US). **Zambales Prov.:** without further data: *Edaño BS 34666* (L). **Quezon Prov.:** Infanta, *Forestry Student PNH 35266* (CANB); Pagbilao [?Pagbalao], *Merrill 1903* (K, US), *1906* (K, US); Salakan, Guinayangan, *Mendoza PNH 97948* (A, K, L, PNH); Polillo Island, *Robinson BS 10221* (US), *Salvoza FB 29734* (UC); Alabat Island, Sitio Canumay, Villa Norte Barangay Bacung, *Gaerlan & Romero PPI 23632* (BRIT, K). **Bataan Prov.:** Limay, *McGregor BS 44524* (UC); Mt. Mariveles, *Borden FB 1951* (K, NY, US), *FB 2360* (K, NY), *Meyer FB 2319* (K, NY, US), *Williams 375* (NY). **Rizal Prov.:** Manila, *Andersson s.n.* (NY, S), *Gaudichaud 215* (A), *Merrill 560* (K, NY), *Merrill 3436* (BM, K), *Ramos BS 12164* (BM, US), *Wilkes Expedition s.n.* (NY); Rio Tansá (Manila), *Loher 1506* (K); Antipolo, *Merrill 597* (A, BM, GH, K, L, MO, NY, US, W), *Ramos BS 343* (US), *BS 560* (US), *Vidal 1455bis* (A); Mt. Angilog, *Ramos BS 49776* (L, P, US); Mt. Irid, *Ramos & Edaño BS 48537* (K, UC), *Ramos & Edaño BS 48552* (UC); Montalban, *Manayon 70* (L), *Robinson BS 6144* (L); Mt. Mapatag, San Andales, *Edaño BS 48673* (UC), *BS 48745* (BM, S, UC, W), *BS 48833* (NY, S, UC); Balara, Quezon City, *Del Mundo 11* (L), *Mati 25* (L); Novaliches, Quezon City, *Ramos BS 40866* (BM, PNH); San Francisco del Monte, *Mendoza PNH 3071* (MO); Mt. Susong-Dalaga, *Ramos & Edaño BS 29356* (GH, US); Susong, *Loher 1505* (K); Bosoboso, *Ahern's Collector FB 1893* (K, P, RSA), *Merrill 1869* (K, NY, US), *Vidal 403* (A, L); Mt. Rinagbayanan, San Andales, *Edaño BS 48844* (BM, NY, UC); Pasia, *Vergara 82* (L); Kay Ungulan, Teresa, *Bartlett 15360* (A). **Rizal Prov.:** without further data: *Loher 14265* (UC), *Loher s.n.* (UC), *Ramos BS 16687* (MO, S), *Reillo BS 19176* (K). **Cavite Prov.:** Mt. Palay, Palay National Park, Ternate, *Reynoso, Sagcal & Fuentes PPI 14970* (BRIT); *Reynoso, Sagcal & Fuentes PPI 17007* (BRIT, K); Palicpican River, Mt. Palay Palay National Park, Ternate, *Reynoso, Sagcal & Fuentes PPI 17089* (BRIT); Pico de Loro, Mt. Palay Palay, Ternate, *Reynoso, Sagcal & Fuentes PPI 17156* (BRIT). **Laguna Prov.:** Los Banos, *Elmer 8200* (NY), *Hallier 4038* (L), *Sulit PNH 7084* (A); Mt. Makiling, *Alombro PNH 35430* (LAE), *Amihan PNH 34316* (BM, K), *Collado PNH 35492* (BM, K), *Curran FB 13159* (L), *Elmer 18156* (A, BISH, BM, GH, K, L, NY, MO, S, UC), *Esteve PNH 35352* (BISH, CANB), *Foxworthy 32* (US), *Lamanilao PNH 34252* (BM), *Landingin PNH 35452* (US), *Lazo PNH 33403* (L, US), *Loher 6368* (K), *Lorenzo PNH 35288* (CANB), *Narciso PNH 35310* (BISH, CANB, L), *Ordon PNH 33509* (L, US), *Quisumbing s.n.* (UC), *Robinson BS 17224* (A, US), *Servinas BS 16918* (BM, L, NY, US), *Sohmer 11967* (BISH, K), *Whitford FB 19714* (BM); Santa Maria, *Curran FB 8866* (NY); Calauan, *McGregor BS 12346* (US); Pagsanjan, *Steiner PNH 22946* (L, US). **Batangas Prov.:** Mt. Maculot, Cuenca, *Castro, Barbon & Garcia PPI 22085* (BRIT); *Castro, Barbon & Garcia PPI 22115* (BRIT); Mataasn Nakahoy, *Aravez 35* (L); Inland from Malabrigo Point, *Ridsdale 1875* (K). **Camarines Sur Prov.:** Pili, *Convocar PNH 2928* (A, L, MO). **Albay Prov.:** Castillo [?Castilla], *Vidal 3062* (K), *Vidal 3067* (K); Daraga, *Quisumbing PNH 22992* (L). **Sorsogon Prov.:** Mt. Bulusan, Irosin, *Elmer 15214* (A, BISH, BM, GH, K, L, MO, NY, P, S, UC, US), *15438* (A, BISH, BM, GH, K, L, MO, NY, S, UC, US), *15523* (A, BISH, BM, K, L, MO, NY, S, UC, US). **Luzon. Without further data:** *Vidal 1455c* (A), *Loher 1507* (K). **Mardinique. Mardinique Prov.:** Hawillian, So. Maulawin, Barangay Dampulan Torrijos, *Romero & Chavez PPI 29080* (BRIT, K). **Mindoro. Mindoro Oriental Prov.:** Subaan River, inland from San Teodoro, *Coode 5450* (K); Mt. Minolo, Puerto Galera, *Britton PNH 19454* (L), *Kienholz BS 15215* (UC), *BS 15425* (UC); Mahabang Parang Cove, Puerto Galera Bay, *Santos 5164* (L, US), *5256* (L, US); Pinamalayan, *Ramos BS 40937* (A); Mt. Halcon, *Merrill 6150* (L). **Mindoro Occidental Prov.:** S (?) Alatin, Abra de Ilog, *Sulit PNH 13760* (A, L, PNH); Iriron, *Merritt FB 8844* (MO). **Mindoro:** without further data: *Amarillas FB 19932* (BM, L, NY, PNH, US). **Masbate. Masbate Prov.:** Masbate Island, Barangay Tugbo, Barbon, *Garcia & Fernando PPI 12578* (BRIT). **Masbate Prov.:** without further data: *Merrill 3395* (BM, K, NY, US), *Rosenbluth FB 12563* (NY). **Panay. Iliolo Prov.:** Miagao, *Vidal 1481bis* (A); Guimaras Island, Buenavista, *Santos 4657* (L, US), *Sulit PNH 11687* (A, L). **Iloilo Prov.:** without further data: *Robinson BS 18068* (K). **Antique Prov.:** Pungtod, Semirara Island, Caluya, *Romero & Majaducon PPI 29712* (BRIT), *Romero & Majaducon PPI 29716* (BRIT, K); Belawarte, Semirara Island, Caluya, *Romero & Majaducon PPI 29487* (BRIT, K); Palawan, Barangay Semirara Island, Caluya, *Romero & Majaducon PPI 29501* (K). **Antique Prov.:** without further data: *McGregor BS 32459* (L). **Negros. Negros Oriental Prov.:**

San Jose, Calo, *Kondo & Edaño PNH 36610* (BISH, PNH), Dumaguete, Cuernos Mountains, *Elmer 10117* (A, BM, K, L, MO, NY, W), *Vidal 403c* (A). **Negros Occidental Prov.:** Luzuriaga, *Robinson BS 9956* (K). **Palawan. Palawan Prov.:** Lipuun Island, Lipuun Peak, *Mendoza & Cordero NHM 91360* (K); Malapackun Island, *Ridsdale SMHI 454* (K). **Palawan. Without further data:** *Foxworthy BS 561* (A, NY). **PHILIPPINES. Without further data:** *Comision Flora Forestal 1455* (K, L), *Cuming 518* (BM, K, L, W), *Hallier 237* (NY).

Distribution.—Widely distributed throughout the Philippines.

Ecology.—Forest, including primary, disturbed, and secondary forest; recorded from limestone and well-drained substrates; sea level to nearly 2000 m but mostly below 700 m.

Phenology records.—Flowering: all year round; fruiting: January to June, September to December.

Preliminary conservation assessment.—Least Concern (**LC**). Last collected in 1996.

Distinguishing characteristics.—Leaf blades elliptic, narrow, often less than 4.5 cm wide, tending to dry green or brownish green; infloresences trichotomous, fruit tending to be diplococcus; pyrenes smooth on outer (dorsal) surface.

Remarks.—This is one of the most widely distributed species of Philippine *Psychotria*, perhaps the most widely distributed. *Psychotria luzoniensis*was first described by Chamisso and Schlechtendal (1829) in the genus *Coffea* and later transferred to *Psychotria* by Celestino Fernandes-Villar (1880), where it rightfully belongs.

Some of the variation is unusual. The collection by Vidal (*Vidal 1481*) from Panay, for example, has flowers that are 8-merous. There also seems to be intergradation with other species. For example, *Ramos BS 41956* shares features of *P. pinnatinervia* Elmer: it has a globose rather than diplococcus fruit with pyrenes that are smooth on the outer (dorsal) surface, not ridged as they would be in *P. pinnatinervia*.

93. Psychotria bayombongensis Sohmer & A.P. Davis, sp. nov. (**Fig. 109**). TYPE: PHILIPPINES. LUZON: Nueva Vizcaya Prov.: Bayombong, Mar 1886, *Vidal 3068* (HOLOTYPE: K; ISOTYPE: A).

Frutices vel arbores parvae 2–7 m alti sunt, stipulis 0.8–1 cm longis imbricatis glabris, laminis 8 × 17–15 × 31.5 cm latis et longis crasse coriaceis glabris ovatis ad elliptici-obovatis quando siccis viridibus ad subbadio-viridibus quoque nervis tertis non evidentibus, petiolis (3–)4–6 cm plerumque plus que 4 cm longis, inflorescentia/infructescentia trichotoma vel pentachotoma in flore 8–10 cm longa, pedunculis (1.5–)3–5.5 cm longis axibus glabris, nodis primariis cum ramis tertiis, floribus 5-meris, pedicellis 2–4 mm longis, calycibus 4 mm longis lobis obscuris corollis albis glabris tubo (longistylis) 5–6 mm longo, lobis 5–6 mm longis, fructibus maturis 5 mm longis glabris costatis vel rugosis, pyrenis in dorso costatis, endospermo ruminato.

Shrub or small tree, 2–7 m high, young stems glabrous. Stipules imbricate, lanceolate, 0.8–1 cm long, glabrous; apex acute. Leaves: petioles (3–)4–6 cm long, rather stout (3–5 mm wide at mid-point), glabrous; leaf blades ovate to elliptic-obovate, 17–31.5 × 8–15 cm, distinctly coriaceous, drying green to brownish green, glabrous above and below; secondary veins (11)15–20 pairs; tertiary venation weak to obscure, but rather prominent on the upper surface; base acute; apex broadly acute to obtuse. Inflorescence/infructescence trichotomous or pentachotomous, 8–10 cm long, puberulous or glabrous; principal axis (1.5–)3–5.5 cm long, each axis with 2–3 primary nodes and verticillate branching at each node, branches terminating in rather dense cymules; bracts and bracteoles unknown (?absent). Flowers 5-merous, pedicels 2–4 mm long; calyx (incl. hypanthium) ca. 4 mm long, puberulous; calyx lobes obscure; corolla: white; corolla tube 5–6 mm long (long styled), external surface glabrous, internal surface glabrous; corolla lobes 5–6 mm long; stamens: ca. 2 mm long (long styled); anthers unknown; style 3–4 mm long (long styled), lobes ca. 1 mm long. Fruit turbinate to obovoid, ca. 5 mm long, somewhat ribbed or wrinkled, red, glabrous; calyx limb and lobes persistent; pedicels 2–4.9 mm long; pyrenes ribbed on outer surface; endosperm ruminate.

PARATYPES: **Luzon. Cayagan Prov.:** Claveria: *McLean & Catalan s.n.* (field #130) (UC). **Kalinga Prov.:** Balbalan, *Celestino PNH 7818* (A).

Distribution.—North central Luzon.

Ecology.—Primary forest; elevation unknown.

Phenology records.—Flowering May; fruiting: unknown.

Preliminary conservation assessment.—Critically Endangered (**CR B1ab(iii)**) or possibly Extinct (**EX**). Last collected in 1948.

Distinguishing characteristics.—Leaf blades unusually large, distinctly coriaceous (particularly in the holotype), petioles 3–5 mm in diam; inflorescences trichotomous or pentachotomous, robust; pyrenes ribbed (not ridged) on outer (dorsal) surface.

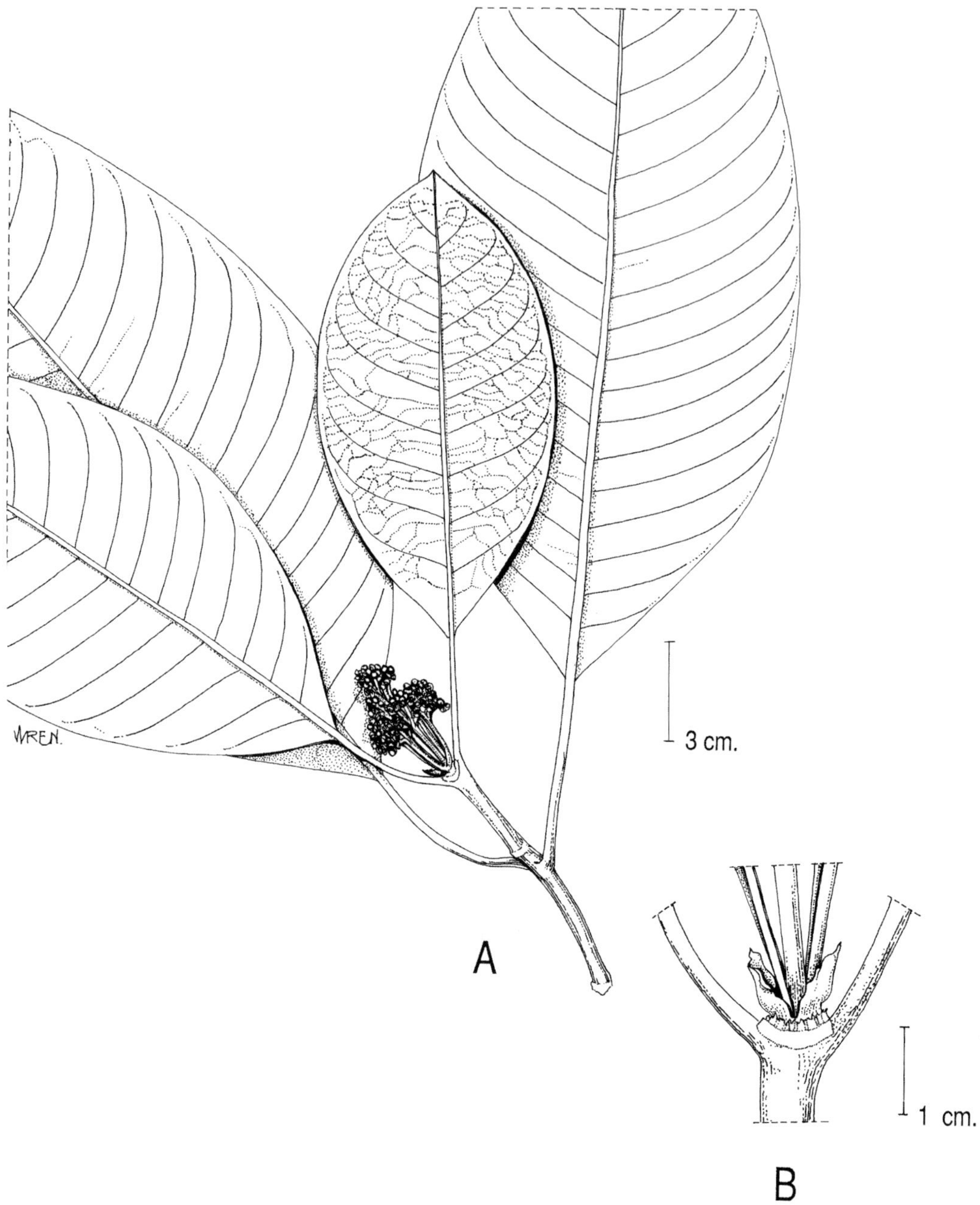

FIG. 109. *Psychotria bayombongensis* (*Vidal 3068*) (K, holotype). **A,** habit of fruiting branch; **B,** stipule.

Remarks.—It would be very useful to see this taxon in its natural environment. The large leaves would surely make this a distinctive, easily recognizable field characteristic. Hopefully, future fieldwork will add to our knowledge of this species.

94. Psychotria negrosensis Elmer, Leafl. Philipp. Bot. 2:520. 1908; Merr., Enum. Philipp. Fl. Pl. 3:560. 1923. (**Fig. 110**). TYPE: PHILIPPINES. NEGROS: Negros Oriental Prov.: Dumaguete, Cuernos Mountains, steep wooded slope at 1060 m, Mar 1908, *Elmer 9686* (LECTOTYPE: L, designated here; ISOLECTOTYPES: A, BM, K, MO, W).

Shrub, 2–3 m high, young stems glabrous. Stipules imbricate, ± narrowly elliptic to ligulate, 1.3–2.4

FIG. 110. *Psychotria negrosensis* (*Elmer 9686*) (L, lectotype).

cm long, glabrous; apex acute. Leaves: petioles 1.5–3.2 cm long, glabrous; leaf blades elliptic, elliptic-oblong to subobovate, 8–16 × 3.4–6 cm, subcoriaceous to coriaceous, drying yellow-green above, grayish brown below or chocolate-brown above and below, glabrous above and below; secondary veins 9–14 pairs; tertiary venation prominent below and above; base obtuse to acute; apex abruptly acute to abruptly acuminate. Inflorescence/infructescence trichotomous, 4–9.7 cm long, glabrous; principal axis 1.5–6 cm long, each axis with 1–2(–3) primary nodes and opposite branching at each node, branches terminating in cymules; bracts and bracteoles unknown (?absent). Flowers 5-merous, pedicels 1–3 mm long; calyx (incl. hypanthium) 2–3.5 mm long, glabrous; calyx lobes obscure or shallowly triangular, ca. 0.3 mm long; corolla: color unknown; corolla tube 5–7 mm long, external surface glabrous, villous at throat within; corolla lobes 2.5–3.5 mm long; stamens: filaments ca. 1 mm long (long styled); anthers ca. 1 mm; style ca. 5 mm long, lobes ca. 1 mm. Fruit ovoid, 7–10 mm long, somewhat ribbed, probably red but reported as yellow, glabrous; calyx limb and lobes persistent, obvious or obscure; pedicels 1.3–5 mm long; pyrenes ribbed-ridged on outer surface; endosperm ruminate.

Other specimens examined: **Negros. Negros Oriental Prov.:** Lake Danao, Sibulan, *Edaño PNH 6620* (A, L, SING), *PNH 6640* (A); Lake Balinsasayao, Tanjay, *Edaño PNH 6696* (A, K, L); Sibulan, Kabalinan, Lake Balinsasayao, *Reynoso, Fuentes & Garcia PPI 946* (BRIT, K); Mt. Silay, *Whitford 1494* (K).

Distribution.—Negros Oriental Province, Negros.

Ecology.—Forested slopes, including forest with *Agathis*; 976 - 1060 m.

Phenology records.—Flowering March; fruiting: May, September.

Preliminary conservation assessment.—Endangered (**EN B1ab(iii)**). Last collected in 1991.

Distinguishing characteristics.—Leaf blades drying green to grayish brown (particularly the type collection), tertiary venation very prominent; inflorescences lax; fruits ovoid, pyrenes ribbed but not ridged on outer (dorsal) surface.

Remarks.—We are not entirely certain of the status of this species but, not being able to match it with any other, we have retained it.

95. Psychotria patulinervia Merr. & Quisumb. ex Sohmer & A.P. Davis, sp. nov. (**Fig. 111**). TYPE: PHILIPPINES. MINDANAO: Davao Prov.: Mt. Apo, Miran River, in forest along stream at 1916 m, 9 Nov 1946. *Edaño PNH 1420* (HOLOTYPE: L).

Arbores parvae vel frutices 1–3 m alti sunt, stipulis (0.5–)1.2–2.2 cm longis imbricatis, laminis 4.6 × 13.8–6.5 × 16.3 cm latis et longis obovatis vel obovati-ellipticis apice abrupte acuto basi acuta chartaceis vel semicoriaceis glabris in sicco obscure badiis nervis lateralibus 11–17 in dimidio quoque, petiolis 1.7–2.2 cm longis, inflorescencia trichotoma glabra 4–8.8 cm longa, floribus 10–15, pedicellis (0.7–)1–5 mm longis, floribus 5-meris hypanthio 2–3.8 mm longo, lobis calycis obscuris glabris vel pubescentibus, corollis albis (brevistylis) glabris tubo (2.9–)4–9 mm longo, lobis 2–5 mm longis, fructibus 7–9.8 mm longis costatis glabris ut dicunt badiis vesimiliter obscure rubris vel nigris pyrenis in dorso costatis, endospermo ruminato.

Small tree or shrub, 1–3 m high, young stems glabrous. Stipules imbricate, lanceolate to narrowly lanceolate, (0.5–)1.2–2.2 cm long, glabrous; apex acute. Leaves: petioles 1.7–2.2 cm long, glabrous; leaf blades obovate to obovate-elliptic, 13.8–16.3 × 4.6–6.5 cm, chartaceous to subcoriaceous, drying dark brown, glabrous above and below; secondary veins 11–17 pairs; tertiary venation prominent; base acute; apex abruptly acute. Inflorescence/infructescence trichotomous, 4–8.8 cm long, glabrous to puberulous; principal axis 2.4–6.1 cm long, each axis with 1–3 primary nodes and opposite branching at each node, branches terminating in condensed cymules; bracts narrowly triangular to acuminate, ca. 1.5 mm long; bracteoles similar to the bracts, 0.6–1.5 mm long. Flowers 5-merous, pedicels (0.7–)1–5 mm long; calyx (incl. hypanthium) 2–3.8 mm long, glabrous to puberulous; calyx lobes ± deltate, 0.4–1.2 mm long, or obscure; corolla: white; corolla tube (2.9–)4–9 mm long, external surface glabrous, internal surface glabrous (at least short styled flowers); corolla lobes 2–5 mm long; stamens: filaments ca. 2 mm long (short styled); anthers ca. 1.5 mm long; style unknown. Fruit turbinate, 7–9.8 mm long, distinctly keeled-ridged, probably dark red or black but reported brown, glabrous; calyx limb and lobes not persistent or rarely persistent; pedicels (2–)4.4–7.2 mm long; pyrenes ribbed-ridged on outer surface; endosperm ruminate.

PARATYPES: **Mindanao. Davao Prov.:** Todayo (Mt. Apo), *Elmer 11462* (A, BM, K, L, MO, US, W); Mt. Apo, Lake Lanao, *Edaño PNH 1379* (K, L); Mt Apo, Miran River, *Edaño PNH 1445* (K, L). **Bukidnon Prov.:** Tangculan: *Ramos & Edaño BS 39078* (A, K), *BS 39100* (A, K).

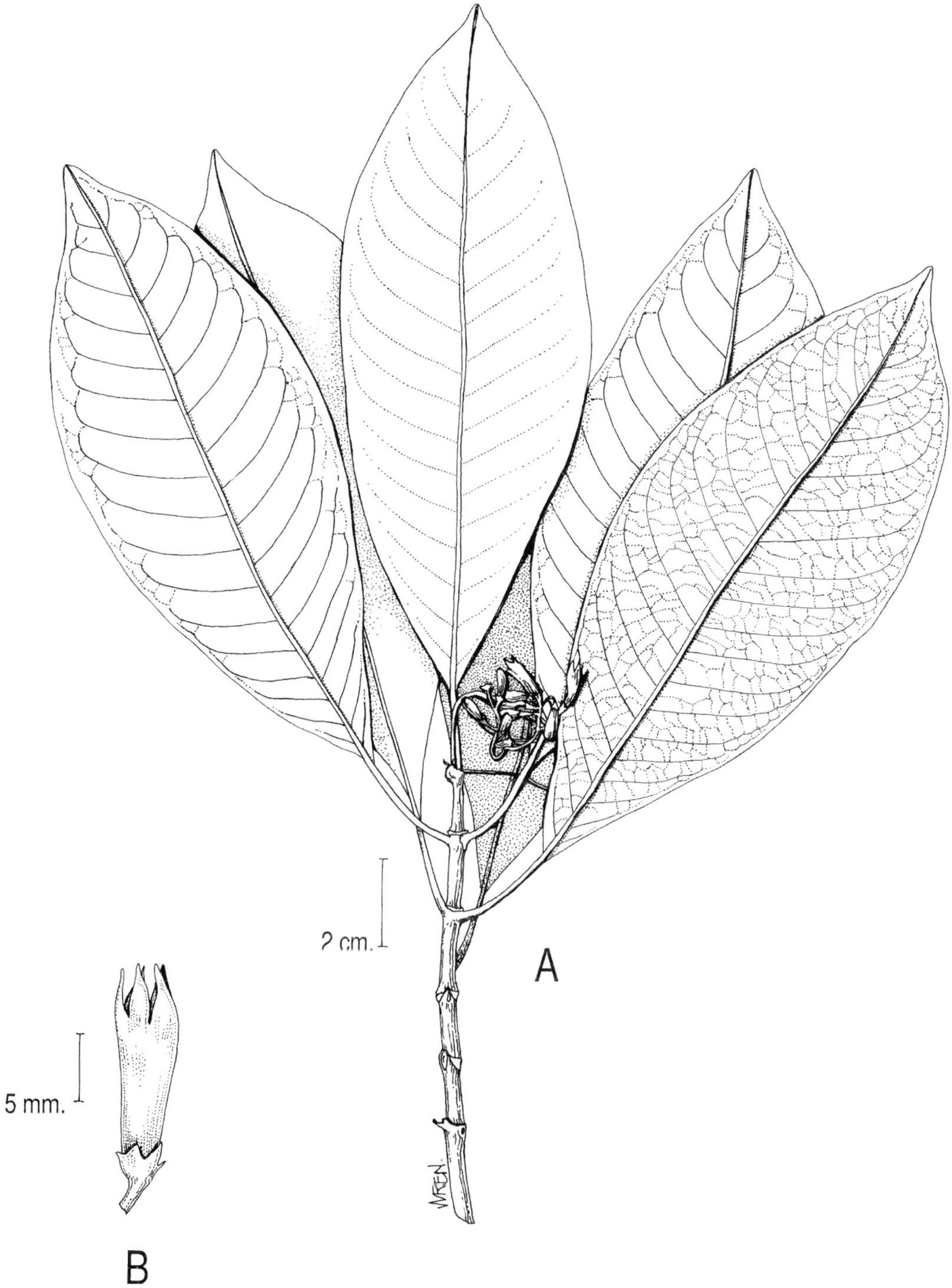

Fig. 111. *Psychotria patulinervia* (*Edaño PNH 1420*) (L, lectotype). **A,** habit of flowering branch; **B,** flower.

Distribution.—Known only from Mindanao (Mt. Apo and the vicinity of Tangculan).

Ecology.—Forest, and streamside forest; high elevation on Mt. Apo (1916–2123 m).

Phenology records.—Flowering: August to November; fruiting: May to July, October.

Preliminary conservation assessment.—Critically Endangered (**CR B1ab(iii)**). Last collected in 1946.

Distinguishing characteristics.—Leaves: lateral veins forming a very prominent network as each one loops to join the adjacent one; inflorescences trichotomous, glabrous; corolla tube ca. 9 mm long.

Remarks.—*Psychotria patulinervia* was first recognized in the herbarium by Merrill and Quisumbing, as evidenced by annotations to herbarium labels (including some of the specimens listed above). It is unclear why Merrill and Quisumbing did not publish this rather distinct taxon; we take this opportunity of formally describing *P. patulinervia*.

96. Psychotria casiguraensis Sohmer & A.P. Davis, sp. nov. (**Fig. 112**). TYPE: PHILIPPINES. LUZON: Quezon Prov.: Casiguaran, along stream in forest, 4 May 1925, *Ramos & Edaño BS 45291* (HOLOTYPE: BM; ISOTYPES: K, NY, P, UC).

Frutices 1 m alti sunt, laminis (7 × 16–)9.4 × 20–12.5 × 25.5(–15 × 22) cm latis et longis glabris quando siccis subviridi-badiis ad khakiis late ellipticis ad elliptici-oblongis apice abrupte acuto basi subrotunda nervis lateralibus 15–20 in dimidio quoque et infra prominentioribus nervis tertiis debilibus, petiolis (3–)4–6 cm longis glabris, infructescentia 4–7 cm longa trichotoma nonnihil condensata, pedunculis 1.5–3 cm longis glabris, nodis principalibus 2–3 plerimque cum ramis oppositis, fructibus 9–11 mm longis globosi-obovoideis laevibus glabris vesimiliter rubris et plerumque cum capitis tubulosis calycinis, pyrenis in dorso laevibus vel cum costis, endospermo ruminato.

Shrub, ca. 1 m high, young stems glabrous. Stipules unknown. Leaves: petioles (3–)4–6 cm long, glabrous; leaf blades broadly elliptic to elliptic-oblong, (16–)20–25.5 × (7–)9.4–12.5(–15) cm, subcoriaceous, drying greenish brown to khaki, glabrous above and below; secondary veins 15–20 pairs, very prominent below; tertiary venation manifest to rather prominent; base obtuse to nearly round; apex abruptly acute or broadly acute. Inflorescence/infructescence trichotomous, somewhat condensed, 4–7 cm long, glabrous; principal axis 1.5–3 cm long, each axis with 2–3 primary nodes and with opposite or sometimes verticillate branching at each node, branches terminating in few-flowered/fruited cymules; bracts and bracteoles unknown (?absent). Flowers unknown. Fruit globose-obovoid, 9–11 mm long, smooth, probably red, glabrous; calyx limb and lobes usually persistent as a tubular cap; pedicels ca. 1 mm long; pyrenes smooth on outer surface or ribbed; endosperm ruminate.

PARATYPES: **Luzon. Quezon Prov.:** Infanta-Siniloan Trail, *Ramos & Edaño BS 29225* (A).

Distribution.—In Quezon Province.

Ecology.—Moist forest, including streamsides; low elevations.

Phenology records.—Flowering: unknown; fruiting: May, June.

Preliminary conservation assessment.—Extinct (**EX**). Last collected in 1925.

Distinguishing characteristics.—Leaf blades broadly ovate, relatively large, 20–25.5 cm long, lateral veins 15–20, prominent below; infructescences trichotomous; fruits large, 9–11 mm long; pyrenes usually smooth on outer (dorsal) surface.

Remarks.—Judging from the specimen localities for this species, which are from the Quezon Province, we assume that this species is extinct. Herbarium specimens are remarkably distinctive.

97. Psychotria mariguidonensis Sohmer & A.P. Davis, nom. nov. (**Fig. 113**). *Psychotria longipetiolata* Merr., Philipp. J. Sci. 17:451. 1921. Non *Psychotria longipetiolata* Thwaites, Enum Fl Zeyl. 149. 1849. TYPE: PHILIPPINES. CATANDUANES: Ilocos Norte Prov.: Santo Domingo River and Mt. Mariguidon, on forested slopes, Dec 1917, *Ramos BS 30418* (LECTOTYPE: K, designated here).

Small tree or shrub, 2–4 m high, young stems glabrous. Stipules unknown. Leaves: petioles 4.3–4.5(–6) cm long, glabrous; leaf blades ovate, 10–21(–25) × 7–10 cm, coriaceous, drying reddish brown, glabrous above and below; secondary veins 15–20 pairs, very conspicuous below; tertiary venation very prominent below and above; broadly acute; acute to abruptly acute. Inflorescence/infructescence trichotomous, 7–12 cm long, glabrous to minutely puberulous; principal axis 2.9–4 cm long, each axis with 1–3 primary nodes and verticillate branching at each node, inflorescence spreading, with probably more than 50 fruit flowers/fruits; bracts absent; bracteoles triangular, ca. 0.3 mm long, often absent (caducous). Flowers unknown. Fruit globose to globose-obovoid, 5–7 mm long, smooth to

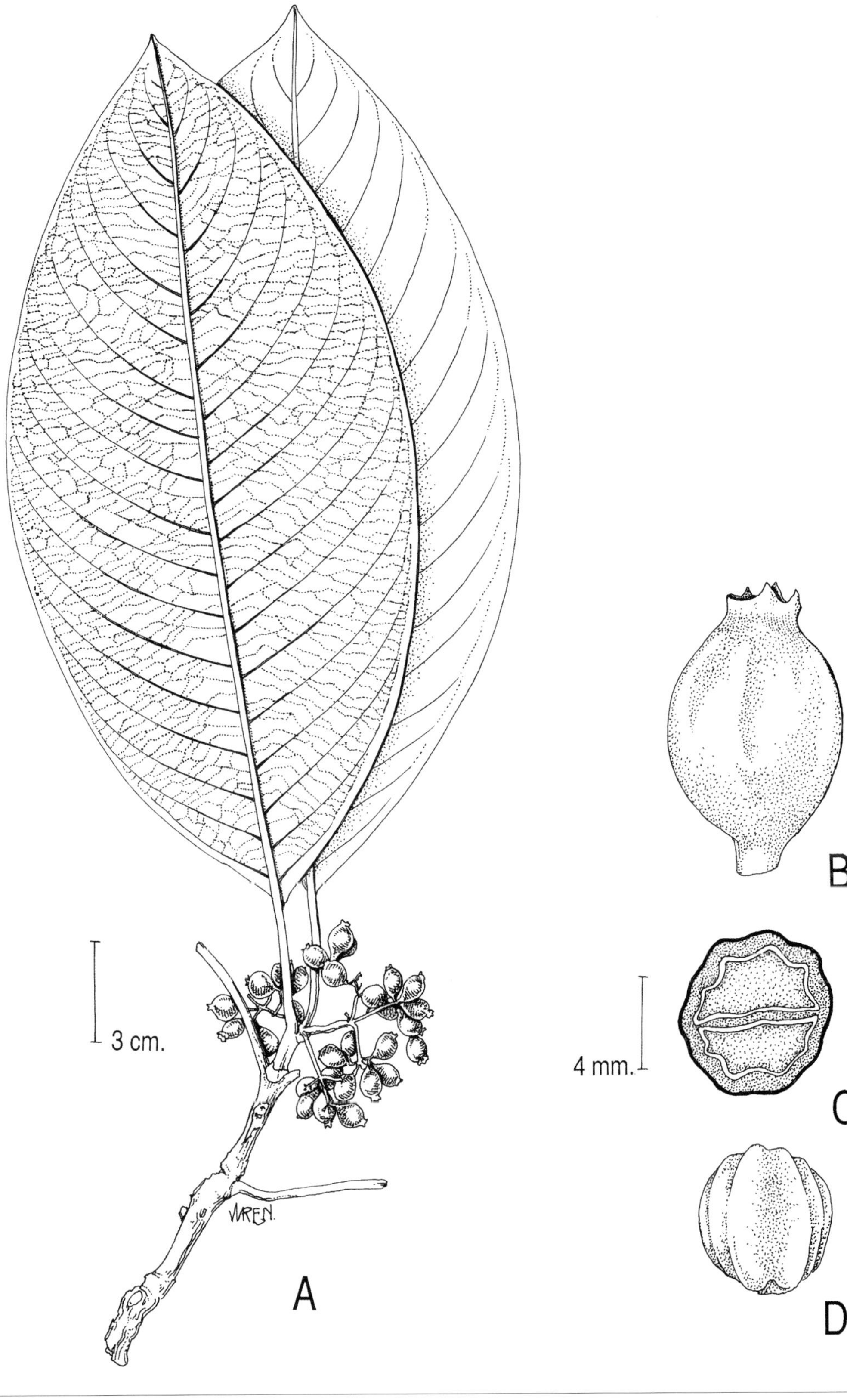

FIG. 112. *Psychotria casiguraensis* (*Ramos & Edaño BS 45291*) (BM, holotype). **A,** habit of fruiting branch; **B,** fruit; **C,** fruit in cross section; **D,** pyrene.

FIG. 113. *Psychotria mariguidonensis* (*Ramos BS 30418*) (K, lectotype).

obscurely ribbed (when dry), reported red, glabrous; calyx limb and lobes persistent but obscure; pedicels 2–3 mm long; pyrenes smooth or obsurely ribbed on outer surface; endosperm ruminate.

Other specimens examined: **Luzon. Mountain Prov.:** Mt. Taggutu, Banue, *Conklin & del Rosario PNH 72673* (K, L). **Isabela Prov.:** Palanan, Dibungko (Marikit), *Gaerlan et al. PPI 2794* (K).

Distribution.—Known from Mountain, Isabela and Catanduanes provinces of Luzon.

Ecology.—Forest and forested slopes; ca. 180 m.

Phenology records.—Flowering: unknown; fruiting: April, August to December.

Preliminary conservation assessment.—Endangered (**EN B1ab(iii)**). Last collected in 1991.

Distinguishing characteristics.—Leaf blades relatively large, distinctly coriaceous, lateral veins 15–20 per side, tertiary venation very prominent; petioles over 4 cm long; infructescences spreading; fruits numerous, each fruit less than 7 mm long; pyrenes smooth on outer (dorsal) surface.

Remarks.—Merrill (1921) suggest that *P. longipetiolata* (now *P. mariguidonensis*) shares a close relationship to *P. plumeriaefolia* Elmer, but we believe that there is only a superficial resemblance between these two taxa.

98. Psychotria plumeriaefolia Elmer, Leafl. Philipp. Bot. 3:1026. 1911; Merr., Enum. Philipp. Fl. Pl. 3:561. 1923. (**Fig. 114**). *Grumilea plumeriaefolia* (Elmer) Merr., Philipp. J. Sci. 10:31. 1915. TYPE: PHILIPPINES. SIBUYAN: Romblon Prov.: Magallanes (Mt. Giting-giting), along wooded banks at about 250 m, Apr 1910, *Elmer 12250* (LECTOTYPE: US, designated here; ISOLECTOTYPES: A, BISH, BM, K, L, MO, US, W).

Shrub, 3–4 m high, young stems glabrous. Stipules imbricate, lanceolate to lanceolate triangular, 0.6–1.4 cm long, glabrous; apex acute. Leaves: petioles 2.5–5.5 cm long, glabrous; leaf blades broadly obovate to subobovate, 17.5–24.5 × 7–12.1 cm, subcoriaceous, drying grayish green above, lighter below, glabrous above and below; secondary veins 13–17 pairs; tertiary venation weak to obscure; base attenuate; apex abruptly acute to abruptly acuminate. Inflorescence/infructescence trichotomous, 3.5–5.5(–10) cm long, glabrous; principal axis 1.7–5.1 cm long, each axis with 1–2 primary nodes and opposite branching at each node, branches terminating in cymules; bracts and bracteoles unknown (?absent). Flowers poorly known, 5-merous, pedicels ca. 1 mm long; calyx (incl. hypanthium), ca. 2 mm long, glabrous; calyx lobes ± absent (calyx limb truncate); corolla: color unknown; corolla tube ca. 2 mm long, external surface glabrous, pubescent at throat within; corolla lobes ca. 2 mm long; stamens: filaments ca. 0.3 mm long; anthers unknown; style ca. 2 mm long (?long styled), lobes ca. 1 mm long. Fruit globose, 5–7.5 mm long, ribbed, color when ripe unknown but presumed red or black, glabrous; calyx limb and lobes persistent but rather inconspicuous; pedicels 2–4 mm long or pedicels absent; pyrenes ribbed on outer surface; endosperm ruminate

Other specimen examined: **Mindoro.** Mindoro Oriental Prov.: Banus River, *Reed PNH 91880* (PNH).

Distribution.—Sibuyan and Mindoro.

Ecology.—Forest, including forest along streams; 100–250 m.

Phenology records.—Flowering: unknown; fruiting: April.

Preliminary conservation assessment.—Critically Endangered (**CR B1ab(iii)**). Last collected in 1964.

Distinguishing characteristics.—Leaf blades tending to be broadly obovate, broadest above the middle, base attenuate, tertiary venation not discernable with a 10x lens, drying grayish green; inflorescences trichotomous; pyrenes assummed to be smooth or with irregular ribs on outer (dorsal) surface.

Remarks.—Even less is known about *P. plumeriaefolia* now than when it was first described because more specimens were available then, before the Philippine National Herbarium was destroyed. Elmer considered his new species to be similar to *P. banahaensis* but distinguishable from it by leaf blades widest above the middle. It shares with *P. banahaensis* the lack of a discernable tertiary venation pattern on the lower surface of the leaf blade. However, *P. plumeriaefolia* does not have pyrenes with the distinct sharp ridges on the outer (dorsal) surface, which are a distinct feature of *P. banahaensis*.

99. Psychotria nitens (Merr.) Merr., Enum Philipp. Fl. Pl. 3:560. 1923. (**Fig. 115**). *Grumilea nitens* Merr., Philipp. J. Sci. 10:137. 1915. TYPE: PHILIPPINES. LUZON: Isabela Prov.: Palanan Bay, 25 Jun 1913, *Escritor BS 21157* (LECTOTYPE: K, designated here).

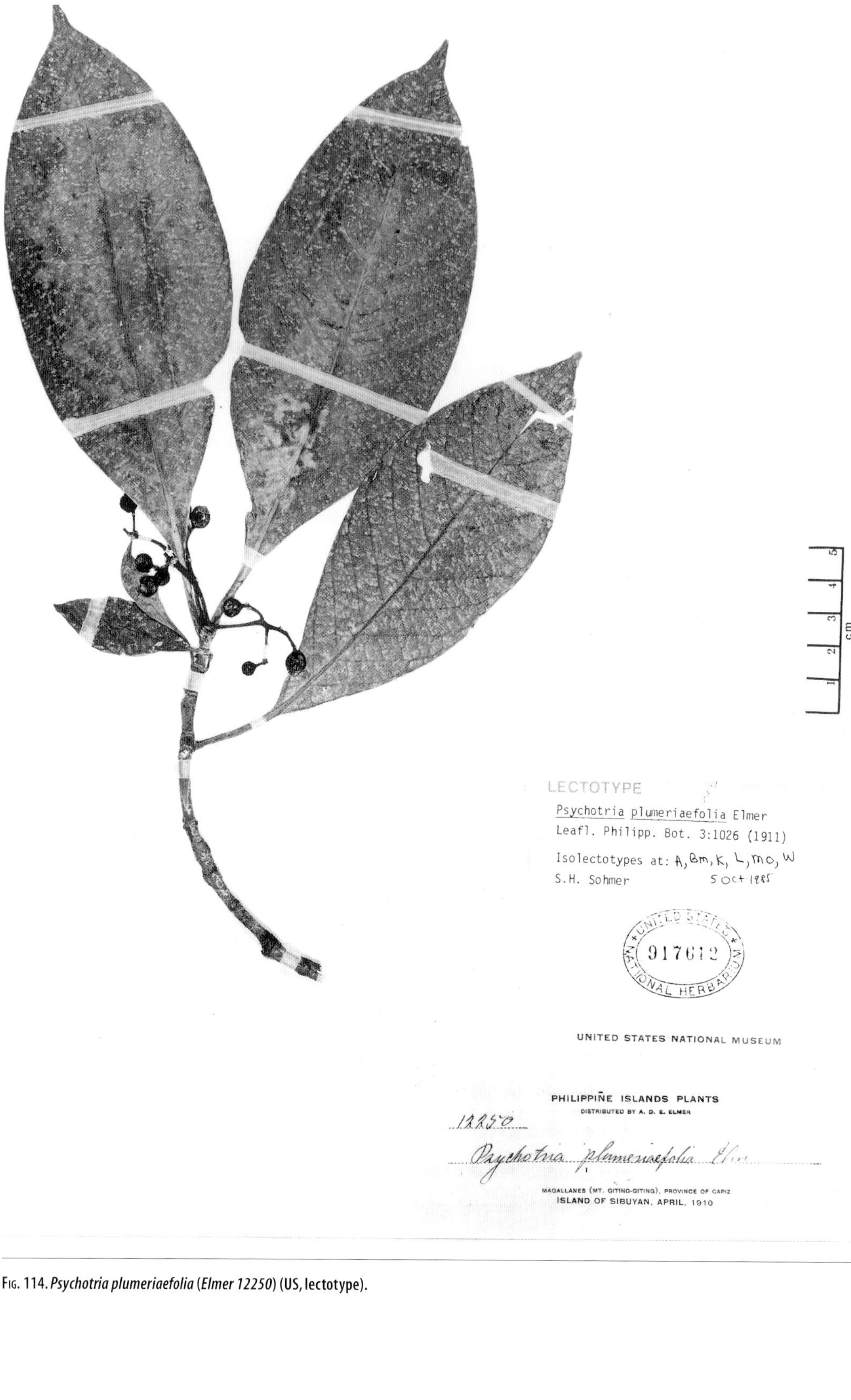

Fig. 114. *Psychotria plumeriaefolia* (*Elmer 12250*) (US, lectotype).

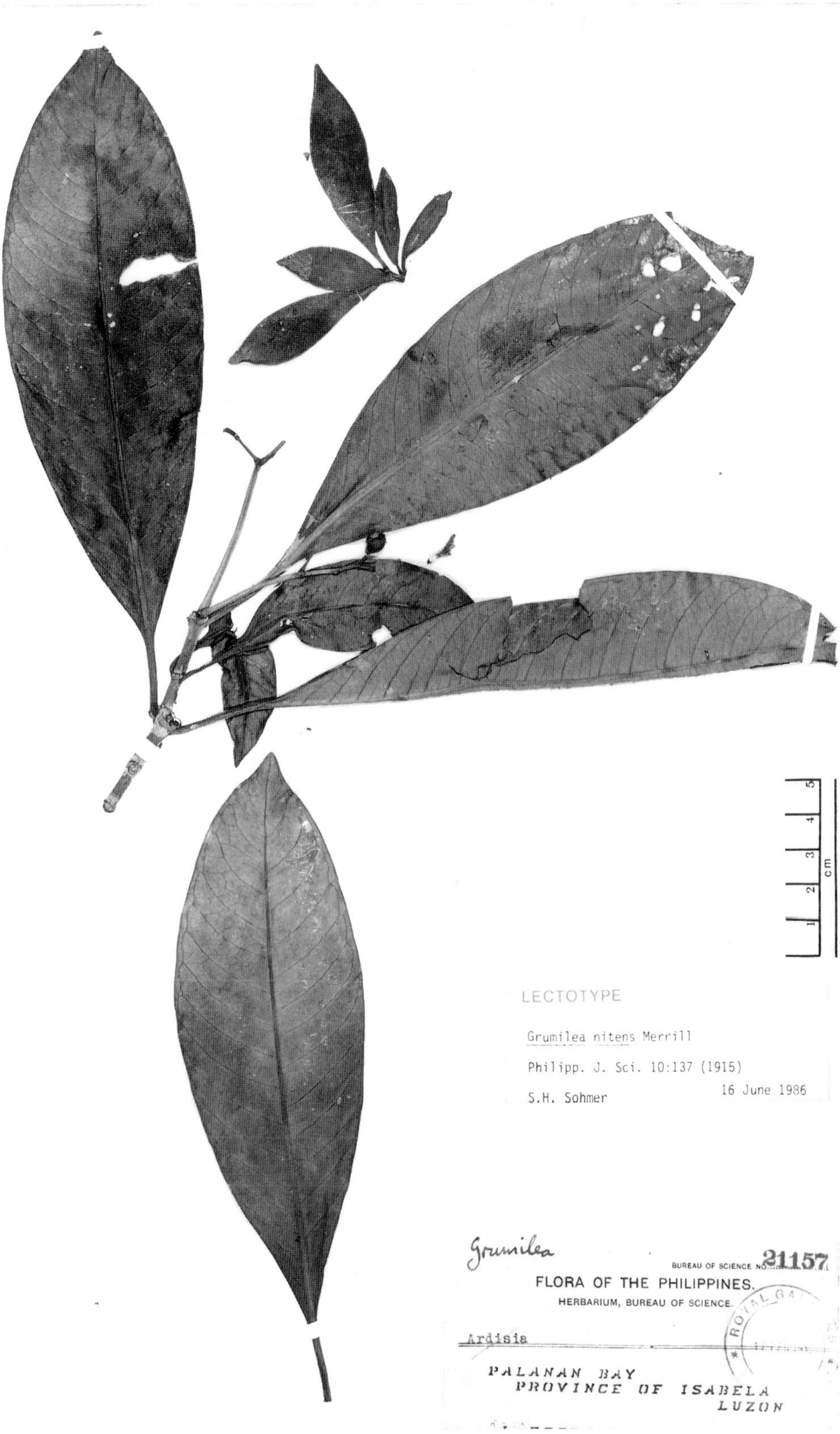

Fig. 115. *Psychotria nitens* (*Escritor BS 21157*) (K, lectotype of *Grumilea nitens*).

Shrub, 1–2.5 m high, glabrous throughout. Stipules imbricate, ovate to lanceolate, or ligulate, 0.5–1.9 cm long, glabrous; apex acute to acuminate. Leaves: petioles 1–2.5 cm long, glabrous; leaf blades ovate to oblanceolate, 7.5–17.5 × 3.1–5.8 cm, coriaceous, drying shiny yellow-brown to orange or reddish brown, glabrous above and below; secondary veins (10–)13–16 pairs; tertiary venation manifest to very prominent but usually prominent; base acute; apex usually abruptly acute to obtuse. Inflorescence/infructescence trichotomous, 3.8–9.5 cm long, glabrous; principal axis 1.2–7.5 cm long, each axis with 1–2 primary nodes and verticillate or opposite branching at each node, branches terminating in 1–3 flowers or few flowered cymules; bracts and bracteoles unknown (?absent). Flowers unknown. Fruit obovoid, 7–8 mm long, usually smooth, reported yellow but probably ripening red; calyx limb and lobes persistent but inconspicuous; pedicels 1–2 mm long; pyrenes smooth or sometimes ribbed on outer surface; endosperm ruminate.

Other specimens examined: **Luzon. Cagayan Prov.:** ridge north of Baguio Cove, *Allen 394-81* (BISH); Penablanca, *Ramos BS 76747* (A); Mt. Babatngin, *Edaño BS 79423* (A). **Isabela Prov.:** Mt. Moises, *Ramos & Edaño BS 47265* (K, NY, P, UC), *BS 47316* (UC); Sierra Madre Mountains, Palanan, *Gutierrez PNH 78240* (A, PNH); 20 km south of Maconacon, *Allen 51-81* (BISH). **Benguet Prov.:** Mt. Nangaota, above road km 61, *Sulit PNH 7702* (A, PNH), above road km 65, *Sulit PNH 7683* (A, L, PNH). **Quezon Prov.:** Mt. Binuang, *Ramos & Edaño BS 28615* (A, K). **Samar. Western Samar Prov.:** Mt. Concord, *Gutierrez et al. PNH 117178* (L).

Distribution.—Found throughout Luzon.

Ecology.—Mossy forests and on forested slope; 732–1099 (and possibly up to 2400) m.

Phenology records.—Flowering: March; fruiting: May, June.

Preliminary conservation assessment.—Near Threatened (**NT**). Last collected in 1981.

Distinguishing characteristics.—Leaf blades drying shiny yellowish brown, orange-brown or reddish brown, often thickly coriaceous; pyrenes usually smooth on outer (dorsal) surface; inflorescences trichotomous.

Remarks.—*Psychotria nitens*, as defined by the type collection, is fairly easy to recognize. If the fruits of some of the collections referred to it, however, were sessile and pubescent, such collections would fit well with the concept of *P. crispipila*.

100. Psychotria mendozii Sohmer & A.P. Davis, sp. nov. (**Fig. 116**). TYPE: PHILIPPINES. MINDANAO: Agusan del Norte Prov.: Mt. Hilong-hilong, mossy forest slopes at 1333 m, 21 Apr 1949, *Mendoza & Convocar PNH 10789* (HOLOTYPE: A; ISOTYPES: K, L).

Arbores vel frutices 3 m alti sunt, stipulis 1–1.9 cm longis imbricatis acuminatis glabris 3–4-plo longioribus quam latis, laminis 3 × 7–7 × 12 cm latis et longis coriaceis glabris lute ellipticis obtusis vel breve acuti in sicco olivacei-viridibus nervis lateralibus 10–15 in dimidio quoque in paginis rubis prominentibus nervis tertiis prominentibus petiolis 1–2.8 cm longis glabris, inforescentia trichotoma axibus glabris vel puberulis, axi centrali 4–9 cm longo, axibus omnibus cum 1–3 nodis cum ramis verticillatis, ramulis paucis vel multis cymulis ferrentilus, bracteis minimis inconspicuis, floribus dimorphicis 5-meris pedicellatis, hypanthio 0.9–1.2 mm longo lobis acutis vel obscuris glabris, corollis albis tubo extra glabro in fauce piloso 2.5–5 mm longo lobis pariter longis reflexis.

Small tree or shrub, ca. 3 m high, young stems glabrous. Stipules imbricate, narrowly lanceolate, 1–1.9 cm long, glabrous; apex narrowly acute to acuminate. Leaves: petioles 1–2.8 cm long, glabrous; leaf blades broadly elliptic, 7–12 × 3–7 cm, subcoriaceous to coriaceous, drying dark olive-green, glabrous; secondary veins 10–15 pairs, prominent above and below; tertiary venation prominent; base broadly acute to acute; apex obtuse to abruptly acute. Inflorescence/infructescence trichotomous, 7.4–12.6 cm long, glabrous or puberulous; principal axis 4–9 cm long, each axis with 1–3 nodes (situated towards the apex) and verticillate branching at each node, the main branches terminating in several to many cymules; bracts deltate to triangular, ca. 0.3–0.6 mm long; bracteoles similar to the bracts, 0.3–0.4 mm long. Flowers 5-merous, pedicels 0.8–3.8 mm long; calyx (incl. hypanthium) 0.9–1.2 mm long, glabrous; calyx lobes deltate to broadly ovate, ca. 0.4 mm long; corolla: white; corolla tube 2.5–5 mm long, external surface glabrous, pubescent at throat within; corolla lobes 2.5–4 mm long; stamens: filaments 1–2 mm long; anthers ca. 1 mm long; style 1–2 mm long, lobes ca. 1 mm long. Fruit unknown.

PARATYPE: **BILIRAN. Leyte Prov.:** Mt. Suiro, north slope, *Sulit PNH 21689* (L, PNH).

Distribution.—Only known from Biliran Island (Leyte Province) and northern Mindanao.

Ecology.—Forest, including mossy forest, and forested slopes; mid to higher elevations (e.g. ca. 1333 m).

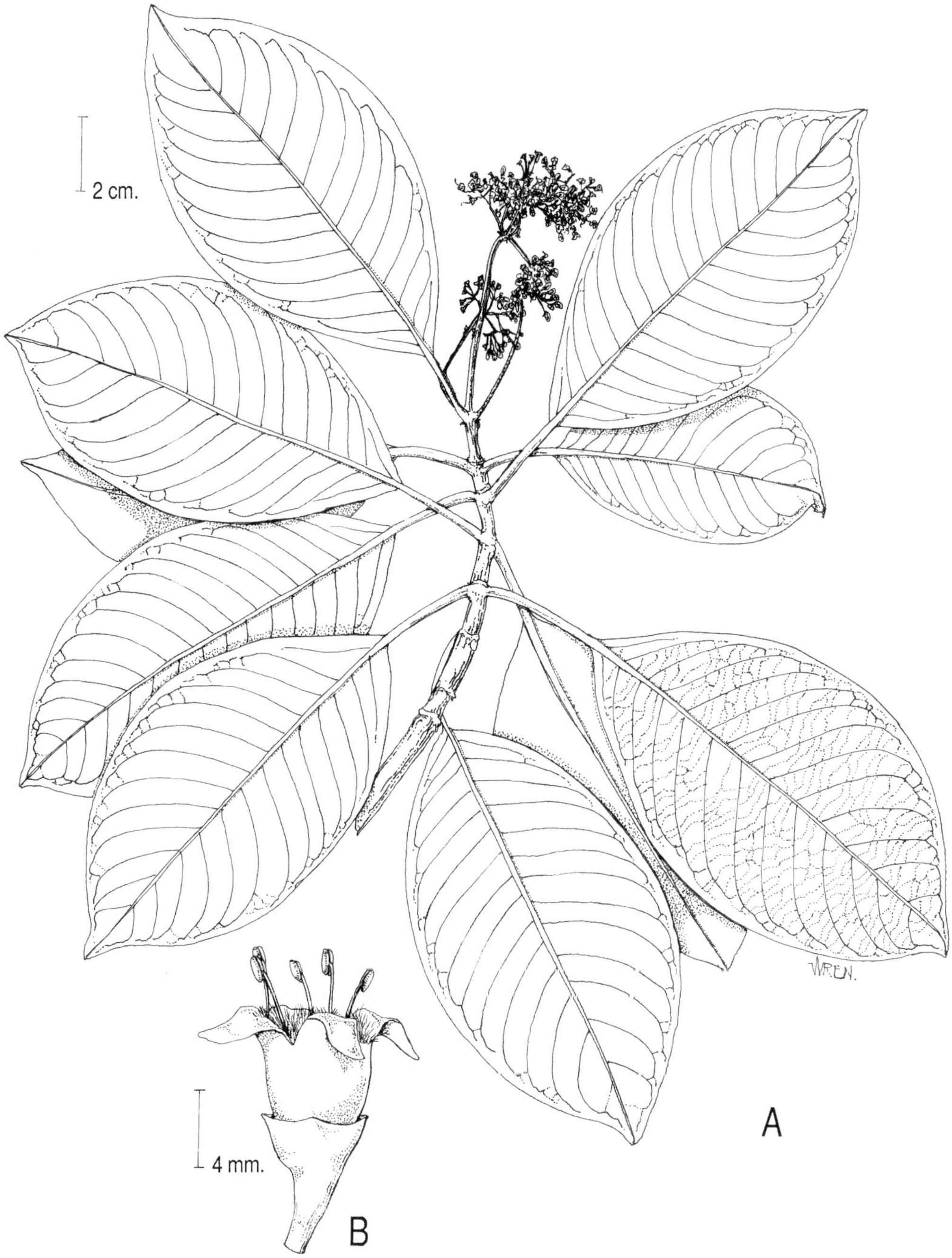

Fig. 116. *Psychotria mendozii* (*Mendoza & Convocar PNH 10789*) (A, holotype). **A,** habit of flowering branch; **B,** flower.

Phenology records.—Flowering: April and May; fruiting: unknown.

Preliminary conservation assessment.—Endangered (**EN B1ab(iii)**). Last collected in 1954.

Distinguishing characteristics.—Leaf blades broadly elliptic, lateral veins prominent; inflorescences trichotomous.

Remarks.—Herbarium collections of this species stand out quite clearly from all Philippine *Psychotria*, due mainly to leaf shape and prominent lateral venation.

FIG. 117. *Psychotria yatesii* (*Yates BS 25518*) (K, lectotype of *Grumilea yatesii*).

101. Psychotria yatesii (Merr.) Merr., Enum. Philipp. Fl. Pl. 3:564. 1923. (**Fig. 117**). *Grumilea yatesii* Merr., Philipp. J. Sci. 12:172. 1917. TYPE: PHILIPPINES. LUZON: Quezon Prov.: Mt Cadig, low elevational forests, Dec 1916, *Yates BS 25518* (LECTOTYPE: K, designated here; ISOLECTOTYPE: P).

Shrub or small tree, 1–4 m high, glabrous throughout. Stipules unknown. Leaves: petioles 1.3–4 cm long, glabrous; leaf blades broadly elliptic to oblong-elliptic or somewhat obovate, 15.5–20(–25) × (5–)7–10.4(–11) cm, subcoriaceous to coriaceous, drying reddish brown, glabrous above and below; secondary veins 12–16 pairs, prominent below, and often drying conspicuously lighter than surrounding leaf tissue; tertiary venation prominent or sometimes weak; base acute to attenuate; apex abruptly acuminate to abruptly acute. Inflorescence/infructescence trichotomous, conspicuously condensed, 2–5 cm long (3–4 cm wide), glabrous; principal axis 0.9–2 cm long, each axis with 2–3 primary nodes and verticillate branching at each node, branches terminating in cymules, with 40–70 flowers per Inflorescence; bracts and bracteoles unknown, probably absent. Flowers 5-merous, pedicels (0.3–)0.9–2 mm long; calyx (incl. hypanthium) 0.8–1.5 mm long, glabrous; calyx lobes very broadly ovate, ca. 0.2 mm long, or limb with a sinuate margin; corolla: white; corolla tube 2.5–3 mm long (short styled), external surface glabrous, villous at throat within; corolla lobes ca. 1.5–2 mm long; stamens: filaments ca. 2 mm long; anthers ca. 0.8 mm long (short styled); style ca. 1 mm long, lobes ca. 0.5 mm long (short styled). Fruit globose to globose-obovoid, 5–8 mm long, smooth or sometimes wrinkled when dry, orange or dark red to nearly black, glabrous; calyx limb and lobes not persistent; pedicels 1.7–3 mm long; pyrenes smooth to very obscurely ribbed-striated on outer surface; endosperm ruminate.

Other specimens examined: **Luzon. Isabela Prov.:** Palanan Area, Digallorin, Divinsia camp site, *Ridsdale & Baquiran et al. ISU 363* (K). **Quezon Prov.:** Mt. Binuang, *Ramos & Edaño BS 28498* (NY); Mt. Cadig, *Yates BS 25519* (K); Polillo Island, Santa Maria, *Salvoza FB 29726* (NY, UC). **Laguna Prov.:** San Antonio, *Ramos BS 10948* (BM). **Camarines Sur Prov.:** Sarapan, *Edaño BS 76186* (NY); Mt. Bagacay, without further data, *Ramos & Edaño BS 33869* (K). **Camarines Sur Prov.:** without further data: *Alvarez FB 23787* (A, K, NY, P, US). **Sorsogon Prov.:** Mt. Bulusan, *Elmer 16160* (A, BM, K, MO, NY, P). **Samar. Eastern Samar Prov.:** Oras, *Castro & Anonuevo PNH 5760* (K, L). **Western Samar Prov.:** Bagacay, *Sulit PNH 6279* (L), *Sulit & Conese PNH 6375* (K, L).

Distribution.—Isabela, Quezon and Camarines Provinces of Luzon and also on Samar and Polillo Islands.

Ecology.—Forest; one recorded on steep sea cliffs, 50–200 m.

Phenology records.—Flowering: March to May; fruiting: April, December.

Preliminary conservation assessment.—Near Threatened (**NT**). Last collected in 1992.

Distinguishing characteristics.—Leaf blades broadly elliptic to oblong-elliptic, usually drying reddish brown, lateral venation prominent; inflorescences trichotomous and compact; pyrenes usually smooth on outer (dorsal) surface.

Remarks.—A number of collections were seen during the course of this revision that possess the characteristics listed above. We have little doubt that the collections cited above belong together with the type collection here designated, although the latter has much more broadly elliptic leaf blades than any of the other specimens. Merrill (1917) thought that *P. yatesii* resembled *P. plumeriaefolia*, but we see little or no such resemblance. The range of this species has been much diminished due to the loss of lowland forests. This species resembles *P. talasensis* Sohmer from New Britain and northeast New Guinea, particularly in the structure of its compact inflorescences.

102. Psychotria weberi Merr., Philipp. J. Sci. 8:54. 1913; Merr., Enum. Philipp. Fl. Pl. 3:564. 1923. (**Fig. 118**). *Grumilea weberi* (Merr.) Merr., Philipp. J. Sci. 10:133. 1915. TYPE: PHILIPPINES. LUZON: Cagayan Prov.: Abulug River, rocky hillsides at ca. 30 m, Jan 1912, *Weber 1573* (LECTOTYPE: US, designated here; ISOLECTOTYPES: A, GH, K, P, UC).

Shrub and small tree, 2–4 m high, young stems glabrous. Stipules imbricate, triangular-lanceolate, 1.1–2.4 cm long, glabrous; apex acute. Leaves: petioles 1–3.5 cm long, glabrous; leaf blades oblong-elliptic to elliptic-lanceolate, 10.6–22 × 4.8–8.6 cm, subcoriaceous, usually drying yellowish green, or grayish green, sometimes reddish brown, glabrous above and below or sparsely pubescent below; secondary veins 11–15(–19) pairs; tertiary venation manifest to prominent; base acute; apex obtuse to abruptly acuminate. Inflorescence/infructescence trichotomous, usually condensed, 1.1–2.6(–6.5) cm long, glabrous; principal axis 0.7–1.1(–2.2) cm long, each axis with 1–2 primary nodes and verticillate or opposite branching at each node, branches terminating in few-flowered cymules (e.g. 3 flowers per cymule); bracts and bracteoles unknown, probably absent. Flowers 5-merous, pedicels 0.5–2 mm long, or ± absent; calyx (incl. hypanthium) ca. 3 mm long, glabrous; calyx lobes obscure, or calyx

Fig. 118. ***Psychotria weberi*** **(*Weber 1573*) (US, lectotype).**

limb truncate; corolla: white; corolla tube 1–2 mm long, external surface glabrous, villous at throat within; corolla lobes ca. 4 mm long (short styled). stamens: filaments 1.5–2 mm long (short styled); anthers ca. 1 mm long; style ca. 1 mm long, lobes ca. 0.6 mm long (short styled). Fruit globose-ovoid, 4.5–6 mm long, smooth or somewhat ribbed, color unknown but presumed red, glabrous; calyx limb and lobes persistent but inconspicuous, sometimes chartaceous; pedicels 0.6–4(–6) mm long; pyrenes ribbed on outer surface; endosperm ruminate.

Other specimens examined: **Luzon. Cagayan Prov.:** Logum, near Penablanca, *Adduru 274* (A, K, MO, P). **Isabela Prov.:** Mt. Cresta, *Ramos BS 76963* (A), *BS 76979* (A); Mt. Moises, *Clemens 16842* (UC), *16843* (UC), *17023* (UC). **Quezon Prov.:** Real, Lalawinan, Tipuan, NBG, *Barbon, Alvarez & Garcia PPI 2204* (BRIT, K); Mamator R., Casiguran, *Ramos & Edaño BS 45385* (A, K, P, UC); Casiguran, *Ramos & Edaño BS 45379* (K); mountain west of Philipp. National Botanic Garden, *Sohmer 12008* (BISH); forest near garden and Lalawinan River, *Sohmer 12018* (BISH). **Camarines Norte Prov.:** Basud, Tuaca, Mt. Nilisan, *Reynoso, Romero & Fuentes PPI 1282* (BRIT), *Reynoso, Romero & Fuentes PPI 1215* (BRIT, K). **Camarines Sur Prov.:** Mt. Labo, *Quisumbing BS 76631* (NY); Kinaclan River, *Edaño BS 76344* (NY); Ponawa River, *Edaño BS 76397* (NY).

Distribution.—Luzon.

Ecology.—Humid forest, including dipterocarp forest and secondary forest; along streams or on slopes; 30–610 m.

Phenology records.—Flowering: April to June; fruiting: May, June, September.

Preliminary conservation assessment.—Endangered (**EN B1ab(iii)**). Last collected in 1991.

Distinguishing characteristics.—Leaf blades drying yellowish green or grayish green; inflorescences trichotomous, small, condensed; corolla lobes twice as long as corolla tube; pyrenes irregularly ribbed on outer (dorsal) surface.

Remarks.—According to Merrill (1913) *P. weberi* is similar to *P. banahaensis* Elmer, but there are many significant differences between them. *Psychotria weberi* does not have pyrenes with pronounced, sharp ridges on the outer (dorsal) surface (visible through the dried fruit wall) as characteristic of *P. banahaensis*, but it does have a discernable pattern (with ×10 lens) of tertiary venation whereas *P. banahaensis* does not. *Psychotria weberi* is one of several taxa with smooth or irregularly ribbed pyrenes and as such provides evidence of a closer morphological relationship with *P. luzoniensis*, rather than *P. banahaensis*. If pyrene shape (e.g. ribbed vs. smooth) is found to be of negligible taxonomic value, then the relationship of this taxon might be with *P. pallidifolia* Merr., based on morphological similarity.

IX.—Subalpina Species Group

Shrubs, or rarely trees. Stipules valvate, valvate-imbricate, or rarely imbricate. ***Leaf blades usually less than 3 cm wide, sometimes broader, never more than 15 cm long;*** chartaceous to subcoriaceous, or coriaceous. ***Inflorescences with axes and branches, trichotomous, sometimes pentachotomous or with an indefinite number of axes.*** Flowers pedicellate. Fruits usually red, sometimes yellow or orange; pyrene inner (adaxial) surface flat; seed producing a red color when placed in 70% ethanol, due to a soluble seed coat pigment (SCP positive), rarely SCP negative; endosperm ruminate, usually dark. Comprising ten species.

Note.—Text in ***bold italics*** denotes features useful for the placement of taxa within the respective species group.

KEY TO THE SPECIES OF THE SUBALPINA SPECIES GROUP

1. Corolla tube 4 mm long; calyx (incl. hypanthium) 3–3.5 mm long ____ **103. P. hilonghilongensis**
1. Corolla tube 2–3 mm long, usually ca. 2 mm long; calyx (incl. hypanthium) 2–3 mm long.
  2. Leaf blades broadly obovate or oblong-elliptic to obovate-elliptic, drying brownish to orange-brown ____ **104. P. ilocana**
  2. Leaf blades usually narrowly elliptic or obovate, never drying orange-brown.
    3. Inflorescences un-branched or few-branched, flowers and fruit attached ± directly to main axes or secondary branches ____ **105. P. macgregorii**
    3. Inflorescences branched or many-branched, flowers and fruit attached to tertiary branches.
      4. Fruits ellipsoid to narrowly ellipsoid, 1–1.3 cm long ____ **106. P. palimlimensis**
      4. Fruits globose to turbinate, ellipsoid-globose, or obovoid, 0.5–0.7 cm long.
        5. Pedicels very slender, usually < 0.3 mm in diam. when dry ____ **107. P. gracilipes**
        5. Pedicels not slender, usually > 0.5 mm in diam. when dry.
          6. Leaf blades narrowly elliptic to narrowly lanceolate, usually < 2 cm wide, apex acuminate, lateral veins (8–)10–14 pairs.
            7. Inflorescence axes each terminating in a single node with a single cymule at each ____ **108. P. acuminatissima**
            7. Inflorescence axes each terminating in a single node with 5–15 cymules at each ____ **109. P. sohotonensis**
          6. Leaf blades usually elliptic to narrowly elliptic, or oblong-elliptic, or oblanceolate to obovate, usually >2 cm wide, lateral veins usually < 10 pairs.

8. Fruit strongly ribbed on back when dry; leaf base narrowly attenuate ______ **110. P. subalpina**
8. Fruit smooth on back when dry; leaf base acute to narrowly acute, or acute to attenuate.
  9. Flowers/fruit 5–10 per inflorescence/infructescence. ______ **111. P. pauciflora**
  9. Flowers/fruit usually >10 per inflorescence/infructescence. ______ **112. P. loheri**

103. Psychotria hilonghilongensis Sohmer & A.P. Davis, sp. nov. (**Fig. 119**). TYPE: PHILIPPINES. MINDANAO: Agusan Prov.: Mt. Hilong-hilong, Cabadbaran, mossy forest ridge at 1050 m, 29 Apr 1949, *Mendoza & Convocar PNH 10760* (HOLOTYPE: A; ISOTYPE: L).

Arbores parvae ca. 4 m. altae glabrae sunt, stipulis valvatis subimbricatis ca. 0.5 cm longis acutis, laminis crasse chartaceis vel subcoriaceis anguste ellipticis vel anguste lanceolatis 1.6 × 5.2–2.2 × 8.8 cm latis et longis, nervis lateralibus 7–9 cm dimidio quoque apice acuminato obtuso basi acuminata subbrunneo-viridibus, petiolis 0.7–1 cm longis, inflorescentia trichotoma rhachidi principali 0.5–2.2 cm longo, rhachidibus 1–2 nodis ramis oppositis floribus paucis bracteis nullis vel minimus et caducis, floribus 4–5-meris dimorphicis, pedicellis 1–3 mm longis, tubo calycis et hypanthio 3–3.5 mm longis lobis reductis vel nullis, corollis extra glabris intra in fauce villosis tubo 4 mm longo, lobis 2.5 mm longis.

Small tree, ca. 4 m high, glabrous throughout. Stipules valvate to slightly imbricate, ± deltate, ca. 0.5 cm long, glabrous; apex broadly acute to acute. Leaves: petioles 0.7–1 cm long, glabrous; leaf blades narrowly elliptic to narrowly lanceolate or oblanceolate, 5.2–8.8 × 1.6–2.2 cm, chartaceous to subcoriaceous, brownish green to brownish when dry, glabrous above and below; secondary veins 7–9 pairs; tertiary venation manifest to weak; base narrowly cuneate; apex shortly acuminate. Inflorescence/infructescence trichotomous, 1.1–3.6 cm long, glabrous; principal axis 0.5–2.2 cm long, each axis with 1–2 nodes with opposite branching at each node, branches terminating in 1 or 2 flowers; bracts and bracteoles not seen (caducous). Flowers 4- or 5-merous, pedicels 1–3 mm long; calyx (incl. hypanthium) 3–3.5 mm long, glabrous; calyx lobes reduced or broadly ovate to broadly triangular, ca. 0.5 mm long; corolla: probably white; corolla tube ca. 4 mm long, external surface glabrous, villous at throat within; corolla lobes ca. 2.5 mm long; stamens: filaments ca. 1 mm long (long styled); anthers ca. 1 mm long; style ca. 4 mm long (long styled), lobes ca. 1 mm long. Fruits unknown.

PARATYPES: None—known only from the type collection.

Distribution.—Known only from Mt. Hilong-hilong, Mindanao.

Ecology.—Mossy forest; recorded at 1050 m.

Phenology records.—Flowering: April; fruiting: unknown.

Preliminary conservation assessment.—Critically Endangered (**CR B1ab(iii)**). Only collected once in 1949.

Distinguishing characteristics.—Relatively large flowers, corolla tubes ca. 4 mm long.

104. Psychotria ilocana (Merr.) Merr., Enum. Philipp. Fl. Pl. 3:556. 1923. (**Fig. 120**). *Grumilea ilocana* Merr., Philipp. J. Sci 12:173. 1917. TYPE: PHILIPPINES. LUZON: Ilocos Norte Prov.: Bangui, low elevation forest, Feb 1917, *Ramos BS 27456* (LECTOTYPE: K, designated here).

Shrub, 1.5–3 m high; glabrous throughout. Stipules valvate-imbricate, ± ovate to ovate, 0.4–1.5(–2) cm long, glabrous; apex acute to acuminate. Leaves: petioles 0.5–2.3 cm long, glabrous; leaf blades broadly obovate, or oblong-elliptic to obovate-elliptic, 4.9–6.2 × 1.9–2.4(–3.1) cm, subcoriaceous, drying brownish to orange-brown, glabrous above and below; secondary veins 8–12(–14) pairs; tertiary venation weak to obscure; base cuneate to acute; apex round to obtuse with a very short, blunt tip. Inflorescence/infructescence trichotomous, 2.4–4.6 cm long, glabrous; principal axis 0.1–4(–7) cm long, each axis with 1–2 nodes and opposite branching at each node, branches terminating in cymules; bracts and bracteoles unknown (?caducous or absent). Flowers unknown. Fruit obovoid, 6–8 mm long, smooth to obscurely ribbed, color unknown but probably red, glabrous; calyx limb and lobes not persistent; pedicels 1–5 mm long; pyrenes smooth on outer surface; endosperm ruminate.

Other specimens examined: **Luzon. Ilocos Norte Prov.:** Bangui, *Ramos BS 27544* (K, NY); Burgos, *Ramos BS 32802* (NY). **Zambales Prov.:** without further data: *Ramos BS 4704* (NY).

Distribution.—Luzon, principally in Ilocos Norte province except for one collection cited from Zambales province.

Ecology.—Dry forest; low elevation.

Phenology records.—Flowering: unknown; fruiting: February, March.

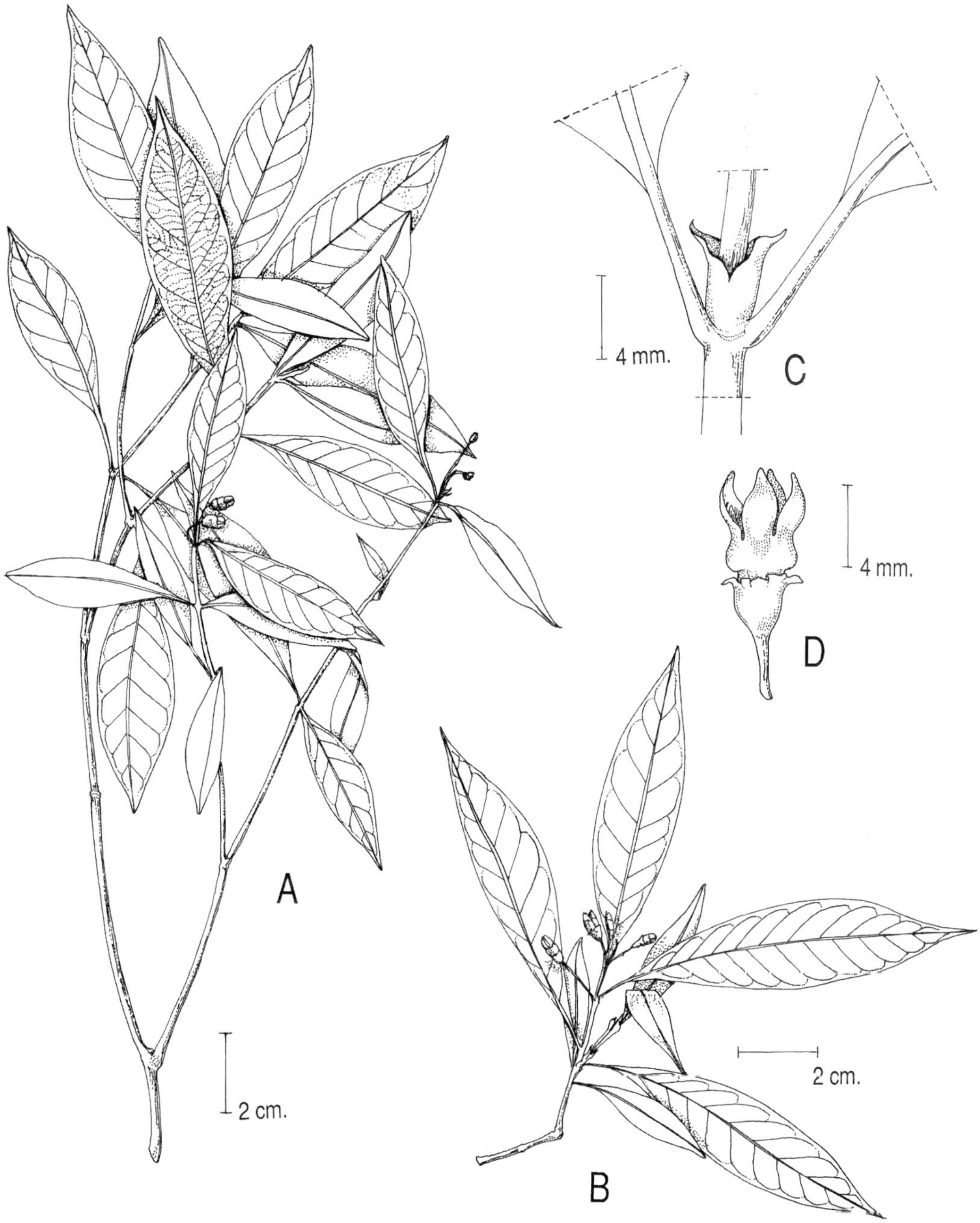

FIG. 119. *Psychotria hilonghilongensis* (*Mendoza & Convocar PNH 10760*) (A, holotype). **A,** habit of branch; **B,** habit of branch; **C,** stipule; **D,** flower (*Mendoza & Convocar PNH 10760*) (L, isotype).

Preliminary conservation assessment.—Critically Endangered (**CR B1ab(iii)**) or possibly extinct (**EX**). Last collected in 1918.

Distinguishing characteristics.—Leaf blades broadly obovate to oblong-elliptic, drying orange-brown.

Remarks.—We suspect that this species might be a variant of *P. luzoniensis*, although fieldwork is needed to verify this idea.

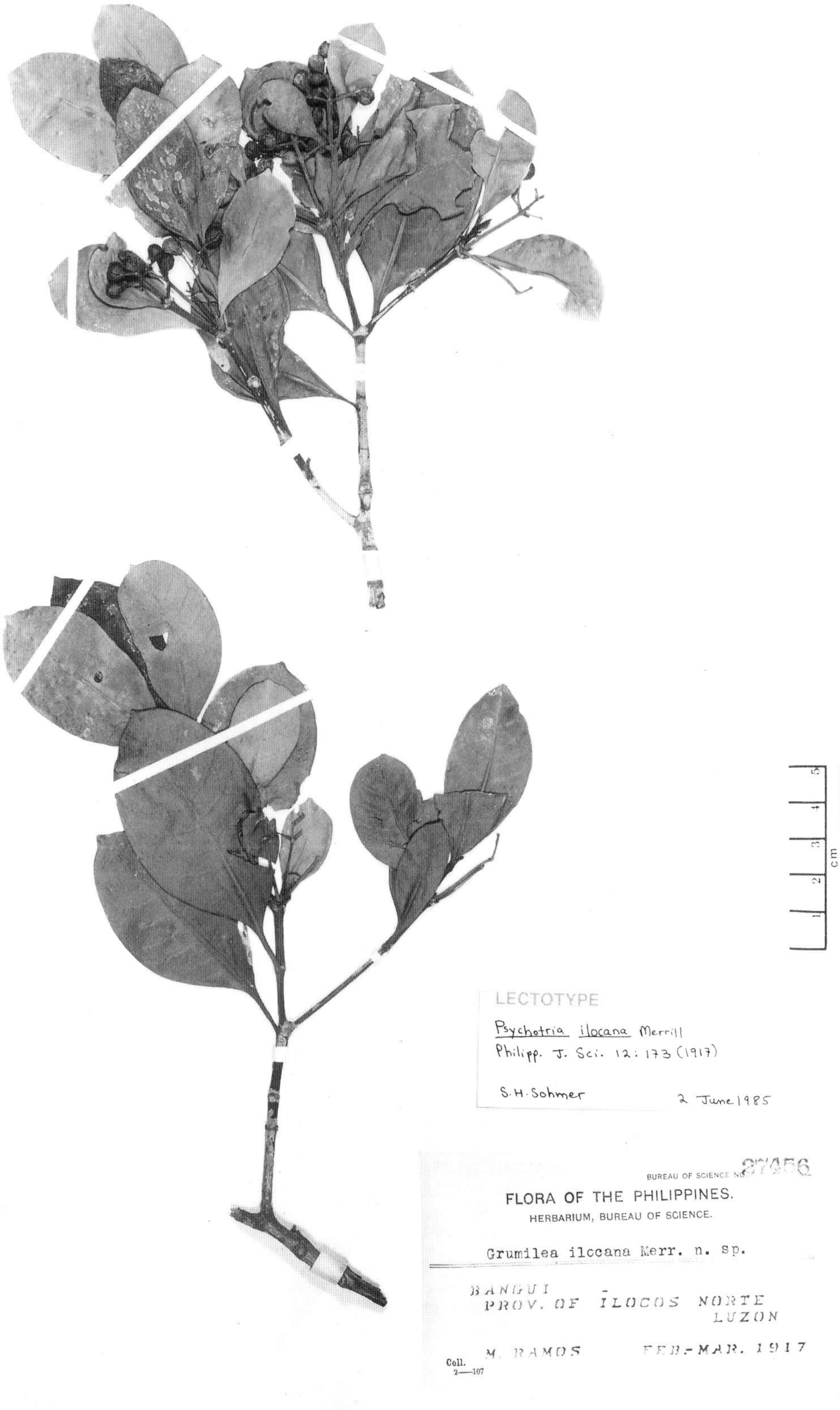

FIG. 120. *Psychotria ilocana* (*Ramos BS 27456*) (K, lectotype).

105. Psychotria macgregorii Merr., Philipp. J. Sci. 5:389. 1910; Merr., Enum. Philipp. Fl. Pl. 3:558. 1923. (**Fig. 121**). *Grumilea macgregorii* (Merr.) Merr., Philipp. J. Sci. 10:133. 1915. TYPE: PHILIPPINES. LUZON: Benguet Prov.: Mt. Pulog, mossy forest above 2400 m, May, 1909, *Merrill 6495* (published as *8495*) (LECTOTYPE: US, designated here; ISOLECTOTYPE: NY).

Shrub or small tree, ca. 4 m high, glabrous throughout. Stipules valvate, narrowly triangular to narrowly lanceolate, 0.6–1.3 cm long, glabrous; apex acute to acuminate. Leaves: petioles 0.5–1.5 cm long, glabrous; leaf blades elliptic to obovate, 3–8.5 × 1.6–3.3 cm, subcoriaceous to coriaceous, drying olivaceous to dark brown, glabrous above and below; secondary veins 8–10(–12) pairs; tertiary venation weak to prominent; base acute to narrowly acute; apex acute to shortly and abruptly acuminate, usually curved. Inflorescence/infructescence trichotomous, 2.2–5.2 cm long, glabrous; principal axis 1–3.5 cm long, each axis with 1–3 nodes (situated towards the apex) and verticillate branching, nodes and short branches terminating in small clusters of flowers/fruits; bracts narrowly triangular, 0.7–1.5 mm long, but often caducous; bracteoles triangular to narrowly triangular, ca. 0.6 mm long. Flowers 4- or 5-merous, pedicels absent or infrequently 0.1–0.7 mm long; calyx (incl. hypanthium) 1.5–3 mm long, glabrous; calyx lobes obscure to very broadly triangular-ovate, ca. 0.3 mm long; corolla: color unknown but probably white; corolla tube ca. 3 mm long, external surface glabrous, villous at throat within; corolla lobes 2–3 mm long; stamens: filaments ca. 1 mm long; anthers ca. 1 mm long; style ca. 2 mm long, lobes ca. 0.5 mm long (short styled). Fruit obovoid, 6–8 mm long, smooth or somewhat wrinkled when dry, red/orange, glabrous; calyx limb and lobes persistent and obvious (usually lighter in color than fruit body when dry); sessile or with pedicels, 0.1–0.7 mm long; pyrenes smooth on outer surface; endosperm ruminate.

Other specimens examined: **Luzon. Mountain Prov.:** Mt. Santo Tomas, *Van Steenis PNH 18846* (PNH); Barangay Garling, Mt. Amuyao, Barlig, *Reynoso, Garcia & Sagcal PPI 14563* (BRIT, K); Mt. Polis, *McGregor 19825* (K). **Ifugao Prov.:** Mt. Data, *Ramos & Edaño BS 40193* (A, K,UC, US); Mt. Tabayoc, *Jacobs 7449* (L), *7474* (CANB, K, L, PNH); Sumigar, Mt. Pimmage, *Barbon & Pedersen PPI 12870* (BRIT). **Benguet Prov.:** Mt. Pulog, Lusod, Kabayan, NE of ranger station, summit, *Gaerlan, Romero & Conran PPI 26530* (BRIT, K); *Gaerlan, Romero & Conran PPI 26533* (BRIT, K); Mt. Pulog National Park, Barangay Lusod, Bokod, *Garcia PPI 25083* (BRIT); Mt. Pulog, *Celestino et al. 0–275* (PNH), *Curran et al. FB 18077* (US), *Argent et al. PPI 9528* (K), *Jacobs 7055* (K), *7154* (K), *s.n.* (L), *Ramos & Edaño BS 40367* (A, K, L, P, US), *BS 44959* (NY, P, UC); Mt. Pauai (A), *Quisumbing & Sulit BS 82372* (A); Tonglon, *Loher 5091* (K). **Benguet/Nueva Vizcaya Prov.:** without further data: *Curran & Merritt FB 15841* (US). **Nueva Ecija Prov./Nueva Vizcaya Prov.:** Mt. Umingan, *Ramos & Edaño BS 26333* (A, US). **Province unknown:** Montiquiran, *Loher 1515* (K). **Palawan. Palawan Prov.:** Mt. Mantalingahan, *Edaño PNH 89* (K, L); Brooke's point, Mt. Mantalingahan, *Argent & Romero PPI 9580* (BRIT); Mt. Manalsal, *Edaño BS 77755* (PNH). **Mindanao. Surigao del Norte Prov.:** Dinagat Island, KROMINCO, Mt. Redondo, *Gaerlan, Sagcal & Fernando PPI 4717* (BRIT).

Distribution.—Found on the islands of Luzon, Palawan and Mindanao.

Ecology.—Montane forest, including mossy forest, with one record from grassland; (?700–)1500–2800 m.

Phenology records.—Flowering: January, March to June; fruiting: January to June, September.

Preliminary conservation assessment.—Near Threatened (**NT**). Last collected in 1996.

Distinguishing characteristics.—Leaves elliptic to obovate, small, subcoriaceous to coriaceous; flowers sessile; inflorescences trichotomous, compact.

Remarks.—*Psychotria macgregorii* is one of several *Psychotria* species of the upper montane or "mossy" forests and may be closely related to taxa in adjacent parts of Malesia in similar habitats, such as the *P. boloboensis* complex in New Guinea (Sohmer 1988). *Psychotria macgregorii* also merges morphologically with *P. loheri* Elmer, but the sessile flowers and fruit separate it from the latter. Hybridization may occur between these two species. The collections from Palawan are slightly different from other populations throughout the distribution range, but otherwise conform to this species.

106. Psychotria palimlimensis Sohmer & A.P. Davis, sp. nov. (**Fig. 122**). TYPE: PHILIPPINES. LUZON: Ilocos Norte Prov.: Mt. Palimlim, Aug 1918, *Ramos BS 33364* (HOLOTYPE: A; ISOTYPES: K, P).

Frutex ca. 2 m altus est, stipulis imbricatis glabris 0.3–1.1 cm longis, laminis chartaceis glabris ellipicis vel subobovatis 1.4 × 5.6–2.2 × 9 cm latis et longis nervis lateralibus 9–10 in dimidio quoque apice acuminato basi acuta vel acuminata in sicco obscure olivaceis, petiolis glabris 0.9–1.8 cm longis, infructescentia trichotoma interdum pentachotoma vel irregulariter quoque cum nodo terminali et 1–3 fructibus pedicellatis, rhachidi principali 0.8–1 cm longo, pedicellis saepe longioribus, fructibus et dicunt aurantiacis sed licet rubris valde sulcatis et costatis ellipsoideis 1–1.3 cm longis, pyrenis in dorso valde costatis, endospermo non ruminato.

FIG. 121. *Psychotria macgregorii* (*Merrill 6495*) (US, lectotype).

Fig. 122. *Psychotria palimlimensis* (*Ramos BS 33364*) (A, holotype). **A,** habit of fruiting branch; **B,** stipule; **C,** cross section of fruit; **D,** pyrene; **E,** fruit.

Shrub, ca. 3 m high, glabrous throughout. Stipules imbricate, narrowly triangular to triangular-lanceolate, 0.3–1.1 cm long, glabrous; apex acute to narrowly acute. Leaves: petioles 0.9–1.8 cm long, glabrous; leaf blades elliptic to sub-obovate, 5.6–9 × 1.4–2.2 cm, chartaceous, drying dark oliveaceous green, glabrous above and below; secondary veins 9–10 pairs; tertiary venation manifest; base acute to narrowly acute; apex acuminate. Inflorescence/infructescence trichotomous, sometimes pentachotomous, or with an indefinite number of axes, ca. 2.5 cm long, glabrous; principal axis 0.8–1 cm long, each axis with 1 terminal node bearing 1–3 flowers/fruits; bracts and bracteoles unknown

(?caducous or absent). Flowers unknown. Fruit ellipsoid to narrowly ellipsoid, 1–1.3 cm long, prominently ribbed, reported orange but probably red at maturity, glabrous; calyx limb and lobes persistent and rather obvious (distinctly lobed); pedicels 3–12 mm long; pyrenes ribbed on outer surface; endosperm entire.

PARATYPES: **Luzon. Ilocos Norte Prov.:** Mt. Nagapatan, *Ramos BS 33138* (A, K, L, NY, P, UC, US); Mt. Pico de Loro, *Edaño PNH 17975* (BM, CANB, K, L); Mt. Darna, *Edaño PNH 18132* (L).

Distribution.—Apparently restricted to the mountains in Ilocos Norte Province.

Ecology.—Forest; ca. 1000 m, or less.

Phenology records.—Flowering: unknown; fruiting: February, March, August.

Preliminary conservation assessment.—Critically Endangered (**CR B1ab(iii)**). Last collected in 1953.

Distinguishing characteristics.—Leaves relatively small, 5.6–9 cm long; fruits ellipsoid to narrowly ellipsoid, 1–1.3 cm long, prominently ribbed, pedicels relatively long, calyx persistent, with distinct, triangular lobes, pyrenes long and narrow, with non ruminate (entire) endosperm.

Remarks.—In *P. palimlimensis* the nature of the inflorescence is not clearly understood. More data based on new collections would be useful. If the inflorescence is generally trichotomous then this taxon belongs with the small-leaved Subalpina Species Group. If it was discovered that the inflorescence was greatly reduced (i.e. lacking axes and branches) it might be easily placed in the Linearis Species Group.

107. Psychotria gracilipes Merr., Philipp. J. Sci. 8:52. 1913; Merr., Enum. Philipp. Fl. Pl. 3:556. 1923. (**Fig. 123**). *Grumilea gracilipes* (Merr.) Merr., Philipp. J. Sci. 10:132. 1915. TYPE: PHILIPPINES. LUZON: Cagayan Prov.: Abulug River, rocky, forested hillside on limestone, elevation unknown, 5 Feb 1912, *Weber 1572* (LECTOTYPE: A, designated here; ISOLECTOTYPES: GH, K, P, US).

Psychotia isarogensis (Merr.) Merr., Enum. Philipp. Fl. Pl. 3:556. 1923. *Grumilea isarogensis* Merr., Philipp. J. Sci. 10:136. 1915. TYPE: PHILIPPINES. LUZON: Camarines Sur Prov.: Mt. Isarog, mossy forest, elevation unknown, 29 Nov 1913, *Ramos BS 1553* (LECTOTYPE: BM, designated here; ISOLECTOTYPES: A, L, MO, NY, P).

Shrub, 2–4 m high, young stems glabrous. Stipules valvate-imbricate, twisted in bud, 0.5–1.8 cm long, glabrous or rarely puberulous; apex narrowly acute to acuminate. Leaves: petioles 0.5–1.9 cm long, glabrous; leaf blades narrowly elliptic to subobovate, 4.5–11 × 1.2–3.2 cm, thickly chartaceous to subcoriaceous, drying olive-brown, grayish brown, or blackish brown, glabrous above and below; secondary veins 6–9(–11) pairs; tertiary venation manifest to rather prominent; base acute to narrowly acute; apex acuminate to abruptly acuminate, the tip sharp or blunt. Inflorescence/infructescence trichotomous, 1.4–3.2 cm long, glabrous; principal axis 0.6–3.5 cm long, each axis with 1–2 nodes and opposite branching at each node, branches usually terminating in 1–3-flowers/fruits; bracts narrowly triangular, ca. 1.6 mm long, but often caducous; bracteoles triangular, ca. 0.3 mm long, but often caducous. Flowers 4- or 5-merous, pedicels 1–3.2 mm long (very slender, less than 0.3 mm in diam. when dry); calyx (incl. hypanthium) 0.9–2 mm long, glabrous; calyx lobes very broadly ovate, ca. 0.3 mm long, or obsolete; corolla: white; corolla tube ca. 2 mm long, external surface glabrous, villous at throat within; corolla lobes ca. 2 mm long; stamens: filaments ca. 1 mm long (short styled); anthers ca. 0.5 mm long; style ca. 1.5 mm long (short styled), lobes ca. 0.5 mm long. Fruit obovoid, 5–7 mm long, smooth, reported red, orange and yellow when ripe, probably ripening red in all cases, glabrous; calyx limb and lobes not persistent; pedicels 2.7–5 mm long (less than 0.3 mm in diam. when dry); pyrenes smooth on outer surface; endosperm ruminate.

Other specimens examined: **Luzon. Cagayan Prov.:** without further data: *Darling FB 14776* (US), *Ramos BS 14507* (BM). **Nueva Vizcaya Prov.:** Mt. Alzapan, *Ramos & Edaño BS 45699* (NY, P, UC). **Camarines Sur Prov.:** Mt. Isarog, *Edaño BS 76192* (A, K, NY), *Edaño BS 76193* (A, NY), *BS 76227* (NY), *BS 76238* (NY), *Ramos BS 22037* (BM, K, L, NY, P, SING); *Utzurrum RCBU 210* (K). **Sorsogon Prov.:** without further data: *Ramos BS 23700* (A, BM, K, L, NY, P, SING). **Leyte. Leyte Prov.:** Biliran Island, Mt. Suiro, *Sulit PNH 21628* (L, PNH), *PNH 21632* (L). **Mindanao. Camiguin Prov.:** Camiguin Island, *Ramos BS 14658* (K). **Davao Prov.:** Mt. Mayo, *Edaño PNH 11359* (K, L, PNH), *PNH 11371* (A, L, PNH), *Ramos & Edaño BS 49476* (NY, UC), *BS 49490* (NY, UC); Mt. Kampalili, *Edaño PNH 11584* (A, L, PNH). **Agusan Prov.:** Mt. Urdaneta, Cabadbaran, *Elmer 14162* (A, BM, K, L, MO, NY, P, UC, US, W).

Distribution.—Reported from the islands of Luzon, Biliran, and Mindanao.

Ecology.—Moist to wet forest; from ca. 700–1800 m.

FIG. 123. *Psychotria gracilipes* (*Weber 1572*) (A, lectotype).

Phenology records.—Flowering: March, April; fruiting: January, April to August, October.

Preliminary conservation assessment.—Near Threatened (**NT**). Last collected in 1988.

Distinguishing characteristics.—Leaves small, less than 10 cm long and less than 3 cm wide; inflorescences trichotomous; pedicels very slender, less than 0.3 mm in diameter.

Remarks.—*Psychotria gracilipes* is another apparently common or once common shrub of mid to upper elevation forests. It is probably closely related to such species as *P. loheri*, *P. acuminatissima*, *P. macgregorii*, and *P. subalpina*. It appears morphologically very close to *P. subalpina*, but the smooth fruit (pyrenes smooth on outer (dorsal) surface) separates it from the latter if the delicate pedicels fail to do so.

108. Psychotria acuminatissima Elmer, Leafl. Philipp. Bot. 1:353. 1908; Merr., Enum. Philipp. Fl. Pl. 3:552. 1923. (**Fig. 124**). *Grumilea acuminatissima* (Elmer) Merr., Philipp. J. Sci. 10:132. 1915. TYPE: PHILIPPINES. LUZON: Benguet Prov.: limestone shrubbery below Antop barrio, about 1000 m, Mar 1907, *Elmer 8752* (LECTOTYPE: US, designated here; ISOLECTOTYPES: A, K, MO, NY, W).

Shrub, ca. 2 m high, glabrous throughout. Stipules valvate (?), narrolwy triangular to ± ligulate, 0.6–1.6 cm long, glabrous; apex acute. Leaves: petioles 0.5–1.2 cm long, glabrous; leaf blades narrowly elliptic to narrowly lanceolate, 6.5–11 × 1.1–1.9 cm, chartaceous, drying dark brown to dark oliveaceous green, glabrous above and below; secondary veins 8–13 pairs; tertiary venation rather prominent (tightly reticulate tertiary venation); base attenuate; apex acuminate, the tip sharp and curved. Inflorescence/Infructescence trichotomous, 1.5–2.5 cm long, glabrous; principal axis 0.6–1.1 cm long, each axis with 1(–2) nodes, with a single cymule at each node; bracts and bracteoles unknown (?caducous or absent). Flowers unknown. Fruit obovoid, 5–7 mm long, smooth or wrinkled, color unknown, glabrous; calyx limb and lobes not persistent, apex of fruit usually chartaceous; pedicels 2–4(–5) mm long (more than 0.5 mm in diam. when dry); pyrenes smooth on outer surface; endosperm ruminate.

Other specimens examined: **Luzon. Mountain Prov.:** road to Asin, *Steiner PNH 39791* (PNH). **Benguet Prov.:** Baguio, head of gully behind Easter School, *Clemens 16465* (UC); Baguio, *Merrill 11663* (A, BM, K, P, PNH, W), *Ramos & Edaño BS 45051* (L, NY, UC); Mountain Trail, *Ramos & Edaño BS 45051* (L, UC). **Quezon Prov.:** Mt. Alzapan, *Ramos & Edaño BS 45671* (A, NY, P, UC). **Luzon. Nueva Vizcaya Prov.:** Mt. Caraballo, *Vidal 3045* (K). **Rizal Prov.:** Montalban, *Loher 12133* (UC), *9256* (UC).

Distribution.—In the Mountain, Quezon and Benguet provinces of Luzon.

Ecology.—Shrubby forest, including those on limestone outcrops; ca. 1000 m.

Phenology records.—Flowering: unknown; fruiting: January, March, December.

Preliminary conservation assessment.—Endangered (**EN B1ab(iii)**). Last collected in 1925.

Distinguishing characteristics.—Inflorescence trichotomous; leaf blades very narrow, 1.1–1.9 cm wide, chartaceous, with curved acuminate tips; fruits relatively smooth, obovoid, pedicellate.

Remarks.—According to Elmer (1908) *P. acuminatissima* is close to *P. subalpina*. *Psychotria acuminatissima* is distinguished from *P. subalpina* by the unribbed, relatively smooth fruits, and from *P. loheri* by the more acuminate leaf apex. It is similar in leaf shape to *P. sohotonensis* but can be separated from this species by the smaller number of cymules per inflorescence. However, with so few collections of *P. sohotonensis* at our disposal a definitive statement concerning relationships are not possible. Some of the collections of *P. acuminatissima* resemble *P. luzoniensis*.

109. Psychotria sohotonensis Sohmer & A.P. Davis, sp. nov. (**Fig. 125**). TYPE: PHILIPPINES. SAMAR: Western Samar Prov.: Mt. Sohoton, Babasian Na Gatte, 12 km, Apr 1970, *Madulid et al. PNH 117798* (HOLOTYPE: L).

Frutex 2 m altus est, stipulis valvatis imbricatis glabris 0.7–1.8 cm longis, laminis chartaceis glabris anguste ellipticis 1 × 7–2.2 × 14.2 cm latis et longis nervis lateralibus 10–14 in dimidio quoque apice acuminata basi acuto vel acuminata obscura badiis vel obscure olivaceis quando siccis, petiolii glabris 0.7–1.8 mm longis, inflorescentia trichotoma rhachidibus 1.5–3.5 cm longis quoque in nodis cum 5–15 cymulis terminantibus, bracteis nullis vel caducis, floribus 5-meris, pedicellis 1 mm longis vel longioribus, calyce et hypanthio 1.5–2 mm longo lobis fere nullis, tubo corollae glabro 1.5–2 mm longo, lobis fere tam grandibus quam tubam, fructibus versimiliter rubris glabris laevibus ellipsoideo-globosis 8–9 mm longis, pyrenis in dorso laevibus vel paulum costatis, endospermo non ruminato.

Shrub, ca. 2 m high, glabrous throughout. Stipules valvate-imbricate, lanceolate or ± narrolwy triangular, 0.7–1.8 cm long, glabrous; apex acute to slightly acuminate. Leaves: petioles 0.7–1.8 cm long,

Fig. 124. *Psychotria acuminatissima* (*Elmer 8752*) (US, lectotype).

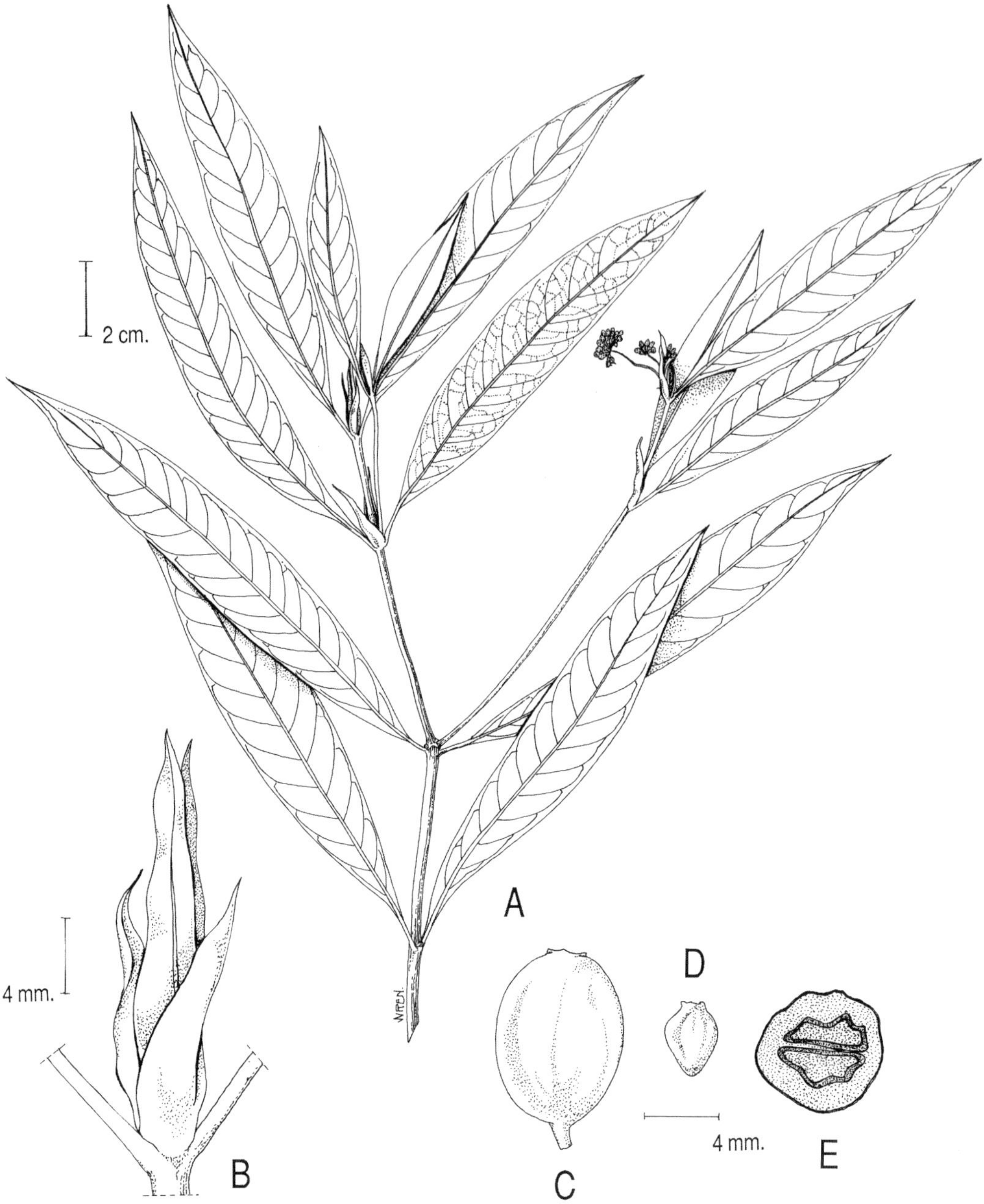

Fig. 125. ***Psychotria sohotonensis*** **(*Madulid et al. PNH 117798*) (L, holotype). A, habit of fruiting branch; B, stipule; C, fruit; D, pyrene; E, fruit in cross section.**

glabrous; leaf blades narrowly elliptic, 7–14.2 × 1–2.2 cm, chartaceous, dark brown to dark oliveaceous brown when dry, glabrous above and below; secondary veins 10–14 pairs; tertiary venation obscure; base acute to narrowly acute; apex acuminate. Inflorescence/Infructescence trichotomous, 2.8–4 cm long, glabrous; principal axis 1.5–3.5 cm long, each axis terminating in a node with 5–15 shortly branched cymules; bracts and bracteoles absent. Flowers 5-merous, pedicels ca. 1 mm long; calyx (incl. hypanthium) 1.5–2 mm long, glabrous; calyx lobes obsure to tooth-like; corolla: greenish white; corolla tube 1.5–2 mm long, external surface glabrous, internal surface glabrous; corolla ca. 1.5 mm long; stamens: filaments ca. 1 mm long (?short styled); anthers ca. 0.6 mm long; style unknown. Fruit ellipsoid-globose, 8–9 mm long, smooth, probably red, glabrous; calyx limb and lobes persistent but obscure; pedicels ca. 1 mm long; pyrenes smooth or slightly ribbed on outer surface; endosperm entire.

PARATYPES: **Samar. Western Samar Prov.:** Mt. Sohoton, *Gutierrez et al. PNH 117579* (L).

Distribution.—Restricted to Mt. Sohoton, Western Samar Province, Samar.

Ecology.—Moist forest; altitude not known but probably at low to mid elevations.

Phenology records.—Flowering: April; fruiting:unknown.

Preliminary conservation assessment.—Critically Endangered (**CR B1ab(iii)**) or possibly Extinct (**EX**). Last collected in 1970.

Distinguishing characteristics.—Leaf blades very narrowly elliptic, relatively long for this group, 7–14.2 cm long; inflorescences unusual, with only one node terminating the inflorescence axes from which 5–15 cymules arise.

110. Psychotria subalpina Elmer, Leafl. Philipp. Bot. 1:64. 1906; Merr., Enum. Philipp. Fl. Pl. 3:563. 1923. (**Fig. 126**). *Grumilea subalpina* (Elmer) Merr., Philipp. J. Sci. 10:131. 1915. TYPE: PHILIPPINES. LUZON: Quezon Prov.: Mt. Banahão, wet forest between 1500 and 2000 m, May 1906, *Elmer 7942* (LECTOTYPE: K, designated here; ISOLECTOTYPES: A, L, US).

Shrub, 3–5 m high, glabrous throughout. Stipules valvate, twisted in bud, triangular-lanceolate, or ± narrowly elliptic, 0.6–1.1 cm long, glabrous; apex acute to narrowly acute. Leaves: petioles 0.5–1.8 cm long, glabrous; leaf blades narrowly ellpitic to obovate, 4–11 × 1–3.5 cm, chartaceous to subcoriaceous, drying dark oliveaceous green to dark brown, glabrous above and below; secondary veins 6–11 pairs; tertiary venation rather weak to rather prominent; base narrowly attenuate; apex acuminate to abruptly acuminate. Inflorescence/infructescence trichotomous, 2.1–6.4 cm long, glabrous; principal axis 1.2–3.8 cm long, each axis (or at least primary axis) with 2 nodes and opposite or verticillate branching at each node, branches terminating in cymules; bracts ovate to triangular, ca. 0.8 mm long, but usually absent (?caducous or absent); bracteoles ± triangular, ca. 0.5 mm long, usually absent (?caducous). Flowers poorly known (immature), 4- or 5-merous, pedicels 0.9–3.1 mm long; calyx (incl. hypanthium) 1–1.5 mm long, glabrous; calyx lobes broadly ovate to ± triangular, 0.3–0.7 mm long, or ± obsolete; corolla: color unknown; corolla tube ca. 3 mm long (short styled), external surface glabrous, pubescent at throat within; corolla lobes ca. 1.5 mm long; stamens and style unknown. Fruit globose to subobovoid, rarely ellipsoid, 5–9 mm long, conspicuously ribbed, reported yellow but probably maturing red-orange, glabrous; calyx limb and lobes persistent and usually obvious (limb lobes or ± truncate); pedicels 1–3.2 mm long; pyrenes ribbed on outer surface; endosperm ruminate.

Other specimens examined: **Luzon. Ilocos Norte Prov.:** Mt. Palimlim, *Ramos BS 33270* (NY, SING). Bangui to Claveria, *Ramos BS 32994* (A, BM, K, P, US), *BS 32997* (L), *BS 33058* (A, K, P). **Ifugao Prov.:** Mt. Polis, *Sandkuhl 305* (K); Bontoc Sub-province, Mt. Polis, *Ramos & Edaño BS 37642* (K). **Laguna Prov.:** Mt. Banahao, *Curran & Merritt FB 7855* (NY), *Ocampo BS 27916* (A, P), *Ramos BS 19568* (BM, K). **Quezon Prov.:** Lucban, *Elmer 7582* (A, K, L, MO, US, W), *7596* (A, K, L, NY), *7672* (A, K, L, MO, NY, W); Kinatakutan, *Oro FB 30648* (NY); Mt. Alzapan, *Ramos & Edaño BS 45683* (BM, NY, UC). **Nueva Vizcaya Prov.:** Caraballo mountain, *Loher 13645* (UC); S. Caraballo (Mt.), *Vidal 400* (K). **Camarines Sur Prov.:** Naga, West slope of Mt. Isarog, Barangay Panicuason, *Argent, Romero & Fuentes PPI 20143* (K). **Mindoro. Mindoro Oriental Prov.:** Mt. Halcon, *Merrill 6159* (NY); NE Mt. Halcon (Parabatugan area) following Paitan access trail, Naujan, *Mandia 497* (K). **Sibuyan. Romblon Prov.:** Mayos Peak, *Stone, Reynoso & Fernando PPI 6934* (BRIT). **Panay. Aklan Prov.:** Tibulong, So. Aklan, Barangay Dalagasaan, Libacao (Mt. Baloy), *Garcia & Romero PPI 15932* (BRIT, K); Laguiron, So. Aklan, Barangay Dalagsaan, Libacas (Mt Baloy), eastern side of Mt. Baloy, *Garcia & Romero PPI 15922* (BRIT). **Capiz Prov.:** Mt. Madiaas, *Ramos & Edaño BS 30713* (A, K, P, UC, US). **Negros. Negros Occidental Prov.:** Mt. Canlaon, *Edaño PNH 21950* (K). **Mindanao. Miasamis Occidental Prov.:** Mt. Malindang, Barangay Lake Duminagat to Liboron, *Gaerlan, Sagcal & Romero PPI 10820* (BRIT, K). **Bukidnon Prov.:** Mt. Lipa, *Ramos & Edaño BS 38517* (A, K, P, US); Invatas, Impasug-ong, Mt. Kitanglad, *Madulid et al. PPI 3144* (BRIT, K). **Camaguin Prov.:** without further data: *Ramos BS 14627* (K). **Davao Prov.:** Mt. MacKinley, *Kanehira 2717* (NY); *Edaño PNH 1565* (K).

Distribution.—Luzon, Sibuyan, Panay, Negros, Mindoro, and Mindanao.

Ecology.—Humid forest, primary and disturbed, including 'mossy' and 'elfin' forest; 1200–2100 m.

Phenology records.—Flowering: March to July, September; fruiting: January to October.

Preliminary conservation assessment.—Near Threatened (**NT**). Last collected in 1995.

Distinguishing characteristics.—Leaves narrowly elliptic or subobovate; fruits (pyrenes) ribbed, pedicels stout.

Fig. 126. *Psychotria subalpina* (*Elmer 7942*) (K, lectotype).

Remarks.—Elmer (1906) noted that this species was rather common in the subalpine vegetation in Quezon province on Luzon, but relatively few herbarium collections are known after the 1930s. The strongly ribbed fruits (when dry) separate this taxon from the presumed other closely related species such as *P. gracilipes*, *P. acuminatissima*, *P. pauciflora* and *P. loheri*.

111. Psychotria pauciflora Bartl. ex DC., Prodr. 4:522. 1830; Merr., Philipp. J. Sci. 10:129. 1915; Merr., Enum. Philipp. Fl. Pl. 3:560. 1923. (**Fig. 127**). TYPE: PHILIPPINES. LUZON: without further data, *Haenke* s.n. (HOLOTYPE: PR).

Shrub, ca. 3 m high, glabrous throughout. Stipules valvate, narrowly ovate, 0.5–2.2 cm long, glabrous; apex acuminate. Leaves: petioles 0.6–2 cm long, glabrous; leaf blades narrowly elliptic to oblong-elliptic, 4.1–13 × 1.1–4.2 cm, chartaceous, drying grayish brown to chocolate-brown, glabrous above

Fig. 127. *Psychotria pauciflora* (*Haenke s.n.*) (PR, holotype).

and below; secondary veins 8–10(–13) pairs; tertiary venation weak; base acute to narrowly acute; apex acuminate to abruptly acuminate. Inflorescence/infructescence trichotomous, although secondary axes much smaller and sometimes not developing, (0.5–)1.1–4.6 cm long, glabrous; principal axis (0.3–)0.6–4.7 cm long, each axis with 1 node, node terminating in few flowered cymules, with 5–10 flowers/fruits per inflorescence; bracts and bracteoles unknown. Flowers unknown. Fruit globose, 7–10 mm long, smooth or ribbed, reported orange but probably red or black at maturity, glabrous; calyx limb and lobes not persistent; pedicels 0.1–4 mm long, or pedicels absent; pyrenes smooth on outer surface; endosperm ruminate.

Other specimens examined: **Luzon. Cagayan Prov.:** Bauan, Mt. Tabuan, *Ramos BS 77116* (A). **Kalinga-Apayao Prov.:** Mt. Masingit, Lubuagan, *Ramos & Edaño BS 37554* (A, K, L). **Benguet Prov.:** without locality, *Loher 1511* (K). **Quezon Prov.:** Mt. W. of Philippine Nat. Bot. Garden (Mt. Bataan Mt.), *Sohmer 11998* (BISH, K). **Bulacan Prov.:** Mt. Biak na Bato, San Miguel, *Garcia et al. PPI 15078* (BRIT, K); *Garcia et al. PPI 15044* (BRIT, K); Angat Watershed Area, Norzagaray, *Barbon & Garcia PPI 22396* (BRIT, K), *Barbon & Garcia PPI 22428* (BRIT); *Barbon & Garcia PPI 22461* (BRIT). **Rizal Prov.:** Sitio Bayucan, Barangay Samploc, Tanay, *Romero & Chavez PPI 29395* (BRIT). **Mindoro. Mindoro Oriental Prov.:** Mt. Halcon, *Ramos & Edaño BS 40737* (A).

Distribution.—On Luzon and Mindoro.

Ecology.—Forest, including primary and secondary forest; 450–500 m.

Phenology records.—Flowering: August, November, December; fruiting: February to August.

Preliminary conservation assessment.—Vulnerable (**VU B1ab(iii)**). Last collected in 1996.

Distinguishing characteristics.—Leaf blades usually less than 3 cm wide, base acute to acuminate; inflorescences trichotomous; fruits less than 5–10 per infructescence, usually pedicellate but sometimes sessile, ribbed or smooth when dry, pyrenes smooth or ribbed on outer (dorsal) surface.

Remarks.—The type specimen of *P. pauciflora* was examined (by S.H. Sohmer) in Prague (PR). Dr. Jiri Sojak, Curator at PR, believes that this was indeed the specimen Bartling had in hand when describing his new species and that it is the holotype. This specimen was also examined by Merrill (1915), and it matches a number of collections cited here. *Psychotria subsessiliflora* and *P. wenzelii* are the taxa most easily confused with *P. pauciflora*. In *P. subsessiliflora*, the inflorescence is often simple (much reduced and compact) and the flowers and fruits nearly always sessile, and in *P. wenzelii* the fruits are more or less sessile. In *P. pauciflora* the inflorescences are trichotomous and the fruits are nearly always pedicellate. Variation within *P. pauciflora* is not fully understood, and good field knowledge is lacking. *Psychotria pauciflora* is difficult to separate from *P. loheri* Elmer, and some of the collections assigned to it here look very much like smaller-leaved representatives of *P. luzoniensis*.

112. Psychotria loheri Elmer, Leafl. Philipp. Bot. 1:356. 1908; Merr., Enum. Philipp. Fl. Pl. 3:557. 1923. (**Fig. 128**). *Grumilea loheri* (Elmer) Merr., Philipp. J. Sci. 10:132. 1915. TYPE: PHILIPPINES. LUZON: Benguet Prov.: Baguio vicinity, Mt. Santo Thomas at 2500 m, Mar 1907, *Elmer 8584* (LECTOTYPE: K, designated here; ISOLECTOTYPES: A, L, NY, US, W).

Psychotria vanoverberghii Merr., Philipp. J. Sci. 9:457. 1915. TYPE: PHILIPPINES. LUZON: Mountain Prov.: Bauco, in forests from 1300–1600 m, Apr 1913, *Vanoverbergh 2818* (LECTOTYPE: US, designated here; ISOLECTOTYPE: K).

Shrub or small tree, 2–5 m high, glabrous throughout. Stipules valvate, twisted in bud, lanceolate-triangular, (0.3–)0.6–2 cm long; glabrous; apex narrowly acute. Leaves: petioles 0.5–1.2 cm long, glabrous; leaf blades elliptic, oblanceolate, or obovate, 4–8.5 × 1.5–2.4 cm, subcoriaceous to coriaceous, drying gray-green, olivaceous to dark chocolate-brown, glabrous above and below; secondary veins 6–10(–13) pairs; tertiary venation obscure to manifest; base acute to attenuate; apex acute to narrowly acuminate with rounded or blunt tip. Inflorescence/infructescence trichotomous, rarely pentachotomous, 1.5–5 cm long, glabrous; principal axis (0.5–)1.1–4 cm long, each axis with 1–2(–3) nodes and opposite branching at each node, branches terminating in cymules, with more than 10 flowers/fruits per inflorescence; bracts deltate to triangular, or ovate, ca. 0.7 mm long, but often absent (?caducous); bracteoles unknown. Flowers 5-merous, pedicels (0.1–)0.4–1.1 mm long; calyx (incl. hypanthium) 1–1.9 mm long, glabrous; calyx lobes broadly ovate, 0.3–0.6 mm long, or obscure to obsolete and then calyx limb truncate; corolla: color unknown; corolla tube 2.1–2.6 mm long, external surface glabrous, villous at throat within; corolla lobes 1.9–2.2 mm long; stamens: filaments ca. 1.2 mm long (short styled); anthers ca. 1.2 mm long; style unknown. Fruit obovoid, 6–7 mm long,

FIG. 128. *Psychotria loheri* (*Elmer 8584*) (K, lectotype).

smooth, color unknown, glabrous; calyx limb and lobes persistent, either ± truncate or lobed; pedicels (0.2–)0.8–1.6 mm long; pyrenes smooth on outer surface; endosperm ruminate.

Other specimens examined: **Luzon. Ilocos Norte Prov.:** Mt. Darna, *Edaño PNH 18141* (K, L). **Abra Prov.:** without further data: *Ramos BS 7062* (US). **Mountain Prov.:** Bontoc subprovince, without further data, *Vanoverbergh 2861* (P). **Ifugao Prov.:** Mt. Data, *Ramos & Edaño BS 40235* (A, L, P). **Benguet Prov.:** Baguio, *Williams 1269* (K, NY); Mt. Santo Tomas, area near satellite and radar station, *Sohmer 11958* (BISH); Mt. Santo Tomas, *Mendoza PNH 20401* (L, PNH), *van Steenis 18571* (L), *Williams 1219* (NY), *1338* (A, GH, K, US); Mt. Tonglon, *Merritt FB 14164* (US), *Ramos BS 5475* (L, US). **Zambales Prov.:** Santa Cruz, Acoje Mine concession area, *Ridsdale & Reynoso 1427* (L, K); Santa Cruz, *Vidal 1461* (K), *1619bis* (K), *3028* (K), *3028bis* (A). **Leyte. Leyte Prov.:** Buenavista, near Jaro, *Wenzel 622* (A, BM, US).

Distribution.—Most collections are from the mountainous provinces of north central Luzon, but also found on Leyte.

Ecology.—Forest. Including mossy forest; 1300–2200(–2500) m.

Phenology records.—Flowering: January to May, November; fruiting: February, March, October.

Preliminary conservation assessment.—Near Threatened (**NT**). Last collected in 1985.

Distinguishing characteristics.—Leaves less than 10 cm long and less than 3 cm wide; inflorescences trichotomous, with more than ten flowers per inflorescence; fruits smooth or wrinkled, not ribbed, pyrenes smooth on outer (dorsal) surface.

Remarks.—In the mid to high elevation forests of the Philippines there are five species within the Subalpina Species Groups that possess small leaves and trichotomous inflorescences. This group comprises *P. loheri*, *P. macgregorii*, *P. subalpina*, *P. gracilipes*, and *P. palimlimensis*. They have not been studied in the field, mainly because many of the original localities and habitats are gone. Based on the characters observed in the herbarium they can be separated from each other, but their true relationships will be elucidated only after fieldwork and further systematic studies have been undertaken. In this treatment we have synonymized *P. vanoverberghii* with *P. loheri*, for the type is the only collection known for the former species, and the only way it differs from *P. loheri* is in the larger number of lateral veins on the leaves. In all other ways it is impossible to distinguish *P. vanovergerghii* from *P. loheri*, and also the habitats and localities appear to be the same. The Wenzel collection from Leyte, a syntype of *P. wenzelii* Merr., is considered by us to represent *P. loheri*.

POTENTIAL NEW SPECIES

The following section comprises potential new species. Insufficient material prevents the formal description of these taxa.

113. Psychotria sp. A

Specimens examined: **Palawan. Palawan Prov.:** Rizal, Ransang, Kumaraka, *Gaerlan, Romero & Sagcal PPI 13205* (BRIT); Barangay Bundog, Municipality of Quezon, *Pipoly & Romero et al. PPI 38209* (BRIT).

Species group.—Probably belonging to the Gitingensis Species Group.

Distribution.—Only known from Palawan (Palawan Provice).

Remarks.—Despite the presence of trichotomous inflorescences, *Psychotria* sp. A apperars to be closely related to *P. rubiginosa*, and especially *P. rubiginosa* var. *jacobsii*. *Psychotria* sp. A has the general appearance of *P. rubiginosa*, for example, large oblanceolate leaves with more than 11 pairs of lateral veins and large stipules with 'shoulders' at the base. It is glabrous, like *P. rubiginosa* var. *jacobsii*. It differs from *P. rubiginosa* on the basis of its trichotomous inflorescences, and somewhat larger auriculate stipules with an entire apex (as opposed to stipules divided at the apex, in *P. rubiginosa*). The stipules are persistent, like *P. rubiginosa var. jacobsii*; the stipules of *P. rubiginosa* are deciduous to semi-persistent. The pyrenes of *P.* sp. A are similar to *P. rubiginosa*, as they both have pyrenes of about the same shape and size, distinctly ribbed or ridged on the outer (abaxial) surface. However, the seeds of *P.* sp. A have entire endosperm, whereas the endosperm of *P. rubiginosa* is ruminate. Lack of good flowering material prevents the formal description of *P.* sp. A.

114. Psychotria sp. B

Specimens examined: **Palawan. Palawan Prov.:** Brookes point, Pulot, ca. 50 km N. of town, *Podzorski SMHI 2160* (K, L); Pulot, Massin River, 12 km N. of Brookes Point, *Ridsdale 1000* (K, L).

Species group.—Probably belonging to the Gitingensis Species Group.

Distribution.—Only known from Palawan (Palawan Provice).

Remarks.—*Psychotria* sp. B is a pubescent plant with monochotomous inflorescences. We are unsure as to its affinity with other Philippine *Psychotria*, as there is only one specimen in bud (*Ridsdale 1000*) and one other with immature fruits (*Podzorski SMHI 2160*).

115. Psychotria sp. C

Specimens examined: **Luzon. Cagayan Prov.:** Claveria, Barangay, Kilkiling, Taggat Watershead Area, *Garcia, Fuentes & Romero PPI 18318* (BRIT, K); Barangay San Juan, Santa Praxedes, *Garcia, Fuentes & Romero PPI 18492* (BRIT).

Species group.—Belonging to the Luzoniensis Species Group.

Distribution.—Only known from Cagayan Province, northern Luzon.

Remarks.—*Psychotria* sp. C appears close to *P. banahaensis* but differs by its smaller (c. 0.5 cm long), valvate stipules and obovoid fruits. The stipules of P. banahaensis are usually 1.4–2.6 cm long, and the fruits distinctly obovoid-turbinate.

116. Psychotria sp. D

Specimens examined: **Luzon. Mountain Prov.:** Mt. Pukis, *Ramos & Edaño BS 37842* (A, K); Barangay Garling, Mt. Amuyao, Barlig, *Reynoso, Garcia & Sagcal PPI 14588* (BRIT, K).

Species group.—Belonging to the Luzoniensis Species Group.

Distribution.—Restricted to the Mountain Province of nortern Luzon.

Remarks.—*Psychotria* sp. D can not be matched with any other spcies of Philippine *Psychotria* on the basis of its subcoriaceous to coriaceous leaves, which have prominent tertiaty venation on both surfaces, trichotomous infloresences with verticillate branching, and smooth walled pyrenes with very short preformmed germination slits (PGSs). It is known only from two fruiting specimens and flowers are unkown. *Psychotria* sp. D appears to be a high altitude entity; the only specimen with field notes (*Reynoso, Garcia & Sagcal PPI 14588*, BRIT, K) gives the altitude as 2700 m, and the habitat as submossy forest.

117. Psychotria sp. E

Specimens examined: **Ilocos Norte Prov.:** Mt. Palimlin, *Ramos BS 33291* (A, K, UC). **Nueva Vizcaya Prov.:** Mt. Alzapan, *Ramos & Edaño BS 45591* (K, NY, P, UC); *Ramos & Edaño BS 45592* (A); *Ramos & Edaño BS 45622* (A, K, NY, P, UC).

Species group.—Belonging to the Luzoniensis Species Group.

Distribution.—Restricted to the provinces of Ilocos Norte and Nueva Vizcaya, in nortern Luzon.

Remarks.—*Psychotria* sp. E is very similar to *Psychotria* sp. D, on the basis of its subcoriaceous to coriaceous leaves, with prominent tertiaty venation on both surfaces, and trichotomous infloresences. It differs from *Psychotria* sp. D by its opposite or rarely verticillate branching, more slender inflorescence axes, weakly two or three ridged pyrenes (outer surface) and long preformmed germination slits (PGSs). Further material is required to ascertain whether the three specimens cited here are conspecific.

UNKNOWN SPECIES—SPECIES INCOGNITAE

This part contains Philippine *Psychotria* names that are uknown to us because the type, or other original material, is either missing (e.g. destroyed) or has been misplaced within herbaria. We are reluctatnt to assign names to taxa on the basis of descriptions alone, particularly where there are several options for comparisons with taxa from the same location, and have not done so in the treatment presented here.

118. Psychotria cagayanensis Merr., Philipp. J. Sci. 8:51. 1913; Merr., Enum. Philipp. Fl. Pl. 3:553. 1923.
Grumilea cagayanensis (Merr) Merr., Philipp. J. Sci. 10:133. 1915. TYPE: PHILIPPINES. LUZON: Cagayan Prov.: Casambalangan, Apr 1910, in forests at an altitufr of ca. 10m, *Bernado FB 15491* (not seen).

Notes.—According to Merrill (1913) this taxon resembles *P. pinnatinervia* and *P. luzoniensis*. Merrill also stated that the pyrenes were not ridged or grooved. Probably belonging to the Luzoniensis Species Group.

119. Psychotria irosinensis Elmer, Leafl. Philipp. Bot. 9:3262. 1934. TYPE: PHILIPPINES. LUZON: Sorsogon Prov.: Sorsogon Prov.: Irosin (Mt. Bulusan), Sep 1916, *Elmer 17356* (not seen).

Notes.—According to Elmer (1934) *P. irosinensis* has unbranched inflorescences but is otherwise similar to *P. manillensis* and *P. pallidiflora*. Species group unkown.

120. Psychotria samarensis Merr., Philipp. J. Sci. 12:174. 1917; Merr., Enum. Philipp. Fl. Pl. 3:562. 1923. TYPE: PHILIPPINES. SAMAR: Province unknown: Salcedo, on ridges, ca. 200 m, 11 Mar 1915, *Lasquety FB 23576* (not seen).

Notes.—This species seems to be distinctive on the basis of its brown indumentum, smooth fruits and entire (non ruminate endosperm) (Merrill 1917). Species group unknown.

121. Psychotria voluta Elmer, Leafl. Philipp Bot. 4:1347. 1912. TYPE: PHILIPPINES. PALAWAN: Palawan Prov.: Puerto Princesa (Mt. Pulgar), Apr 1911, *Elmer 13253* (not seen).

Notes.—*Psychotria voluta* is clearly a member of the Ixoroides Species Group but we are unsure of its identity or relationships with other climbing species.

INDETERMINATE SPECIMENS

122. Psychotria sp. indet.

Cagayan Prov.: Calayan Island, Barangay Cabudadan Centro 2, *Fuentes & de la Rosa PPI 38866* (BRIT, K) [Gitingensis Species Group]. **Kalinga-Apayao Prov.:** Mt. Masingit, Lubuagan, *Ramos & Edaño BS 37562* (K) [Gitingensis Species Group]. **Isabela Prov.:** Palanan, Dipud-gen, *Gaerlan, Sagcal & Alvarez PPI 3060* (BRIT, K) [Luzoniensis Species Group]; Palanan, Dipaguidin, nr. Palanan Point, *Ridsdale, Dejan & Baquiran ISU 34* (L, K) [Magnifolia Species Group]. **Cavite Prov.:** Tala, Ternate, *Reynoso, Sagcal & Fuentes PPI 17127* (BRIT) [Tayabensis Species Group]. **Luzon. Without further data:** *Loher 1510* (K) [Subalpina Species Group]. **Panay. Antique Prov.:** Culasi, Hangod Tubig, Mt. Madia-as, *Gaerlan & Fuentes PPI 4955* (BRIT) [Tayabensis Species Group]. **Capiz Prov.:** Mt. Kinablangan, *Edaño BS 46030* (A, K, NY, UC). **Samar. Eastern Samar Prov.:** Pinamgasan, Borongan, *Castro et al. PNH 5850* (K) [species group unknown]. **Palawan. Palawan Prov.:** Taytay, Ibangley, *Stone PPI 350* (BRIT) [*P. rubignosa* cf.; Gitingensis Species Group]. **Mindanao. Surigao del Norte Prov.:** KROMINCO, Mt. Redondo, *Gaerlan, Sagcal & Fernando PPI 4663* (K). **Misamis Occidental Prov.:** Mt Malindang, Barangay Lake Duminagat to Liboron, *Gaerlan, Sagcal & Romero PPI 10823* (K) [*P. rubignosa* cf.; Gitingensis Species Group]. **Zamboanga del Sur Prov.:** Malangas, *Ramos & Edaño BS 37363* (K) [species group unkown]. **PHILIPPINES. Without further data:** *Loher 6445* (K); *Loher 6445* (K). **Tawi-tawi. Tawi-tawi Prov.:** Seratang, Languyan, *Gaerlan & Sagcal PPI 10217* (BRIT, K) [Ixoroides Species Group].

EXCLUDED SPECIES

Psychotria philippensis Cham. & Schltdl. in Linnaea, 4:21. 1829. TYPE: unknown.

Notes.—This is obviously not a spcies of *Psychotria* as the description for this species gives a plant with paired axillary to supraxillary inflorescences. At the present time we are unable to assign this taxon to genus but believe that it is very likely to be Rubiaceae.

ACKNOWLEDGMENTS

The support of the National Science Foundation (grant # BSR-8306982) is gratefully acknowledged. It is difficult to perform comprehensive revisions, like the one presented here, without such support; the field work and research assistance provided by the grant made this work possible. Mr. and Mrs. Watson Smith (both now deceased) provided support towards maintaining a research assistant for S.H. Sohmer, and are gratefully acknowledged; Mrs. Smith, or Lucy Cranwell as her friends all knew her, also provided funds for the publication of this revision, before she passed away.

Christa Russell was the primary assistant during much of this project and to her we express appreciation for the meticulousness and care with which she worked during the formative stages of this work. Fieldwork (for S.H. Sohmer) in the Philippines was strongly supported by the staff of the National Museum in Manila. S.H. Sohmer would particularly like to thank his colleague Dr. Domingo Madulid, who made available the resources of the Division of Botany. Mr. Estong Reynoso, now deceased, accompanied S.H. Sohmer in the field on nearly all occasions; without his help the fieldwork would not have been as successful as it was.

We thank the staffs of the following herbaria: Harvard University (A), the Royal Botanic Gardens, Kew (K), Muséum National d'Histoire Naturelle, Paris (P), the Rijksherbarium, Leiden (L) and the Naturhistorisches Museum, Vienna (W), for their help and courtesy during visits made to those institutions. S.H. Sohmer would particularly thank the staff of the Rijksherbarium, especially the late Kees Kalkman, Colin Ridsdale, Jan Frits Veldkamp, Max van Balgooy, and the support staff such as Leo Vogelenzang, Andre Groenewegen, and Frans Waterschoot. Colin Ridsdale was extremely helpful due to his interest in and knowledge of the Rubiaceae and the Philippines. At the Royal Botanic Gardens, Kew we are indebted to Diane Bridson and Sally Dawson who supported this work in various ways; Timothy Utteridge provided valuable assistance in reading through parts of the manuscript, Hazel Wilks produced the illustration for Figure 2, and Patricia Malcolm-Tompkins provided invaluable assistance with database management.

We thank Peter O'Connor for organizing loans at the Bishop Museum, and Warren L. Wagner for technical discussions. Harold St. John, Research Associate at the Bishop Museum, provided Latin diagnoses for the new taxa. The illustrations were undertaken by Wren [sic] and we appreciate her care and conscientiousness. Katie Anderson, the Bishop Museum Botany Department Secretary, spent a great amount of time on this manuscript as well.

Without the support of the Board of Trustees of the Botanical Research Institute of Texas, for a short sabbatical at Kew, S.H. Sohmer could not have completed the original manuscript. Peter Crane, the present Director of Kew, was instrumental in arranging the appointment as Honorary Research Associate for S.H. Sohmer, which provided the opportunity for the collaboration between S.H. Sohmer and A.P. Davis. The first author especially thanks A.P. Davis for collaborating on this project, and in particular for revising the original manuscript in the light of recent collections and new data. The authors appreciate the support of the editor throughout the long process of bringing this manuscript to publication. There are many others we are grateful to for support during this revision, and we hope that this effort will provide a tangible fulfillment of that support.

REFERENCES

Andersson, L. 2001. *Margaritopsis* (Rubiaceae, Psychotrieae) is a pantropical genus. Syst. & Geogr. Pl. 71:73–85.

Andersson, L. 2002. Relationships and generic circumscriptions in the *Psychotria* complex (Rubiaceae, Psychotrieae). Syst. & Geogr. Pl. 72:167–202.

Bantayan, N.C. 2001. Geomatics-assisted impact assessment of landuse change on the biodiversity of Mt Makiling, Philippines. In: 22nd Asian Conference on Remote Sensing, 5–9 November 2001, Singapore. [http://www.crisp.nus.edu.sg/~acrs2001/pdf/012BANTA.pdf].

Bartling, F.G. 1830. In: A.P. de Candolle, Prodromus systematis naturalis regni vegetabilis Vol 4. Treuttel et Würtz, Paris, France.

Blanco, F.M. 1837. Flora de Filipinas, Segun el sistema sexual de Linneo. En la imprenta de Sto. Thomas por D. Candido Lopez, Manila, Philippines.

Chamisso, L.A. von and D.F.L. von Schlectendal. 1829. De plantis in expeditione speculatoria Romanzoffiana observatis rationem dicunt. Linnaea 4:1–36.

Danielsen, F. and C.G. Treadway. 2004. Priority conservation areas for butterflies (Lepidoptera: Rhopalocera) in the Philippine islands. Anim. Conserv. 7:79–92.

Davis, A.P., D. Bridson, C. Jarvis, and R. Govaerts. 2001. The typification and characterization of the of the genus *Psychotria* L. (Rubiaceae). Bot. J. Linn. Soc. 135:35–42.

Davis, S.D., V.H. Heywood, and A.C. Hamilton (eds.). 1995. Centres of plant diversity: A guide and strategy for their conservation. Volume 2, Asia, Australasia and the Pacific. IUCN Publications, Cambridge, UK.

Elmer, A. D.E. 1906. Additional new species of Rubiaceae. Leafl. Philipp. Bot. 1:63–73.

Elmer, A.D.E. 1908. A century of new planths. Leafl. Philipp. Bot. 1:272–359.

Elmer, A.D.E. 1911. New and noteworthy Rubiaceae. Leafl. Philipp. Bot. 3:971–1045.

Elmer, A.D.E. 1912. Palawan Rubiaceae. Leafl. Philipp. Bot. 4:1327–1362.

Elmer, A.D.E. 1913. Rubiaceae from Mount Urdaneta. Leafl. Philipp. Bot. 5:1855–1905.

Elmer, A.D.E. 1934. New Urticaceae and Rubiaceae. Leafl. Philipp. Bot. 9:3227–3272.

Fernandez-Villar, C. 1880. In: Naves, A., and C. Fernandez-Villar., Flora de Filipinas R.P.Fr. Emmanuëlis Blanco, eds. 3., Vol. 4. Novissima Appendix ad Floram Philippinarum. Apud Plana et socios, Manila, Philippines.

Hamilton, C.W. 1985. Architecture in neotropical *Psychotria* L. (Rubiaceae): dynamics of branching and its taxonomic significance. Amer. J. Bot. 72:1081–1088.

Hickey, L. 1979. Classification of the architecture of dicotyledonous leaves. Amer. J. Bot. 60:17–33.

HOLMGREN, P.K., N.K. HOLMGREN, and L.C. BARRETT. 1990. Index herbariorum. Pt. I. The Herbaria of the World. 8th ed. New York Botanical Garden, Bronx, New York, USA.

IUCN. 1992. Protected areas of the world. A review of national systems, Volume 1: Indomalaya, Oceania, Australia and Antartic. IUCN Publications, Cambridge, UK.

IUCN. 2001. IUCN Red List Categories: Version 3.1. Prepared by the IUCN Species Survival Commission. Gland, Switzerland and Cambridge, UK.

KOOPOWITZ, H., S.H. SOHMER, A. THORNHILL, and G. PEREZ. 1998. Deforestation and plant species extinctions in the Philippines: *Psychotria* as an example. In: C.-I Peng and P.P. Lowry, eds. Rare, threatened and endangered floras of Asia and the Pacific Rim. Academia Sinica Monograph Series No. 16:111–121.

MADULID, D.A. and S.H. SOHMER. 1997. An update on the flora of the Philippines project. In: J. Dransfield, M.J.E. Coode and D.A. Simpson. Plant diversity in Malesia III. Proceeddings of the third international Flora Malesiana Symposium 1995. The Royal Botanic Gardens. Kew, UK.

MERRILL, E.D. 1906a. The flora of the Lamao Forest Reserve. Philipp. J. Sci. 1 (Suppl. 1): 1–139.

MERRILL, E.D. 1906b. New and noteworthy Philippine plants, V. Philipp. J. Sci. 1 (Suppl. 3): 169–246.

MERRILL, E.D. 1909. On a collection of plants from Batanes and Babuyanes. Philipp. J. Sci. (1908) 3:385–442.

MERRILL, E.D. 1910a. New and noteworthy Philippine plants, VII. Philipp. J. Sci. 5:167–257.

MERRILL, E.D. 1910b. The flora of Mt. Pulog. Philipp. J. Sci. 5:371–403.

MERRILL, E.D. 1913. Studies on Philippine Rubiaceae, I. Philipp. J. Sci. 8:31–64.

MERRILL, E.D. 1915a. Sertulum Bontocense: new or interesting plants collected in Bontoc subprovince, Luzon, by Father Morice Vanoverbergh, II. Philipp. J. Sci. 9:433–459.

MERRILL, E.D. 1915b. Studies on Philippine Rubiaceae, II. Philipp. J. Sci. 10:99–144.

MERRILL, E.D. 1917. Studies on Philippine Rubiaceae, III. Philipp. J. Sci. 12:159–176.

MERRILL, E.D. 1921. Studies on Philippine Rubiaceae, IV. Philipp. J. Sci. 17:425–485.

MERRILL, E.D. 1923. An enumeration of Philippine flowering plants. Vol. 3, fasc. 5. Bureau of Printing, Manila, Philippines. Pp. 513–628.

MERRILL, E.D. 1925. Additions to our knowledge of the Philippine flora, I. Philipp. J. Sci. 26:447–496.

MERRILL, E.D. 1926. Additions to our knowledge of the Philippine flora, II. Philipp. J. Sci. 29:475–497.

MYERS, N., R.A. MITTERMEIER, C.G. MITTERMEIER, G.A.B. DA FONSECA, and J. KENT. 2000. Biodiversity hotspots for conservation priorities. Nature 403:853–858.

NEPOKROEFF, M., B. BREMER, and K.J. SYTSMA. 1999. Reorganization of the genus *Psychotria* and tribe Psychotrieae (Rubiaceae) inferred from ITS and rbcL sequence data. Syst. Bot. 24:5–27.

PETIT, E. 1964. Les espèces Africaines du genre *Psychotria* L. (Rubiaceae)—I. Bull. Jard. Bot. État. 34:1–229.

PIESSCHAERT, F. 2001. Carpology and pollen morphology of the Psychotrieae (Rubiaceae-Rubioideae). PhD thesis: Katholieke Universiteit Leuven, Belgium.

QUISUMBING, E. and E.D. MERRILL. 1928. New Philippine plants. Philipp. J. Sci. 37:133–213.

QUISUMBING, E. and E.D. MERRILL. 1930. New or interesting Philippine plants, I. Philipp. J. Sci. 41:315–371.

ROBBRECHT, E. 1988. Tropical woody Rubiaceae. Characteristic features and progressions. Contributions to a new subfamilial classification. Opera Bot. Belg. 1:1–271.

ROBBRECHT, E. 1989. A remarkable new *Chazaliella* (African Psychotrieae), exemplifying the taxonomic value of pyrene characters in the Rubiaceae. Bull. Mus. Nat. Hist. Nat. (Paris), 4e sér. Sect. B, Adansonia 4:341–349.

ROLFE, R.A. 1886. On the flora of the Philippine Islands, and its probable derivation. J. Linn. Soc. Bot. 21:283–316.

SOHMER, S.H. 1988. The nonclimbing species of the genus Psychotria (Rubiaceae) in New Guinea and the Bismarck Archipelago. Bishop Mus. Bull. Bot. 1:1–339. Bishop Museum Press, Honolulu, Hawaii.

SOHMER, S.H. 2001. Conservation lessons from studies in Philippine *Psychotria*. Malayan Nat. J. 55:43–47.

WHITMORE, T.C. 1990. An introduction to tropical rain forests. Oxford University Press Inc., New York.

LIST OF COLLECTIONS EXAMINED

Collections are listed in alphabetical order by the last name of the collector (or of the last name of the first collector, if there are more than one collector), followed by collection number in sequential order. For different collecting series (e.g. BS, PNH, FB) the collector(s) name(s) are repeated. The number in parantheses is the taxon number as given in the *List of species*. The taxon number is used in the main body of text. Names in **boldface** represent new taxa.

List of species

1. **olsenii** Sohmer & A.P. Davis
2. **paracalensis** Sohmer & A.P. Davis
3. *cadigensis*. Merr.
4. *sibuyanensis* Elmer
5. *piperi* Merr.
6. **barkleyi** Sohmer & A.P. Davis
7. *rigidaefolia* (Elmer) Merr.
8. *radicans* (Merr.) Merr.
9. *sarcocarpa* Merr.
10. *iwahigensis* Elmer
11. *paucinervia* Merr.
12. *obscurinervia* Merr.
13. *mindoroensis* Elmer
14. **christarusselliae** Sohmer & A.P. Davis
15. *elmeriana* Hochr.
16. *ixoroides* Bartl. ex DC.
17. *cuernosensis* Elmer
18. *elliptifolia* Elmer
19. *lianoides* Elmer
20. *ramosissima* Elmer
21. *lanaensis* (Merr.) Merr.
22. *diffusa* Merr.
22a. *diffusa* var. *diffusa*
22b. *diffusa* var. *cervina* Elmer
22c. *diffusa* var. *agusanensis* (Elmer) Merr.
23. *linearis* Bartl. ex DC.
24. **madulidii** Sohmer & A.P. Davis
25. *cephalophora* Merr.
26. *longipedicellata* Elmer
27. *cuneata* Elmer
28. **yagawensis** Sohmer & A.P. Davis
29. *papillata* (Merr.) Merr.
30. **yenii** Sohmer & A.P. Davis
31. *subcucullata* Merr.
32. *amaracarpoides* (Merr.) Merr.
33. *urdanetensis* Elmer
34. *nagapatensis* Merr.
35. *subsessiliflora* Elmer
35a. *subsessiliflora* var. *subsessiliflora*
35b. *subsessiliflora* var. **carinata** (Quisumb. & Merr.) Sohmer & A.P. Davis
36. *tayabensis* Elmer
36a. *ayabensis* var. *tayabensis*
36b. *tayabensis* var. **castanea** (Merr.) Sohmer & A.P. Davis
36c. *tayabensis* var. **euphlebia** (Merr.) Sohmer & A.P. Davis
36d. *ayabensis* var. **fasciculiflora** (Merr.) Sohmer & A.P. Davis
36e. *tayabensis* var. **pinagteponanensis** Sohmer & A.P. Davis
37. *scaberula* Merr.
38. *bataanensis* Elmer
39. *elliptilimba* Merr.
40. *pygmaea* Merr.
41. *arborescens* Elmer
42. *wenzelii* (Merr.) Merr.
43. **castroi** Merr. & Quisumb. ex Sohmer & A.P. Davis
44. **surigaoensis** Sohmer & A.P. Davis
45. **sulitii** Merr. & Quisumb. ex Sohmer & A.P. Davis
46. **pulgarensis** Sohmer & A.P. Davis
47. *lancilimba* Merr.
48. *pilosella* Elmer
48a. *pilosella* subsp. *pilosella*
48aa. *pilosella* subsp. *pilosella* var. *pilosella*
48ab. *pilosella* subsp. *pilosella* var. **samarensis** Sohmer & A.P. Davis
48b. *pilosella* subsp. **erythrotricha** (Elmer) Sohmer & A.P. Davis
48ba. *pilosella* subsp. *erythrotricha* var. *erythrotricha*
48bb. *pilosella* subsp. *erythrotricha* var. **camarinensis** (Merr.) Sohmer & A.P. Davis
49. *crispipila* Merr.
50. *balabacensis* Merr.
51. *membranifolia* Bartl. ex DC.
52. **conglomeratiflora** Sohmer & A.P. Davis
53. *magnifolia* Merr.
54. **alzapaniensis** Sohmer & A.P. Davis
55. *amplissima* Merr.
56. *rubiginosa* Elmer
56a. *rubiginosa* var. *rubignosa*
56b. *rubiginosa* var. **jacobsii** Sohmer & A.P. Davis
57. *phanerophlebia* Merr.
58. **boholensis** Merr. ex Sohmer & A.P. Davis
59. **iridensis** Sohmer & A.P. Davis
60. *palawanensis* Elmer
61. **quisumbingiana** Sohmer & A.P. Davis
62. *manillensis* Bartl. ex DC.
63. **alabatensis** Sohmer & A.P. Davis
64. **merrittioides** Sohmer & A.P. Davis
65. **foxworthyi** Sohmer & A.P. Davis
66. *panayensis* Merr.
67. **frakei** Sohmer & A.P. Davis
68. *mindanaensis* Merr.
69. *pyramidata* Elmer
70. *sorsogonensis* Elmer
71. *crassifolia* Miq.
71a. *crassifolia* var. *crassifolia*
71b. *crassifolia* var. **culionensis** Sohmer & A.P. Davis
72. *gitingensis* Elmer
73. *cardiophylla* Merr.
74. *cordatula* Merr.
75. *bontocensis* Merr.
76. *merrittii* Merr.
77. *ramosii* Merr.
78. *capizensis* Merr.
79. **dingalanensis** Sohmer & A.P. Davis
80. *lagunensis* (Merr.) Merr.
81. *velutina* Elmer
82. *fenicis* Merr.
83. *longissima* Quisumb. & Merr.
84. *pinnatinervia* Elmer
85. *pubilimba* Quisumb.
86. **catanduaniensis** Sohmer & A.P. Davis
87. *banahaensis* Elmer
87a. *banahaensis* var. *banahaensis*
87b. *banahaensis* var. **montalbanensis** Sohmer & A.P. Davis
88. **galintanensis** Sohmer & A.P. Davis
89. *pallidifolia* Merr.
90. **reynosoi** Sohmer & A.P. Davis
91. *paloensis* Elmer
91a. *paloensis* var. *paloensis*
91b. *paloensis* var. **bukidnonensis** Sohmer & A.P. Davis
91c. *paloensis* var. **ovatifolia** Sohmer & A.P. Davis
91d. *paloensis* var. **subelliptifolia** Sohmer & A.P. Davis
91e. *paloensis* var. **wenzeliana** Sohmer & A.P. Davis
92. *luconiensis* (Cham. & Schltdl.) F.-Vill.
93. **bayombongensis** Sohmer & A.P. Davis
94. *negrosensis* Elmer
95. **patulinervia** Merr. & Quisumb. ex Sohmer & A.P. Davis
96. **casiguraensis** Sohmer & A.P. Davis
97. **mariguidonensis** Sohmer & A.P. Davis
98. *plumeriaefolia* Elmer
99. *nitens* (Merr.) Merr.
100. **mendozii** Sohmer & A.P. Davis
101. *yatesii* (Merr.) Merr.
102. *weberi* Merr.
103. **hilonghilongensis** Sohmer & A.P. Davis
104. *ilocana* (Merr.) Merr.
105. *macgregorii* Merr.
106. **palimlimensis** Sohmer & A.P. Davis
107. *gracilipes* Merr.
108. *acuminatissima* Elmer
109. **sohotonensis** Sohmer & A.P. Davis
110. *subalpina* Elmer
111. *pauciflora* Bartl. ex DC.
112. *loheri* Elmer
113. sp. A
114. sp. B
115. sp. C
116. sp. D
117. sp. E
118. *cagayanensis* Merr.
119. *irosinensis* Elmer
120. *samarensis* Merr.
121. *voluta* Elmer
122. sp. indet.

LIST OF SPECIMENS EXAMINED, ALPHABETICAL BY COLLECTOR

Number in parentheses after specimen number indicates *Psychotria* species in text. Specimens marked with double asterisk (**) indicate either type specimen of accepted species, or type specimen of Unknown species—*species incognitae*. Specimens marked with a single asterisk (*) indicate type specimen of synonyms.

Adduru 274 (102), Adduru FB 22980 (72); Ahern's collector FB 1508 (38), FB 1893 (92), FB 3375 (23); Alcasid et al. PNH 1615 (92); Alegaen et al. 207 (72); Alhambra & Caulas FB 23223 (82); Allen 44-81 (72), 53-81 (36a), 84-81 (35a), 51-81 (99), 238-81 (85), 337-81 (36a), 382-81 (22a), 393-81 (87a), 394-81 (99); Alombro PNH 35430 (92); Alvarez FB 18410 (92); FB 22189* (38), FB 23787 (101); Amarillas FB 19932 (92); Amihan PNH 34316 (92); Andersson s.n. (48aa), s.n. (92); Anonuevo PNH 13451 (72), PNH 13503 (51); Aravez 35 (92); Argent & Reynoso PPI 899 (87a), PPI 8919 (51), PPI 89130 (51); Argent & Romero PPI 9580 (105); Argent et al. PPI 9528 (105); Argent, Romero & Fuentes PPI 20143 (110);

Baker 2457 (84); Banlugan et al. PNH 72467 (84); Barbon & Garcia PPI 22396 (111), PPI 22428 (111), PPI 22461 (111), PPI 22578 (13), PPI 22656 (72), PPI 22685 (72); Barbon & Pedersen et al. PPI 12894 (36a), Barbon & Pedersen PPI 12806 (22a), PPI 12870 (105), PPI 12940 (59); Barbon et al. PPI 8993 (92); Barbon, Alvarez & Fuentes PPI 8560 (16), PPI 8706 (51); Barbon, Alvarez & Garcia PPI 2204 (102); Barbon, Garcia & Alvarez PPI 2393 (35a); Barbon, Garcia & Farnando PPI 12310 (16), PPI 12189 (22a), PPI 12253 (87a), PPI 12311 (87a), PPI 12724 (62), PPI 12732 (72), PPI 12733 (87a), PPI 9120 (90); Barbon, Garcia & Romero PPI 18837 (51); Barbon, Garcia & Sagcal PPI 6035 (72), PPI 6067 (72); Barbon, Romero & Fuentes PPI 8012 (22a), PPI 8237 (22a), PPI 13019 (90), PPI 13150 (90), PPI 13178 (87a); Bartlett 13737 (16), 14637 (62), 15104 (62), 15360 (92), 15679 (51); Basfeinas 76 (92); Bernado FB 15491** (118); Bolster 180 (92); Borden FB 1771 (62), FB 1951 (92), FB 2360 (92), FB 2078 (38); Britton PNH 19454 (92); Brown BS 17907 (84);

Canicosa 729 (51); Canicosa PNH 9763 (48ba); Castro & Anonuevo PNH 5693 (51), PNH 5730 (51), PNH 5736 (51), PNH 5760 (101), PNH 5795 (51), PNH 5855** (43), PNH 6525 (80), PNH 6548 (13); Castro 5826 (48aa); Castro PNH 5834 (52), Castro PNH 6508 (35a); Castro et al. PNH 5850 (122); Castro, Barbon & Garcia PPI 22085 (92), PPI 22115 (92), PPI 22174 (16); Celestino & Castro PNH 1913 (35a); Celestino et al. 0–275 (105); Celestino PNH 7818 (93), PNH 7843 (92), PNH 7992 (22a); Cenabre FB 29168 (71a); Chamisso s.n.** (92); Clark FB 1727 (26); Clemens 924** (21), 7552 (92), 9226 (49), 16417 (49), 16418 (18), 16465 (108), 16841 (90), 16842 (102), 16843 (102), 16845 (35a), 16846 (36c), 16847 (36c), 17021 (36c), 17022 (84), 17023 (102), 17024 (22a), 17025 (48ba), 17246 (92), 17247 (49), 17248 (49), 17404 (92), 17652 (92), 17843 (49), 18799 (49); Clemens BS 15664 (17); Clemens s.n. (17), Co 3428 (35a); Collado PNH 35492 (92); collector unknown BS 14582** (36c); collector unknown FB 1480 (62); Comision Flora Forestal 1455 (92); Conese PNH 15789 (72); Conklin & Buwaya PNH 79575 (74), PNH 80449 (21), PNH 80449 (22a), PNH 80627 (22a); Conklin & del Rosario PNH 72673 (97); Conklin PNH 17507 (62), PNH 18670 (62), PNH 18687** (28); Convocar PNH 2928 (92); Coode 5334 (76), 5450 (92), 5563 (91c), 5662 (17); Costales FB 29787 (92); Croat 53037 (35a); Cuming 518 (92), 1321 (16), 1526 (72), 1863 (51); Curran FB 3540 (71a), FB 3876 (72), FB 4983 (49), FB 5072 (47), FB 8866 (92), FB 10201 (23), FB 13159 (92), FB 17023 (92), FB 19579 (90); Curran et al. FB 18077 (105); Curran & Merritt FB 7855 (110), FB 7870 (36a), FB 8120** (32), FB 15841 (105);

Danao FB 21589 (71a); Darling FB 14776 (107); Del Mundo 11 (92); del Rosario PNH 79916 (62); Delprete 6455 (84); Denoga FB 30601 (35a), FB 30602 (35a); Dransfield SMHI 1240 (71a), SMHI 1273 (68), SMHI 1279 (68); Duyag BS 76656 (51), BS 76684 (48bb), BS 76685 (36a), BS 76694 (51);

Ebalo 619 (71a); Edaño BS 24804 (36d), BS 24859 (51), BS 26803 (82), BS 26954 (23), BS 26960 (87a), BS 34666 (92), BS 38318 (89), BS 41652 (72), BS 41798 (41), BS 41810 (84), BS 42442 (72), BS 42491 (72), BS 46030 (122), BS 46048 (51), BS 46160 (86), BS 48664 (51), BS 48673 (92), BS 48737 (47), BS 48738 (62), BS 48745 (92), BS 48777 (56a), BS 48794 (51), BS 48833 (92), BS 48844 (92), BS 71919 (48aa), BS 75871 (51), BS 75889 (6), BS 75903 (48aa), BS 75908 (89), BS 75916 (35a), BS 75944 (89), BS 75956 (7), BS 75960 (48aa), BS 75974 (22a), BS 75975 (49), BS 75991 (48bb), BS 76013 (89), BS 76018 (89), BS 76121 (51), BS 76128 (36e), BS 76145 (51), BS 76147 (80), BS 76148 BS (87a), BS 76151 (87a), BS 76155, (35b), BS 76161 (35b), BS 76164 (36c), BS 76186 (101), BS 76190 (48bb), BS 76192 (107), BS 76193 (107), BS 76227 (107), BS 76238 (107), BS 76287 (66), BS 76288 (87a), BS 76297 (36a), BS 76306 (36a), BS 76344 (102), BS 76397 (102), BS 77379 (51), BS 77545 (71a), BS 77556 (24), BS 77563 (91a), BS 77607 (71a), BS 77651 (91a), BS 77656 (91a), BS 77670 (91a), BS 77755 (105), BS 77773 (62), BS 78205 (35b), BS 78207 (35b), BS 78215 (35b), BS 78222 (51), BS 78236 (35b), BS 78265 (35b), BS 78271 (51), BS 78275 (35b), BS 78285 (51), BS 78304 (35b), BS 78343** (36e), BS 78349 (51), BS 78362 (35b), BS 78400 (24), BS 78401 (24), BS 78406 (35b), BS 78408** (24), BS 78413 (90), BS 78427 (24), BS 78435 (84), BS 78475 (92), BS 78518 (90), BS 78521 (84), BS 78526 (90), BS 78530 (35b), BS 78585 (36a), BS 78588 (92), BS 78593 (92), BS 78595 (36c), BS 78596 (92), BS 78606 (35a), BS 78613 (36c), BS 78619 (36c), BS 78620 (35a), BS 78635 (36e), BS 79151 (62), BS 79217 (62), BS 79271 (50), BS 79354 (62), BS 79388 (35a), BS 79397 (51), BS 79418 (35a), BS 79423 (99), BS 79435 (35a), BS 79444 (35a), BS 79446 (35a), BS 79488 (35a), BS 79519 (35a), BS 79524 (35a), BS 79537 (35a), BS 79563 (89); Edaño PNH 6 (69), PNH 28 (16), PNH 31 (16), PNH 46 (91a), PNH 48 (69), PNH 89 (105), PNH 91 (91a), PNH 93 (91a), PNH 144 (71a), PNH 227 (71a), PNH 276 (71a), PNH 389 (71a), PNH 1379 (95), PNH 1420** (95), PNH 1445 (95), PNH 1479 (62), PNH 1565 (110), PNH 3253 (17), PNH 3304 (51), PNH 3306 (17), PNH 3350 (17), PNH 3366 (17), PNH 3576 (66), PNH 4018 (38), PNH 4152 (38), PNH 6620 (94), PNH 6640 (94), PNH 6696 (94), PNH 6756 (36a), PNH 7121 (17), PNH 7188 (17), PNH 7193 (17), PNH 7218 (17), PNH 11359 (107), PNH 11371 (107), PNH 11570 (22c), PNH 11584 (107), PNH 11913** (91d), PNH 12067 (51), PNH 12096 (51), PNH 12152 (22a), PNH 12153 (48bb), PNH 12266 (51), PNH 14119 (51), PNH 14207 (16), PNH 15255 (36d), PNH 15257 (35a), PNH 15297 (36d), PNH 15382 (36d), PNH 15434 (36d), PNH 15439 (36d), PNH 17831 (36a), PNH 17886 (36b), PNH 17918 (92), PNH 17975 (106), PNH 17985 (92), PNH 17990 (92), PNH 17992 (92), PNH 18022 (62), PNH 18032 (92), PNH 18035 (62), PNH 18127 (62), PNH 18132 (106), PNH 18141 (112), PNH 19770 (36a), PNH 19774** (64), PNH 19780 (36c), PNH 19786 (62), PNH 19807 (51), PNH 21950 (110), PNH 21962 (91a), PNH

21969 (91a), PNH 21970 (17), PNH 40163 (35a), PNH 40195 (35a), PNH 40217 (48aa); Edaño & Gutierrez PNH 38482 (35a), PNH 38485 (35a); PNH 38495 (35a), Elmer 1103 (72), 5545 (92), 6258 (22a), 6341 (72), 6549 (49), 6718** (56), 6758 (62), 6811** (22), 6975 (32), 6980** (38), 7048 (72), 7055 (72), 7056** (91), 7112 (16), 7581** (36), 7582 (110), 7596 (110), 7672 (110), 7894 (84), 7905 (84), 7934 (36a), 7942** (110), 7943** (18), 7951 (87a), 7952 (36a), 7953 (22a), 7954 (51), 8061 (18), 8195 (51), 8200 (92), 8347 (49), 8584** (112), 8752** (108), 8794 (49), 8953 (92), 9084 (84), 9118** (7), 9124 (87a), 9178** (20), 9185 (51), 9200 (35a), 9213** (48), 9236 (22a), 9505* (22), 9662 (22a), 9686** (94), 9800** (22b), 10076** (17), 10117 (92), 10191 (22a), 10567* (22c), 10587 (91a), 11263** (41), 11424 (17), 11462 (95), 11503 (17), 11645** (19), 12125 (72), 12165 (75), 12250** (98), 12314* (51), 12355** (4), 12431** (72), 12439* (72), 12532 (7), 12532a** (6), 12532b (13), 12737** (60), 12763* (72), 12868 (46), 12981 (71a), 12991** (46), 13052** (10), 13093 (69), 13183** (69), 13253** (121), 13389 (51), 13486** (48b), 13615** (81), 13667** (22c), 13703** (33), 13740 (91a), 14162 (107), 14199** (15), 14230 (91a), 14431 (12), 14534 (91a), 14977 (35a), 15214 (92), 15242 (13), 15256 (35a), 15438 (92), 15466 (36a), 15507 (22a), 15523 (92), 16160 (101), 16564 (12), 16772** (70), 16805 (35a), 17170** (27), 17356** (119), 17504 (12), 17585 (51), 18109 (84), 18156 (92), 18294 (12), 22302 (51); Elmer FB 28205 (91a), Escritor BS 21157** (99), BS 21181 (16), BS 21196 (72), BS 21534 (71a), BS 21541 (50), BS 21605** (50); Espinas FB 30342 (86); Esteve PNH 35352 (92); Everett FB 7258 (90);

Fenix 144 (92); Fenix BS 3648 (62), BS 4048** (25), BS 12950 (92), BS 12701 (92), BS 24976 (71a), BS 28058 (35a), BS 29914** (82), BS 29967 (82), BS 30001 (16); Fernandez FB 21496 (72); Fontanoza FB 31142 (8); FB 31143 (23); Forestry School FB 20137 (62); Forestry Student PNH 35266 (92); Fox PNH 4599 (92), PNH 4600 (82), PNH 4911 (92), PNH 9101 (62), PNH 9183 (16); Foxworthy 16 (62), 32 (92); Foxworthy BS 561 (92), BS 657** (65), BS 705 (22a), BS 900 (69), BS 1599 (38), BS 2405 (84), BS 8989 (22a); Foxworthy FB 25706 (62); Frake PNH 38033 (51), PNH 38038** (67), PNH 38077 (67), PNH 38081 (13), PNH 38102 (13), PNH 38177 (67), PNH 38192 (35a), PNH 38283 (13); Frohne PNH 35688 (16); Fuentes & de la Rosa PPI 38708 (82), PPI 38866 (122); Fuentes & Fernando PPI 37183 (62), PPI 37473 (87a); Fuentes PPI 38502 (92);

Gaerlan & Chavez PPI 26328 (16), PPI 26347 (48aa), PPI 26352 (72), Gaerlan & Ferdinand PPI 26171 (9), PPI 26177 (13); Gaerlan & Fernando PPI 19126 (4); Gaerlan & Fuentes PPI 4940 (17), PPI 4955 (122), PPI 5000 (49); Gaerlan & Reynoso PPI 20288 (72); Gaerlan & Romero PPI 23618 (63), PPI 23632 (92), PPI 23672 (16); Gaerlan & Sagcal PPI 10087 (72), PPI 10217 (122); Gaerlan et al. PPI 2794 (97), PPI 2953 (87a), PPI 9899 (62); PPI 10394 (62), PPI 10406 (62), PPI 10408 (51), PPI 16887 (5), PPI 16893 (16); Gaerlan, Alvarez & Garcia PPI 2533 (17); Gaerlan, Fernando & Majaducon PPI 19962 (35a); Gaerlan, Fuentes & Majaducon PPI 23152 (5); Gaerlan, Romero & Conran PPI 26530 (105), Gaerlan, Romero & Conran PPI 26533 (105); Gaerlan, Romero & Sagcal PPI 13205 (113), Gaerlan, Romero & Sagcal PPI 13225 (69), Gaerlan, Romero & Sagcal PPI 13298 (16), Gaerlan, Romero & Sagcal PPI 13322 (10); Gaerlan, Sagcal & Alvarez PPI 3060 (122); Gaerlan, Sagcal & Fernando PPI 4538 (72), PPI 4551 (44), PPI 4570 (4), PPI 4583 (44), PPI 4588 (51), PPI 4661 (21), PPI 4663 (122), PPI 4717 (105); Gaerlan, Sagcal & Romero PPI 10491 (16), PPI 10797 (17), PPI 10820 (110), PPI 10823 (122), PPI 10836 (17); Garcia & Fernando PPI 12536 (51), PPI 12578 (92), PPI 22790 (35a), PPI 22809 (35a), PPI 22826 (16), PPI 22916 (51), PPI 25018 (35a), PPI 25221 (92), PPI 25226 (62); Garcia & Romero PPI 15715 (16), PPI 15751 (16), PPI 15811 (34), PPI 15837 (17), PPI 15922 (110), PPI 15932 (110); Garcia PPI 25083 (105); Garcia & Sagcal PPI 6010 (72); Garcia et al. PPI 15044 (111), PPI 15078 (111), PPI 15151 (16), PPI 15167 (62), PPI 15197 (16); Garcia, Fuentes & Romero PPI 18167 (36c), PPI 18318 (115), PPI 18431 (62), PPI 18457 (4), PPI 18464 (36e), PPI 18492 (115), PPI 18724 (4); Gaudichaud 215 (92); Gojar FB 31466 (87a); Gutierrez PNH 78165 (35b), PNH 78198 (22a), PNH 78200 (48ba), PNH 78226 (48ba), PNH 78240 (99), PNH 78244 (48ba), PNH 78259 (85), PNH 78268 (36c); Gutierrez et al. PNH 117759 (55), PNH 117023 (51), PNH 117178 (99), PNH 117192 (48ba), PNH 117218 (48ba), PNH 117439 (43), PNH 117463 (48bb), PNH 117464 (51), PNH 117547 (87a), PNH 117579 (109), PNH 117680 (48ba), PNH 118039 (23);

Haenke s.n.** (111), s.n.** (16), s.n. (16), s.n.** (23), s.n.** (51), s.n.** (62); Hallier 237 (92), 4038 (92), 4237 (82), 4721 (51); Hatusima & Sato 28730 (62); Hermida 6 (36a); Holman 87 (62); Iwatsuki et al. P-258 (36b), P-351 (62), P-1161 (84), P-1188 (84), P-1191 (84), P-1196 (84);

Iwatsuki, Murata & Gutierrez P-82 (92), P-110 (92), P-111 (92), P-169 (92), P-380 (92), P-860 (92);

Jacobs 7055 (105), 7114 (49), 7154 (105), 7449 (105), 7474(105), 7500 (49), 7654** (79), 7846** (87b), 7847** (56b), 7937 (49), s.n. (105);

Kanehira 2553 (91b), 2564 (91b), 2717 (110); Keinholz BS 15191 (72), BS 15215 (92), BS 15425 (92); Kondo & Edaño PNH 36539 (50), PNH 36546 (16), PNH 36560 (71a), PNH 36610 (92), PNH 36638 (87a); Korthals s.n.** (71);

Lamanilao PNH 34252 (92); Landingin PNH 35452 (92); Lasquety FB 23576** (120); Lazo PNH 33403 (92); Leiberg 6008 (62); Leonardo 3428 (87a); Logrinan s.n. (87a); Loher 1501 (22a), 1502 (22a), 1505 (92), 1506 (92), 1507 (92), 1510 (122), 1511 (111), 1513 (49), 1514 (49), 1515 (105), 1516 (56a), 1517 (56a), 1518 (56a), 3073 (38), 5091 (105), 5092 (92), 6327 (84), 6333 (87a), 6381 (87a), 6334 (84), 6345* (38), 6368 (92), 6372 (84), 6382 (22a), 6389 (56a), 6394 (56a), 6400 (62), 6402 (59), 6411 (22a), 6433 (56a), 6434 (72), 6439 (22a), 6443 (72), 6445 (122), 6446 (36a), 6447 (47), 9256 (108), 12054 (22a), 12125 (36a), 12133 (108), 12246 (72), 12751 (84), 12752 (87b), 12774 (80), 12775 (87b), 12991 (84), 13017 (75), 13340 (56a), 13394 (51), 13513 (84), 13515 (84), 13645 (110), 13645 (18), 14201 (72), 14265 (92), 14301 (56a), 14324 (84), 14828 (23), 15011 (53); Loher s.n. (18), s.n. (32), s.n. (51), s.n. (47), s.n. (56a), s.n. (62), s.n. (72), s.n. (72), s.n. (92), s.n. (84), s.n. (84), s.n. (87a); Lomibao 2241 (49); Lorenzo PNH 35288 (92);

Mabesa FB 25380 (51); Madulid 1032 (71a); Madulid & Gaerlan et al. PPI 23843 (62), Madulid & Gaerlan et al. PPI 23878 (62); Madulid & Majaducon PPI 36075 (72); Madulid & Reynoso et al. PPI 27487 (72); Madulid & Reynoso PPI 27372 (16), Madulid & Reynoso PPI 27378 (72); Madulid et al. PNH 11784 (22a), PNH 117798** (109), PNH 118346 (51); Madulid et al. PPI 1509 (62), PPI 1528 (62), PPI 3144 (110), PPI 20189 (51); Madulid & Gaerlan et al. PPI 23761 (62); Madulid, Reynoso & Agoo PPI 7954 (72); Maliwanag 330 (72); Mallonga FB 26267 (16); Manayon 70 (92); Mandia 497 (110); Martelino BS 35467 (72); Martelino & Edaño BS 35307 (25), BS 35385 (16), BS 35437 (51), BS 35521 (36a), BS 35540 (72), BS 35576 (34), BS 35635 (72), BS 35739 (17), BS 35741 (72); Martin PNH 33497 (92); Mati 25 (92); McGregor 312** (13); McGregor BS 10212 (80), BS 10775 (87a), BS 11437 (92), BS 12346 (92), BS 18730 (7), BS 18809 (51), BS 19825 (105), BS 19827 (49), BS 22957 (13), BS 32237 (17), BS 32386 (34), BS 32459 (92), BS

32511 (72), BS 41437 (92), BS 43541 (62), BS 43772 (51), BS 43850 (29), BS 44524 (92), BS 45913 (71a), BS 45925 (71a), BS 45958 (16), BS 47367 (51); McLean & Catalan 141 (36a), s.n. (field # 130) (93); Mearns BS 4427 (49); Mearns & Hutchinson FB 4770* (17); Medecilo, Romero & Agoo PPI 29314 (62); Mendoza & Convocar PNH 10231 (55), PNH 10472 (51), PNH 10551 (17), PNH 10585 (87a), PNH 10616 (51), PNH 10760** (103), PNH 10764 (48ba), PNH 10789** (100), PNH 10818 (17); Mendoza & Cordero NHM 91360 (92); Mendoza PNH 10247 (16), PNH 18275 (35a), PNH 18360 (51), PNH 20401 (112), PNH 3071 (92), PNH 41903 (22c), PNH 42119 (13), PNH 42178 (25), PNH 42262 (91b), PNH 97485 (75), PNH 97728 (16), PNH 97948 (92), PNH 97956 (87a); Merrill 560 (92), 597 (92), 742 (71a), 777 (13), 837 (16), 1364 (92), 1869 (92), 1903 (92), 1906 (92), 3180 (38), 3194 (32), 3253 (16), 3361 (26), 3395 (92), 3436 (92), 3765 (38), 3887 (16), 4000* (23), 4083* (51), 4490** (49), 5416 (51), 5576 (51), 6150 (92), 6159 (110), 6170 (17), 6176 (17), 6495** (105), 8077** (9), 8084 (68), 8495 (105), 9518 (69), 10487* (23), 11621 (72), 11663 (108); Merrill BS 1229 (72), BS 7192 (50), BS 8096** (68); Merritt FB 6883 (36a), FB 8638 (32), FB 8844 (92), FB 11382 (51), FB 12073 (16), FB 14164 (112); Merritt & Darling FB 13972 (92); Meyer FB 2319 (92), FB 2778 (56a), FB 2837 (32); Miranda FB 17986 (72); Miras & Oliveros FB 24524 (75); Moseley s.n. (72);

Narciso PNH 35310 (92);

Ocampo BS 27916 (110), BS 27986 (51), BS 27994 (18); Olsen 220 (60), 246 (14), 758** (1); Ondrada FB 31355 (68); Ordon PNH 33509 (92); Oro FB 30648 (110), FB 30696 (13), FB 30972 (48ba);

Paduada FB 29511 (81); Piper BS 502 (16), BS 513** (5); Pipoly & Romero et al. PPI 37851 (71a), PPI 37983 (72), PPI 38041 (51), PPI 38126 (69), PPI 38135 (51), PPI 38209 (113), PPI 38248 (16), PPI 38268 (16); Podzorski SMHI 558 (71a), SMHI 570 (71a), SMHI 2142 (72), SMHI 2160 (114); Ponce & Abejero FB 30446 (91a);

Quimpo FB 30223 (13); Quisumbing s.n. (92), Quisumbing BS 76631 (102); Quisumbing PNH 2053 (13), PNH 2446 (72), PNH 2481 (72), PNH 22992 (92); Quisumbig et al. PNH 79408 (62); Quisumbing & del Rosario PNH 79811 (72); Quisumbing & Sulit BS 82372 (105);

Rafael & Ponce FB 20734 (72); Ramos BS 335 (23), BS 336 (16), BS 337 (84), BS 343 (92), BS 560 (92), BS 1552** (48bb), BS 1553* (107), BS 1701** (29), BS 2099 (84), BS 2121 (49), BS 2634 (84), BS 2674 (16), BS 4704 (104), BS 5018 (47), BS 5348 (22a), BS 5463 (49), BS 5464 (49), BS 5475 (112), BS 5581 (49), BS 5959 (22a), BS 7061 (18), BS 7062 (112), BS 7404** (57), BS 7499** (77), BS 7640 (51), BS 10948 (101), BS 11072 (72), BS 11076 (72), BS 12108 (36a), BS 12164 (92), BS 13362 (80), BS 13569 (23), BS 13648 (22a), BS 13776 (56a), BS 13957 (62), BS 14507 (107), BS 14627 (110), BS 14658 (107), BS 15077 (35a), BS 15198** (55), BS 15296 (7), BS 15357 (48bb), BS 16687 (92), BS 17519** (36d), BS 17549 (51), BS 19377 (23), BS 19429 (48aa), BS 19555 (84), BS 19568 (110), BS 22037 (107), BS 22098 (80), BS 22255 (84), BS 23386 (35a), BS 23460 (13), BS 23526 (22a), BS 23539 (51), BS 23700 (107), BS 23826** (80), BS 24313 (20), BS 24315 (36d), BS 24445 (55), BS 24498 (20), BS 24574 (51), BS 26996 (32), BS 27038 (49), BS 27055** (11), BS 27456** (104), BS 27459 (92), BS 27544 (104), BS 29632 (72), BS 30320 (89), BS 30347 (62), BS 30418** (97), BS 30446 (89), BS 30510** (89), BS 30517 (29), BS 30522 (35a), BS 30534** (39), BS 30564 (7), BS 30567 (89), BS 32724 (62), BS 32802 (104), BS 32819 (16), BS 32960 (35a), BS 32994 (110), BS 32997 (110), BS 33054 (72), BS 33058 (110), BS 33061 (36b), BS 33077** (12), BS 33090 (51), BS 33138 (106), BS 33193** (34), BS 33212** (36b), BS 33260 (92), BS 33261** (56a), BS 33270 (110), BS 33291 (117), BS 33364** (106), BS 33371 (51), BS 39370 (21), BS 33376 (92), BS 39454 (22a), BS 39547 (16), BS 39564** (91c), BS 39618 (84), BS 39634 (32), BS 40745 (59), BS 40756 (72), BS 40779 (56a), BS 40782 (18), BS 40866 (92), BS 40937 (92), BS 41046 (50), BS 41276 (69), BS 41304 (72), BS 41319** (71b), BS 41582 (41), BS 41610 (51), BS 41611 (51), BS 41879 (59), BS 41949 (23), BS 41956 (84), BS 42175 (72), BS 42250 (68), BS 42679 (36a), BS 43062 (20), BS 43083 (31), BS 43127 (51), BS 43134 (62), BS 43140 (61), BS 43144** (31), BS 43145 (58), BS 43169 (13), BS 43188** (61), BS 43201** (58), BS 43230** (14), BS 43337 (51), BS 43360 (36a), BS 46354 (62), BS 46357 (16), BS 49776 (92), BS 76747 (99), BS 76949 (92), BS 76963 (102), BS 76979 (102), BS 77037 (84), BS 77087 (11), BS 77116 (111), BS 79917 (62), BS 80090 (62), BS 80669 (62); Ramos & Convocar BS 83745 (55), BS 83791 (48ba), BS 83831 (68), BS 83886 (51), BS 83888 (48ba), BS 83912 (37), BS 83977 (16), BS 84101 (44); Ramos & Edaño BS 26232* (49), BS 26256* (35), BS 26300 (18), BS 26313 (16), BS 26323 (38), BS 26333 (105), BS 26343 (62), BS 26424 (51), BS 26452 (23), BS 26456 (87a), BS 26503 (38), BS 26568 (87a), BS 26570* (62), BS 26594 (26), BS 26623 (35b), BS 28491 (51), BS 28498 (101), BS 28524 (87a), BS 28568 (90), BS 28580 (87a), BS 28587 (48aa), BS 28602 (48aa), BS 28615 (99), BS 28617 (48aa), BS 28680 (80), BS 28773* (48), BS 28789 (48bb), BS 28828** (53), BS 28853 (51), BS 28962 (63), BS 29001 (13), BS 29005 (51), BS 29105 (51), BS 29166** (73), BS 29198 (48aa), BS 29225 (96), BS 29271** (40), BS 29310 (16), BS 29335 (80), BS 29356 (92), BS 29419 (23), BS 29625 (36a), BS 29629 (47), BS 29693 (84), BS 29713** (47), BS 29715 (89), BS 29716* (72), BS 29735 (22a), BS 29738 (80), BS 29767 (84), BS 29793 (84), BS 29794 (38), BS 30646 (17), BS 30657** (78), BS 30659 (22a), BS 30669 (72), BS 30708 (72), BS 30713 (110), BS 30743 (72), BS 30848 (51), BS 30958 (51), BS 30977 (4), BS 31103 (13), BS 31198** (66), BS 31224* (36c), BS 31331 (13), BS 31335 (66), BS 31441 (86), BS 33484 (89), BS 33612 (35a), BS 33615 (48aa), BS 33637 (80), BS 33658 (51), BS 33662 (48aa), BS 33697** (2), BS 33699 (72), BS 33708 (72), BS 33740 (51), BS 33746 (80), BS 33823 (89), BS 33831 (36c), BS 33842 (48aa), BS 33865 (90), BS 33869 (101), BS 34085 (16), BS 36786 (51), BS 36790 (51), BS 37363 (122), BS 37550 (75), BS 37554 (111), BS 37562 (122), BS 37642 (110), BS 37785 (49), BS 37820 (18), BS 37842 (116), BS 37854** (74), BS 37856 (75), BS 37948 (32), BS 38517 (110), BS 38614 (91b), BS 38787 (17), BS 38884 (91b), BS 39078 (95), BS 39099 (22c), BS 39100 (95), BS 40188 (22a), BS 40193 (105), BS 40235 (112), BS 40275 (49), BS 40286 (49), BS 40367 (105), BS 40368 (49), BS 40447 (22a), BS 40565 (35a), BS 40566 (72), BS 40576 (87a), BS 40599** (76), BS 40690 (13), BS 40736 (72), BS 40737 (111), BS 43903 (72), BS 44726 (82), BS 44772 (82), BS 44808** (90), BS 44959 (105), BS 45051 (108), BS 45051 (108), BS 45200 (51), BS 45278** (35b), BS 45291** (96), BS 45325 (35b), BS 45379 (102), BS 45385 (102), BS 45386 (87a), BS 45518 (35b), BS 45550 (51), BS 45583 (80), BS 45591 (117), BS 45592 (117), BS 45606 (48aa), BS 45611 (22a), BS 45612 (48aa), BS 45622 (117), BS 45666 (84), BS 45671 (108), BS 45681** (83), BS 45683 (110), BS 45688** (54), BS 45699 (107), BS 45707 (22a), BS 45742 (90), BS 46621 (92), BS 46832 (87a), BS 47170 (87a), BS 47221** (85), BS 47265 (99), BS 47273 (85), BS 47288 (90), BS 47316 (99), BS 47321 (85), BS 47979 (16), BS 48006** (63), BS 48008 (80), BS 48015 (87a), BS 48075 (35a), BS 48167 (80), BS 48211 (23), BS 48219 (87a), BS 48228 (36c), BS 48313 (62), BS 48361 (63), BS 48390 (72), BS 48404 (62), BS 48417 (72), BS 48432 (84), BS 48437 (22a), BS 48443** (59), BS 48461 (80), BS 48464 (16), BS 48481 (56a), BS 48486 (84), BS 48490 (47), BS 48505 (62), BS 48526 (47), BS 48537 (92), BS 48552 (92), BS 48559 (51), BS 48561 (62), BS 48592 (32), BS 48621 (56a), BS 48629 (62), BS 48895 (71a), BS 48899** (88), BS 48908 (22a), BS

48916 (16), BS 49476 (107), BS 49490 (107), BS 49691 (50), BS 49702 (46), BS 49760 (62), BS 49793 (46), BS 49871 (16), BS 49891 (50), BS 75114 (25), BS 75166 (63), BS 75193 (36a), BS 75216 (7), BS 75249 (22a), BS 75251 (80), BS 75286 (62), BS 75288 (80), BS 75323 (80), BS 75438 (13), BS 75439 (35b), BS 75440 (89), BS 75474 (51), BS 75525 (89), BS 75546 (16), BS 75570 (35b), BS 75589** (86), BS 75625 (29), BS 75719 (51), Ramos & Pascasio BS 34483 (16), BS 34551 (16), BS 34581** (44), BS 34601 (44), BS 34616 (72), BS 34702 (48ba), BS 34864 (51), BS 34920 (16), BS 35049 (48ba), BS 35053 (72), BS 35159** (37), BS 35188 (5), BS 35225 (68), BS 35234 (44), BS 35263 (51), BS 36416 (72); Reed PNH 91880 (98); Reillo BS 16084 (72), BS 19176 (92); Reynoso PNH 87708 (87a), PNH 87734 (71a), PNH 87848 (62), Reynoso & Espirita PPI 118622 (21); Reynoso & Fuentes PPI 24575 (72), PPI 24616 (27), PPI 24705 (72); Reynoso & Gaerlan et al. PPI 21481 (51); Reynoso & Majaducon PPI 21642 (62), PPI 21662 (51), PPI 21737 (62), PPI 21740 (62), PPI 21856 (51), PPI 24066 (5), PPI 24069 (16), PPI 24176 (16), PPI 24292 (10), PPI 24379 (62), PPI 24820 (51), PPI 24831 (48aa), PPI 24947 (48aa), PPI 27210 (62); Reynoso et al. PPI 3872 (51), PPI 3882 (87a), PPI 3887 (13), PPI 7608 (72), Reynoso, Alvarez & Fernando PPI 3932 (56a), PPI 4144 (82), PPI 4163 (82); Reynoso, Alvarez & Fuentes PPI 7784 (34), PPI 7825 (58), PPI 7833 (51), PPI 7866 (51), PPI 7905 (56a), PPI 11160 (69); Reynoso, Fuentes & Garcia & PPI 946 (94); Reynoso, Fuentes & Majaducon PPI 11736 (92), PPI 11916 (48aa), PPI 11932 (62), PPI 11971 (4); Reynoso, Fuentes & Rivers PPI 14010 (51), PPI 14030 (51), PPI 14080 (4), PPI 14087 (6), PPI 14054 (72), PPI 14304 (72); Reynoso, Garcia & Sagcal PPI 14246 (22a), PPI 14492 (74), PPI 14516 (22a), PPI 14563 (105), PPI 14588 (116), PPI 14598 (18), PPI 14490 (22a), Reynoso, Romero & Fuentes PPI 1215 (102), PPI 1217 (35a), PPI 1282 (102), PPI 1459 (16), PPI 3689 (48ba); Reynoso, Sagcal & Fernando PPI 21047 (48ba), PPI 21266 (13), PPI 21373 (22c); Reynoso, Sagcal & Fuentes PPI 14621 (16), PPI 14626 (16), PPI 14680 (62), PPI 14799 (62), PPI 14829 (13), PPI 14970 (92), PPI 17007 (92), PPI 17089 (92), PPI 17116 (62), PPI 17127 (122), PPI 17156 (92), PPI 17236 (82), PPI 17253 (82); Reynoso, Sagcal & Garcia PPI 7030 (74), PPI 7330 (51), PPI 7384 (51), PPI 7450 (72), PPI 7571 (87a); Ridsdale & Baquiran et al. ISU 363 (101), ISU 470 (85); Ridsdale & Reynoso 1427 (112); Ridsdale 842 (13), 843 (51), 919 (91c), 927 (91c), 1000 (114), 1150 (76), 1167 (16), 1228 (91c), 1400 (56a), 1714 (17), 1733 (76), 1738 (76), 1738A (76), 1756 (22a), 1875 (92); Ridsdale SMHI 10 (69), SMHI 102 (72), SMHI 139 (69), SMHI 275 (71a), SMHI 454 (92), SMHI 1879 (71a), SMHI 1648 (62); Ridsdale, Dejan & Baquiran ISU 34 (122, see also under 54), ISU 109 (89), ISU 271 (51); Rivera BS 46512 (92); Rivera & Duyag BS 49953 (80), BS 49971 (48aa), BS 49972 (36a), BS 75018 (51); Robinson BS 6070 (36a), BS 6144 (92), BS 9056 (16), BS 9295 (72), BS 9363 (48aa), BS 9402 (90), BS 9444 (22a), BS 9726 (62), BS 9956 (92), BS 10221 (92), BS 17224 (92), BS 17185 (62), BS 18068 (92), Romero & Chavez PPI 29080 (92), PPI 29193 (51), PPI 29395 (111); Romero & Fuentes PPI 8462 (16); Romero & Majaducon PPI 29487 (92), PPI 29501 (92), PPI 29604 (72), PPI 29712 (92), PPI 29716 (92); Rosenbluth FB 12563 (92), FB 16906 (48bb);

Salvoza FB 28590 (22a), FB 29634 (84), FB 29654 (22a), FB 29692 (16), FB 29726 (101), FB 29734 (92); Sandkuhl 305 (110); Sands & Vendivil SANDS 3269 (4); Santos 4657 (92), 5164 (92), 5256 (92), 5305 (72),6194 (92); Santos BS 31737 (75), BS 31775 (49); Servinas BS 16883 (12), BS 16918 (92); Simeon FB 28744 (23); Sinclair & Edaño 9633 (36a); Soejarto, Fernando & Fernando PPI 7944 (72); Soejarto, Reynoso, Sagcal & Edrada 6507 (72); Soejarto, Reynoso, Sagcal & Rutz 6430 (72); Sohmer 11958 (112), 11967 (92), 11989 (35a), 11998 (111), 12007 (22a), 12008 (102), 12013 (36a), 12014 (22a), 12016 (48aa), 12017 (48aa), 12018 (102), 12021 (35a), 12024 (35a), 12096 (87a), 12097 (87a), 12098 (87a), 12102 (72), 12105 (16), 12236 (87a), 12239 (22c), 12259 (87a), 12260 (87a), 12261 (87a), 12262 (87a), 12292 (35a), 12293 (35a), 12295 (35a), 12297 (15), 12372 (4), 12378 (51), 12379 (51), 12380 (51), 12393 (72), 12394 (72), 12414 (72), 12421 (51), 12428 (51), 12429 (51), 12436 (4), 12441 (51), 12443 (51), 12458 (51), 12466 (72), 12470 (51), 12472 (87a), 12484 (51), 12487 (72), 12490 (72); Sohmer & Reynoso 12078 (35a), 12079 (35a); Steiner 2009 (49); Steiner PNH 22946 (92), PNH 37697 (14), PNH 39556 (87a), PNH 39791 (108); Stern & Rojo 2316 (18); Stone & Madulid PPI 64 (71a); Stone & Madulid et al. PPI 27 (71a); Stone & Reynoso et al. PPI 290 (16); Stone PPI 50 (71a), PPI 55 (71a), PPI 350 (122); Stone, Barbon & Garcia PPI 9438 (16), PPI 9442 (51), PPI 9462 (72), PPI 9463 (72), PPI 12077 (72); Stone, Reynoso & Fernando PPI 6876 (72), PPI 6934 (110); Stone, Reynoso & Sagcal PPI 513 (76); Stone et al. PPI 3713 (6), PPI 3817 (6), PPI 6202 (22a), PPI 6205 (17), PPI 6262 (91a), PPI 6278 (36a); PPI 6366 (56b), PPI 6702 (4), PPI 6733 (87a), PPI 6773 (87a); Sulit & Conese PNH 6375 (101), PNH 6419 (7), PNH 6446 (48bb), Sulit & Conklin PNH 16765 (62), PNH 16963 (62), PNH 17603 (72), PNH 17608 (62), PNH 17680 (62), PNH 17720 (62); Sulit & Gojar PNH 2633 (20), PNH 2683 (36a), PNH 2687 (16); Sulit FB 30050 (51), FB 30064 (84); Sulit PNH 2748 (35a), PNH 6074 (20), PNH 6209 (48ba), PNH 6279 (101), PNH 6306** (52), PNH 6397 (5), PNH 6408** (45), PNH 7084 (92), PNH 7683 (99), PNH 7702 (99), PNH 8491 (84), PNH 8495 (62), PNH 9748 (62), PNH 9846 (22c), PNH 9977** (91b), PNH 10075 (17), PNH 11687 (92), PNH 11727 (72), PNH 12302 (71a), PNH 12356 (72), PNH 12432 (72), PNH 12545 (71a), PNH 12546 (16), PNH 13760 (92), PNH 14346** (48ab), PNH 14469 (48ab), PNH 14496 (55), PNH 14503 (51), PNH 21495 (35a), PNH 21504 (16), PNH 21541 (22a), PNH 21581 (7), PNH 21628 (107), PNH 21632 (107), PNH 21638 (48bb), PNH 21639 (7), PNH 21689 (100), PNH 21713 (41);

Taleon PNH 34017 (72); Tamesis FB 11967 (51); Tashiro s.n.* (62); Topacio FB 19086 (16); Topacio BS 20027 (16),

Univ. San Carlos 209 (51), 245 (51), 442 (48ab); unknown collector 97 (13), 98 (23); unknown collector US 629668 (22a); Utzurrum RCBU 210 (107);

Valderama 10 (62); Van Steenis 17945 (18), 18571 (112); Van Steenis PNH 18846 (105); Vanoverbergh 2610** (75), 2818** (112), 2861 (112), 4366 (75); Velasco FB 26976 (36c); Vendivil & Reynoso PNH 121680 (56b), PNH 121681 (56b); Vergara 82 (92); Vidal 400 (110), 402 (16), 403 (92), 403c (92), 404 (51), 1455bis (92), 1455c (92), 1455e (84), 1461 (112), 1481 (92), 1481bis (92), 1481c (72), 1619bis (112), 3021 (16), 3028 (112), 3028bis (112), 3043 (84), 3045 (108), 3049 (72), 3050 (51), 3051 (72), 3053 (62), 3054 (72), 3062 (92), 3066 (22a), 3067 (92), 3068** (93), 3070 (56a), 3078 (16), 3079 (72); Villamil FB 21393 (62);

Warburg 11897 (51); Weber 1007 (91a), 1008 (51), 1561 (62), 1572** (107), 1573** (102), s.n. (51); Wenzel 279 (13), 280 (51), 530 (48ba), 775 (48ba), 622 (112), 693 (20), 702 (51), 754 (8), 762 (17), 882 (91a), 887 (22a), 1142** (42), 1193** (8), 1200 (51), 1297 (72), 1361 (72), 1376 (72), 1416 (72), 1502 (22a), 1620 (36a), 1628 (55), 1751 (55), 2539 (13), 2630 (51), 2749 (51), 2856 (4), 2901 (4), 3120 (91e), 3196 (51), 3226 (51), 3261** (91e), 3303 (13), 31209 (91e); Whitford 152 (22a), 214 (38), 243 (56a), 253 (16), 625 (87a), 722** (35), 907** (26), 930 (22a), 932** (84), 983** (87), 1193 (32), 1208 (22a), 1494 (94), s.n. (32); Whitford FB 19714 (92); Wilkes Expedition s.n. (87a); s.n. (92), Williams 375 (92), 411 (38), 560 (62), 613 (38), 1155 (22a), 1219 (112), 1338 (112), 1269 (112), 1301 (49), 1474 (92), 1539 (49), 1222 (49); Wright, Sulit & Conese PNH 6154 (5), PNH 6179 (51); PNH 6025 (29);

Yates BS 25437 (16), BS 25513 (36c), BS 25516 (35a), BS 25518** (101), BS 25519 (101), BS 25533** (3), BS 25545 (35a), BS 25553 (87a); Yen 173** (30);
Zwickey 216 (91a), 569 (13).

INDEX OF SCIENTIFIC NAMES

Accepted names are given in Roman type. New names, including new taxa and new combinations, are given in **boldface** Synonyms are printed in *italics*; unknown species (*species incognitae*) are indicated by an asterisk (*); excluded species are marked with a † sign.